Handbook on Augmenting Telehealth Services

Handbook on Augmenting Telehealth Services: Using Artificial Intelligence provides knowledge of AI-empowered telehealth systems for efficient healthcare services. The handbook discusses novel innovations in telehealth using AI techniques and also focuses on emerging tools and techniques in smart health systems. The book highlights important topics such as remote diagnosis of patients and presents e-health data management showcasing smart methods that can be used to improvise healthcare support and services. The handbook also shines a light on future trends in AI-enabled telehealth systems.

Features

- Provides knowledge of AI-empowered telehealth systems for efficient healthcare services
- Discusses novel innovations in telehealth using AI techniques
- Covers emerging tools and techniques in smart health systems
- Highlights remote diagnosis of patients
- Focuses on e-health data management and showcases smart methods used to improvise healthcare support and services
- Shines a light on future trends in AI-enabled telehealth systems

Every individual (patients, doctors, healthcare staff, etc.) is currently getting adapted to this new evolution of healthcare. This handbook is a must-read for students, researchers, academicians, and industry professionals working in the field of artificial intelligence and its uses in the healthcare sector.

Artificial Intelligence in Smart Healthcare Systems

Series Editors: Vishal Jain and Jyotir Moy Chatterjee

The progress of the healthcare sector is incremental as it learns from associations between data over time through the application of suitable big data and IoT frameworks and patterns. Many healthcare service providers are employing IoT-enabled devices for monitoring patient health care, but their diagnosis and prescriptions are instance-specific only. However, these IoT-enabled healthcare devices are generating volumes of data (Big-IoT Data), that can be analyzed for more accurate diagnosis and prescriptions. A major challenge in the above realm is the effective and accurate learning of unstructured clinical data through the application of precise algorithms. Incorrect input data leading to erroneous outputs with false positives shall be intolerable in healthcare as patient's lives are at stake. This new book series addresses various aspects of how smart healthcare can be used to detect and analyze diseases, the underlying methodologies, and related security concerns. Healthcare is a multidisciplinary field that involves a range of factors like the financial system, social factors, health technologies, and organizational structures that affect the healthcare provided to individuals, families, institutions, organizations, and populations. The goals of healthcare services include patient safety, timeliness, effectiveness, efficiency, and equity. Smart healthcare consists of m-health, e-health, electronic resource management, smart and intelligent home services, and medical devices. The Internet of Things (IoT) is a system comprising real-world things that interact and communicate with each other via networking technologies. The wide range of potential applications of IoT includes healthcare services. IoT-enabled healthcare technologies are suitable for remote health monitoring, including rehabilitation, assisted ambient living, etc. In turn, healthcare analytics can be applied to the data gathered from different areas to improve healthcare at a minimum expense.

This new book series is designed to be a first-choice reference at university libraries, academic institutions, research and development centres, information technology centres, and any institutions interested in using, design, modelling, and analysing intelligent healthcare services. Successful application of deep learning frameworks to enable meaningful, cost-effective personalized healthcare services is the primary aim of the healthcare industry in the present scenario. However, realizing this goal requires effective understanding, application, and amalgamation of IoT, big data, and several other computing technologies to deploy such systems in an effective manner. This series shall help clarify the understanding of certain key mechanisms and technologies helpful in realizing such systems.

Designing Intelligent Healthcare Systems, Products, and Services Using Disruptive Technologies and Health Informatics
Teena Bagga, Kamal Upreti, Nishant Kumar, Amirul Hasan Ansari, and Danish Nadeem

Next Generation Healthcare Systems Using Soft Computing Techniques
D. Rekh Ram Janghel, Rohit Raja, and Korhan Cengiz

Immersive Virtual and Augmented Reality in Healthcare: An IoT and Blockchain Perspective
Rajendra Kumar, Vishal Jain, Garry Han, and Abderezak Touzene

Handbook on Augmenting Telehealth Services: Using Artificial Intelligence
Sonali Vyas, Sunil Gupta, Monit Kapoor, and Samiya Khan

Handbook on Augmenting Telehealth Services
Using Artificial Intelligence

Edited by
Sonali Vyas
Sunil Gupta
Monit Kapoor
Samiya Khan

CRC Press
Taylor & Francis Group
Boca Raton London New York

CRC Press is an imprint of the
Taylor & Francis Group, an **informa** business

Cover design image: Shutterstock Images

First edition published 2024
by CRC Press
2385 Executive Center Drive, Suite 320, Boca Raton, FL 33431

and by CRC Press
4 Park Square, Milton Park, Abingdon, Oxon, OX14 4RN

CRC Press is an imprint of Taylor & Francis Group, LLC

Library of Congress Cataloging-in-Publication Data
Names: Vyas, Sonali, editor. | Gupta, Sunil, 1979- editor. | Kapoor, Monit, editor. | Khan, Samiya, editor.
Title: Handbook on augmenting telehealth services : using artificial intelligence / edited by Sonali Vyas, Sunil Gupta, Monit Kapoor, and Samiya Khan.
Description: First edition. | Boca Raton : CRC Press, 2024. | Includes bibliographical references and index.
Identifiers: LCCN 2023031622 (print) | LCCN 2023031623 (ebook) | ISBN 9781032385464 (hardback) | ISBN 9781032386805 (paperback) | ISBN 9781003346289 (ebook)
Subjects: MESH: Artificial Intelligence | Telemedicine | Medical Informatics Applications | Augmented Reality
Classification: LCC R859.7.A78 (print) | LCC R859.7.A78 (ebook) | NLM W 26.55.A7 | DDC 610.285--dc23/eng/20231027
LC record available at https://lccn.loc.gov/2023031622
LC ebook record available at https://lccn.loc.gov/2023031623

ISBN: 978-1-032-38546-4 (hbk)
ISBN: 978-1-032-38680-5 (pbk)
ISBN: 978-1-003-34628-9 (ebk)

DOI: 10.1201/9781003346289

Typeset in Times New Roman
by MPS Limited, Dehradun

Dedication

This book is dedicated to family and friends ...

Contents

Foreword

Telehealth services have unique advantages of being affordable, ubiquitous, and reliable for such sections of society who seek services via telehealth interfaces. Ever since humanity has faced the extreme challenges of COVID-19, there has been a renewed focus within the scientific community to devise ways and means to provide healthcare to all vulnerable sections of society using digital platforms and telehealth. Inherently, telehealth services are economical and less stressful for patients as has been found in various surveys as well. Telehealth services providers are adopting digital transformation technologies to make the same personalised and precise to the needs of the patients.

As per a report by American Medical Association, around 75% of providers of telehealth said that they were able to provide quality care. Also, around 68% of them said that they were motivated to increase telehealth use to deliver medical services to their patients.

Artificial intelligence has invoked a lot of research interest amongst the scientific community from various domains and disciplines. The scientific community working in the area of healthcare and digital technologies has a unique advantage and motivation to incorporate the power of artificial intelligence into the way telehealth services can be provided. Ethical issues if taken care of in a pragmatic and diligent manner, adoption of AI into telehealth services can become a game changer especially due to shortage of medical professionals that are being witnessed globally. With usage of AI, the offerings of telehealth services are bound to become more accurate, repeatable, and therefore an equally good option comparable to in-person consultation and treatment mechanisms.

The handbook initially touches upon how AI is an enabling technology for healthcare to lay the foundation of understanding for readers. In later chapters, the authors discuss in detail AI applications for healthcare that promise speed and accuracy for various diagnostics that are carried out as part of treatment. Predictive analytics powered by AI are also discussed for early disease detection and effective resource planning. Various challenges like ethical usage, bias, transparency, and data security that impact usage of AI are also discussed.

The *Handbook on Augmenting Telehealth Services: Using Artificial Intelligence* is a good and timely effort by editors and authors to further the cause of development of telehealth services for the masses. The focus on adoption of artificial intelligence while considering various issues of AI adoption like ethical usage, security of data, quality of solutions, and automation is well received in all the chapters. Such efforts need to continue in the same vein and the research community should not stop short of bringing digitization of the healthcare services for consumption of a large mass of the human population.

I congratulate the editors to have worked along with contributing authors of all the chapters to have come up with such unique and novel ideas that are interdisciplinary, embodying digital technologies, telehealth services, and artificial intelligence. I am sure readers of this handbook shall be benefited a lot by the excellent writing done by the authors. I congratulate the editors to have rallied so hard and worked to collect, review, edit, and finally stitch the book in the present form.

I convey them my best wishes and pray to Almighty for a wide acceptability of this book.

Dr Rajnish Sharma
Vice Chancellor
Chitkara University, Himachal Pradesh, India
Secretary, IEEE Delhi Section
Date – 20th June 2023

Preface

The healthcare industry has undergone a significant transition in recent years thanks to the quick development of artificial intelligence (AI) technologies. Modern healthcare systems now depend heavily on telehealth services, which include remote patient monitoring, virtual consultations, and cutting-edge medical equipment. These innovations have completely changed how diseases are identified, predicted, and treated while also improving patient experience and guaranteeing the provision of high-quality healthcare.

The intersection between artificial intelligence and healthcare is thoroughly covered in *The Handbook of Augmenting Telehealth Services Using AI*. This comprehensive collection of 22 chapters examines the different uses of AI in healthcare and considers how it can revolutionize patient monitoring, disease prediction, and disease diagnosis. The book examines the ethical and legal issues that surround the use of AI in healthcare, highlighting the significance of AI ethics and addressing worries about data security and privacy.

The first chapter, "Artificial Intelligence and Healthcare," establishes the context by giving a basic overview of AI and how it is used to the healthcare industry. The following chapters then go into more detailed subjects, like the uses of healthcare items with AI capabilities for disease detection and prognosis. The authors talk about the revolutionary potential of AI-driven technology and offer examples of how they have improved patient outcomes. In a separate chapter, AI ethics and problems in healthcare are highlighted, highlighting the significance of responsible AI development and implementation. The book provides a meta-analysis of remote patient monitoring as the healthcare system develops further, offering light on the future of healthcare delivery and the possible advantages of AI-enabled technologies.

The handbook also discusses the widespread use of automated machine learning in the medical industry for clinical repositories. It investigates the creation and application of intelligent, innovative medical devices based on AI, offering insights into how these technologies might improve patient care and accessibility to healthcare. The scope and applications of virtual consultation in healthcare are thoroughly examined, looking at the advantages and difficulties of providing healthcare remotely. Examining the legal and moral ramifications of augmented reality (AR) in the contexts of telemedicine and telehealth also highlights the necessity for thorough regulations to ensure the appropriate application of AR technology.

The handbook offers in-depth analyses of patient monitoring and the preservation of electronic medical records, two essential components of telehealth services. The book also looks at the possibilities of machine learning methods for anticipating critical health situations and identifying difficult diseases like stomach cancer. It also looks into the usage of interventions for learning difficulties that are based on augmented and virtual reality, demonstrating the potential for AI-powered technology to revolutionize educational healthcare procedures.

Insights, information, and research from leading authorities in the field are collected in *The Handbook of Augmenting Telehealth Services Using Artificial Intelligence*. It is a useful tool for students, researchers, politicians, and healthcare professionals who want to learn more about how AI is being used into telehealth services. The chapters provide readers with a thorough and multifaceted understanding of the subject by fusing theoretical discussions with real-world applications. We believe that this handbook will serve as a catalyst for additional developments and partnerships in the area of AI-driven telehealth services, ultimately improving patient outcomes all across the world. This volume's contents are designed to give readers a thorough grasp of the advantages and disadvantages of enhancing telehealth services with artificial intelligence.

Dr. Sonali Vyas, Dr. Sunil Gupta,
Dr. Monit Kapoor, and Dr. Samiya Khan (Editors)

Editors' Biography

Dr. Sonali Vyas *Ph.D. (Computer Science), SMIEEE*

Dr. Sonali Vyas has served as an academician and researcher for over 13 years. Currently, she is working as an associate professor at the University of Petroleum and Energy Studies, Uttarakhand. She is a professional member of Senior Member-IEEE, ACM-India, CSI, IFERP, IAENG, ISOC, SCRS, and IJERT. She has been awarded the "National Distinguished Educator Award 2021," by the International Institute of organized Research (I2OR) which is a registered MSME, Government of India." She was also awarded the "Best Academician of the Year Award (Female)" in "Global Education and Corporate Leadership (GECL-2018)." Dr. Vyas has authored and edited many books with renowned publishers. She has published many research papers, articles, and chapters in refereed journals, conference proceedings, and patents. She has also been a member of the Organizing Committee, National Advisory Board, and Technical Program Committee for many international and national conferences, as well as chaired sessions. Her research interests include healthcare informatics, blockchain, database virtualization, data mining, and big data analytics.

Prof. (Dr.) Sunil Gupta *B.Tech, M.Tech, PhD (CSE)*

Dr. Sunil Gupta has over more than 20 years of experience in teaching, research, and industry in the field of computer science and engineering. He is an active member of the IEEE Society, Computer Society of India, Member, Computer Science Teacher Association, Life Member, International Association of Engineers, Member, International Association of Computer Science and Information Technology, Member, and Internet Society (ISOC). He has conducted various workshops, conferences, and FDP. He has authored sixty-nine research papers, two textbooks, and has six patents to his credit. He has also acted as a reviewer of international journals and is a member of the Scientific Committee and Editorial Review Board on Engineering and Physical Sciences for the World Academy of Science, Engineering and Technology.

Prof. (Dr.) Monit Kapoor *PhD (Computer Science)*

Dr. Monit Kapoor is an academician and a researcher with a cumulative work experience of twenty-four years and is currently guiding six PhD scholars in various research areas. He has more than thirty journal publications, books, and filed four patents/copyrights in the area of IOMT, blockchain, and healthcare. He currently works as a professor in the Department of CSE at Chitkara University, Punjab, India, and has been voted among the top 50 tech-savvy academicians in India by Ulektz in the year 2019. Dr. Kapoor closely works with the government of India, the Ministry of Electronics, and IT for proposing Mobile Device Security Standards. He has been in TPC of multiple international conferences and serves as a reviewer for SCI and Scopus indexed journals. He has been a session chair in many reputed international conferences and has delivered many invited talks and workshops. His research interests are in the areas of machine learning, ad hoc and sensor networks, UAV networks, speech processing, and artificial intelligence.

Dr. Samiya Khan *PhD (Computer Science)*
Dr. Samiya Khan is an alumna of the University of Delhi, India and received her PhD in computer science from Jamia Millia Islamia, India. She is currently working as a lecturer in computer science at the University of Greenwich, United Kingdom. Previously, she served as a postdoctoral research fellow at the University of Wolverhampton, United Kingdom. She has contributed more than 20 research papers and her publications extend across journal articles and book chapters in high-impact publications of international repute. She has also presented her research at reputed international conferences. She has served as an associate editor for *Scientific India Magazine* and reviewer for many reputed journals and conferences. Besides this, she has authored a book entitled *Big Data and Analytics* and co-edited many books, including *Internet of Things (IoT): Concepts and Applications* published by Springer Nature, Switzerland and *Extended Reality for Healthcare Systems: Recent Advances in Contemporary Research* published by Elsevier. Samiya's expertise spans across data science, artificial intelligence, edge computing, and the Internet of Things (IoT), with experience in the development of heterogeneous systems.

Contributors

Rakhi Ahuja
Delhi Pharmaceutical Sciences &
    Research University
Delhi, India

Puneeta Ajmera
Delhi Pharmaceutical Sciences &
    Research University
Delhi, India

Darpan Anand
Sir Padampat Singhania University
Rajasthan, India

Anwar Ahamed Ansari
Bansal Institute of Science &
    Technology
Madhya Pradesh, India

Pradeep Arya
BML Munjal University
Delhi, India

Susama Bagchi
Chitkara University
Punjab, India

Gaytri Bakshi
UPES University
Uttarakhand, India

Ankit Bansal
Chitkara University
Punjab, India

Harsh Bansal
Sir Padampat Singhania University
Rajasthan, India

Harsh Bansal
Chandigarh University
Punjab, India

Ashima Bhatnagar Bhatia
Vivekananda Institute of Professional
    Studies – Technical Campus
Delhi, India

Vimal Bibhu
Amity University
Uttar Pradesh, India

Mitali Chugh
UPES University
Uttarakhand, India

Neeraj Chugh
UPES University
Uttarakhand, India

Lipsa Das
Amity University
Uttar Pradesh, India

Sanjoy Kumar Debnath
Chitkara University
Punjab, India

Charul Dewan
Galgotias University
Uttar Pradesh, India

Goldie Gabrani
Vivekananda Institute of Professional
    Studies
Delhi, India

Tanya Gera
Chitkara University
Punjab, India

Divya Gupta
Chandigarh University
Punjab, India

Shaurya Gupta
UPES University
Uttarakhand, India

Sunil Gupta
UPES University
Uttarakhand, India

Sunil Gupta
Chitkara University
Punjab, India

Vishan Kumar Gupta
Sir Padampat Singhania University
Rajasthan, India

Danish Jamil
Malaysia University of Science and
    Technology
Selangor, Malaysia

Harsh Joshi
UPES University
Uttarakhand, India

Mukesh Kalla
Sir Padampat Singhania University
Rajasthan, India

Sheetal Kalra
Delhi Pharmaceutical Sciences &
    Research University
Delhi, India

Ridoan Karim
Monash University Malaysia
Selangor, Malaysia

Tejinder Kaur
Chitkara University
Punjab, India

Chetna Kaushal
Chitkara University
Punjab, India

Bhavana Kaushik
UPES University
Uttarakhand, India

Samiya Khan
University of Greenwich
London, United Kingdom

Sourabh Kosey
Gyan Ganga Institute of Technology
    and Science in Jabalpur
Madhya Pradesh, India

T. Ganesh Kumar
Galgotias University
Uttar Pradesh, India

Radhika Mahajan
Vivekananda Institute of Professional
    Studies – Technical Campus
Delhi, India

Shikha Mittal
Chitkara University
Punjab, India

Chetna Monga
Chitkara University
Punjab, India

Ashutosh Negi
UPES University
Uttarakhand, India

Sellappan Palaniappan
Malaysia University of Science and
    Technology
Selangor, Malaysia

Beenkumar R. Prajapati
Gujarat University
Gujarat, India

Bhupendra G. Prajapati
Ganpat University
Gujarat, India

Jigna B. Prajapati
Ganpat University
Gujarat, India

Sunita Rao
Jaipur National University
Rajasthan, India

Nayankumar C. Ratnakar
Gujarat University
Gujarat, India

Shailaja Salagrama
University of Cumberland
Kentucky, USA

Ankit Sharma
Vivekananda Institute of Professional
    Studies – Technical Campus
Delhi, India

Suvarna Sharma
Chitkara University
Punjab, India

Vinod Kumar Shukla
Amity University
Dubai, UAE

Aradhana Kumari Singh
UPES University
Uttarakhand, India

Gurpreet Singh
Sir Padampat Singhania University
Rajasthan, India

Manisha Singh
Sarvottam International School
Uttar Pradesh, India

Rashmi Singh
Bansal Institute of Science &
    Technology
Madhya Pradesh, India

Tanu Singh
UPES University
Uttarakhand, India

Manish Soni
Jaipur National University
Rajasthan, India

Junaid Tantray
NIMS University
Rajasthan, India

Deepak Thakur
Chitkara University
Punjab, India

Harish Tiwari
Sir Padampat Singhania University
Rajasthan, India

Sonali Vyas
UPES University
Uttarakhand, India

Pawan Whig
Vivekananda Institute of Professional
    Studies – Technical Campus
Delhi, India

Priyanka Yadav
UPES University
Uttarakhand, India

Sheetal Yadav
Delhi Pharmaceutical Sciences &
    Research University
Delhi, India

1 Artificial Intelligence and Healthcare

Aradhana Kumari Singh and Manisha Singh

1.1 INTRODUCTION

Artificial intelligence (AI) has the potential to revolutionize healthcare and disease diagnosis. AI-powered systems can analyse vast amounts of medical data, including patient records, medical images, and genomic data, to identify patterns and make predictions. These systems can assist medical professionals in making more accurate diagnoses and treatment decisions, identify potential health risks earlier, and improve patient outcomes. However, the integration of AI into healthcare presents several challenges that need to be addressed for its successful implementation. This research paper will explore the current state of AI in healthcare and disease diagnosis, its potential benefits, and the challenges associated with its adoption. The integration of AI into healthcare and disease diagnosis presents several challenges that need to be addressed. One of the main challenges is data quality and interoperability. The healthcare industry generates vast amounts of data, but this data is often siloed, incomplete, and inconsistent. AI systems rely on high-quality data to make accurate predictions, and therefore, data quality is essential. Another challenge is data privacy and security. Healthcare data is highly sensitive, and patient privacy must be protected. AI systems must be designed to comply with regulations such as HIPAA to ensure patient privacy and data security. Additionally, the lack of standardization in AI development and deployment is also a challenge. AI systems are often developed in isolation and lack interoperability, making it difficult to integrate them into existing healthcare systems. Lastly, the ethical and legal implications of using AI in healthcare are also a challenge. The use of AI in healthcare raises ethical questions regarding bias, transparency, and accountability. The legal framework governing AI use in healthcare is also evolving and needs to be carefully considered.

The standard for storing and transferring medical image data and associated information is the Digital Imaging and Communication in Medicine, or DICOM, standard. In the United States, DICOM is trademarked by the National Electrical Manufacturers Association. The dental, obstetric, cardiology, and cancer communities make extensive use of the norm. DICOM files may be read by systems that are set up to receive medical data and pictures in that format. Images, as well as information about the picture (such as its pixel size and where it was taken), are stored in a digital format called DICOM. It is crucial that patient information and image data remain inseparable, and DICOM files guarantee this. Computer vision, a subfield of AI, models itself after the human visual processing system.

DOI: 10.1201/9781003346289-1

1

The boundaries between AI and more conventional computer vision-based image processing have blurred. However, noise cancellation, picture sharpening, and contrast amplification are examples of low-level processes. Mid-level processes might include visual object identification (image segmentation) and subsequent object categorization. Classified data may be utilized in complex processes and compared to the conclusions reached by a human analyst.

By enhancing diagnostic accuracy, predicting outcomes for patients, and simplifying healthcare processes, artificial intelligence (AI) has the potential to significantly impact the healthcare industry. This study will examine the present state of artificial intelligence in healthcare, the pros and cons of incorporating AI into healthcare, and the prospects for this field in the future. The healthcare industry is already making use of AI to enhance diagnostic precision, forecast patient outcomes, and standardize administrative processes. Artificial intelligence systems can examine medical photos and find anomalies like tumours that may be missed by human doctors. Artificial intelligence may also examine information from patients to predict outcomes like readmission or death. Additionally, AI can help healthcare organizations AI has the potential to improve healthcare outcomes by improving diagnostic accuracy, predicting patient outcomes, and streamlining healthcare operations. Improved diagnostic accuracy can lead to earlier detection of diseases, which can increase the chances of successful treatment. Predicting patient outcomes can help healthcare organizations better allocate resources and provide more targeted care. Streamlining healthcare operations can reduce costs and improve patient satisfaction by reducing wait times and improving the overall patient experience. online operations by automating routine tasks such as appointment scheduling and medication ordering. Despite the potential benefits of AI in healthcare, there are also several challenges that must be addressed before AI can be widely implemented.

One challenge is the lack of data standardization and interoperability. Healthcare data is often stored in disparate systems that don't communicate with each other, which makes it difficult to analyse patient data using AI algorithms. Another challenge is the need for regulatory oversight to ensure that AI algorithms are safe and effective. There is also a risk of bias in AI algorithms, which can result in healthcare disparities if not addressed.

While there is much potential for AI to improve healthcare in the future, several obstacles must be overcome before. One key area of focus is data standardization and interoperability, which will require collaboration among healthcare organizations, technology companies, and regulatory bodies. Another area of focus is addressing bias in AI algorithms, which will require ongoing monitoring and evaluation. Despite these obstacles, AI has great promise for the future of healthcare, and might significantly improve current practises. Using AI to enhance diagnostic accuracy, predict patient outcomes, and streamline healthcare processes has the potential to significantly impact the healthcare industry. Data standardization and interoperability, regulatory monitoring, and bias in AI algorithms are only a few of the obstacles that must be overcome before AI can be broadly applied. The future outlook for AI in healthcare is promising, and ongoing research and collaboration will be necessary to fully realize the potential benefits of AI in healthcare.

1.2 PRE-PROCESSING

Pre-processing steps frequently involve performing picture operations like noise reduction, image normalization, and image enhancement. Image enhancement is the process of modifying an image to make it better suited for viewing or deeper study. Examples in further detail include removing image noise, improving image sharpness, adjusting image intensity, or making object detection easier (image segmentation). There are numerous strategies for achieving these objectives.

Tiny objects in the image can be eliminated, and uneven background illumination can be fixed, using morphological operators as a filtering technique. A structural element with adjustable size and shape may be used to conduct the filtering. Larger objects will be removed by a larger structuring element than by a smaller one. When non-uniform background illumination needs to be fixed, all objects in the picture must be removed so that the background intensity may be measured and deducted from the original picture. One use of median filtering is the elimination of noise from images. The median filtering method operates by setting the value of the output pixel to the median of the adjacent input pixels. With this method, outliers in the pixel values are eliminated since they would be far from the median value of nearby pixel values.

The process of histogram equalization, sometimes referred to as picture normalization, can be used to alter the image's contrast such that the intensity values cover the whole intensity spectrum. Also, the sharper contrasts between dark and bright parts are made possible by the wider range of intensity values. Periodic noise in the photographs, which often results from electrical and/or electromechanical interference that affects image collection, is another possibility in the field of image capture. It is required to ascertain the periodic noise's properties in order to eliminate it. They are normally evaluated by looking at the image's Fourier spectrum. When the frequency spikes produced by this periodic noise become noticeable enough, automated analysis can be used to reduce the difficulty of detecting the input parameters.

1.2.1 SEGMENTATION

Image segmentation plays an important role in several medical imaging applications. The items or areas of interest in a picture are separated using this method. Once a target area or component has been removed, further subdivision or segmentation is no longer necessary. Image segmentation is a challenging process, and its accuracy is crucial to the success of subsequent investigations.

Image segmentation algorithms are frequently based on image intensity value similarities or discontinuities. In the first instance, segmentation divides an image into objects or regions of interest that appear to have comparable intensity levels based on a set of criteria. In the second instance, the segmentation is based on breaking apart the image by paying attention to sharp variations in intensities. There are three primary types of approaches for segmenting images using intensity discontinuities: point, line, and edge detection.

The point and line detection techniques, which entail identifying distinct points or straight lines in an image, respectively, are obviously crucial for image segmentation in general. However, the most popular technique for detecting intensity value discontinuities is edge detection by enlargement. This is because an edge in this context can be thought of as an arbitrary curved line, and notably in medical imaging, it frequently happens that neither points nor straight lines are what we are looking for. First, derivatives are utilized for edge detection because of their qualities, which require that they be nonzero and numerically connected to the degree of intensity change in areas of the image with varied intensities and zero in sections of the image with constant intensity.

Images are often divided into regions with comparable threshold values. Basic thresholding is often used because of its simplicity and utility. In the simplest case, the threshold value is determined via experimentation, with the final result being a number that the user believes adequately fulfils the desired function. Visual inspection of an image's histogram is another option for setting a threshold. Since these two methods rely heavily on the individual using them, it is often recommended to utilize an algorithm that determines a threshold value by considering the necessary picture data.

1.2.2   OBJECT DETECTION

In order to recognize anatomical features in CT images, neural network training methods are often utilized in conjunction with object recognition utilizing bounding boxes (Onieva Onieva et al., 2018). This approach is commonly used to find one or more intriguing spots in a picture. It might be utilized to automatically identify gallstones in patients as well as to find the spline and liver in images (Figure 1.1) (Pang et al., 2019).

1.3   RADIOLOGY'S USE OF ARTIFICIAL INTELLIGENCE AND OVERCOMING ITS CHALLENGES

Some problems are now addressable due to the accessibility of large amounts of data, increased processing power, and process of being developed. Nonetheless, AI in radiography still confronts certain obstacles. One of the biggest obstacles is the need for a large quantity of high-quality training data, an evenly distributed data set that is really representative of all data. In contrast to ImageNet, which contains millions of pictures, medical imaging data sets often only include tens to hundreds of images at most. Particularly for chronic illnesses, there is a scarcity of data from people with a comparable diagnosis, as well as few and costly greater labelled expert sources of data.

The resulting model training is either underfit or routinely overfit, leading to inaccurate classification or the omission of crucial data. False negative findings in radiography should be avoided at all costs since they can have major consequences for both the patient and the practitioner, such as when a benign lesion is overlooked by the model (Erickson et al., 2017; Ker et al., 2017). When a model has too many parameters and not enough training data to properly balance them, overfitting

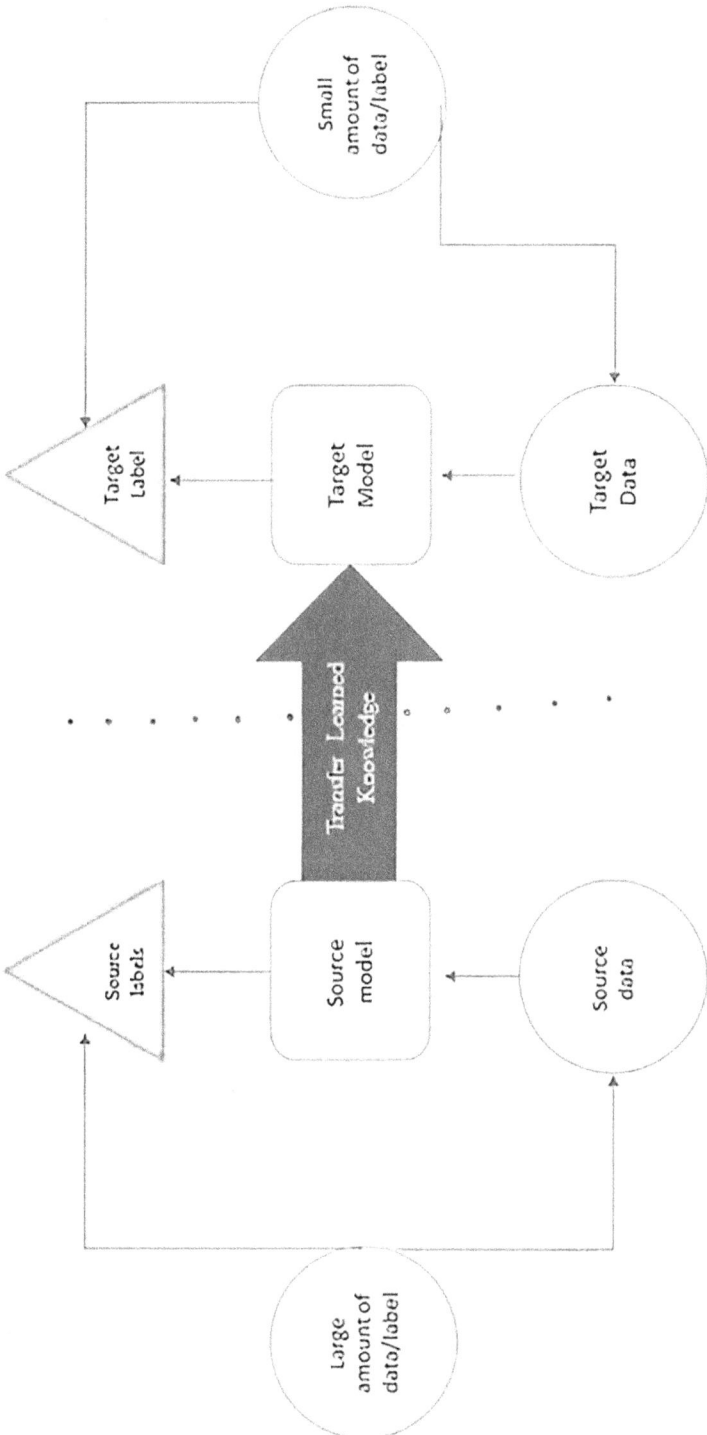

FIGURE 1.1 The process of transfer learning.

occurs. To make up for a deficiency in either training data or labelled data, many methods exist. Fewer than 100 cases in a data set have occasionally generated acceptable results (Ker et al., 2017).

1.3.1 Inadequate Training Data (Data Augmentation)

Data augmentation is the process of adding new pictures to an existing data set by altering its spatial organization, quality, and style without altering the labels. Scanning angles, shape variations, and tissue mobility may all be replicated using spatial transformations, including rotation, flipping, scaling, and deformation. To replicate various scanning techniques and device manufacturers, appearance augmentation affects the statistical features of picture intensities, such as contrast and illumination. Noise level, sharpness, and blurriness are used to alter image quality. The first two are the opposite of one another and are frequently brought on by resolution and motion artefacts in MR and ultrasonography.

Traditional augmentations can produce highly connected images that don't pay enough attention to unusual conditions, which has a limiting effect. Large 3D data sets require expensive computing (Tajbakhsh et al., 2019; Zhang et al., 2019).

Same-domain and cross-domain image synthesis methods, in which labelled data is generated in the target domain or transferred from another domain, may be used to improve synthetic data. Domain adaptation falls under the latter kind, which is discussed in more detail below. In order to generate synthetic pictures that mimic the characteristics of the input data set, it is proposed that generative adversarial networks (GANs) be used. GANs may be used to generate other picture variants or characteristics, hence expanding the data set for further training (Bowles et al., 2018; Tajbakhsh et al., 2019).

1.3.2 Poorly Balanced Training Data (Weighting of Data)

Overfitting may be avoided with the use of regularization in the method. This is the procedure that is used to modify the weights between the two linking layers. The standard method is weight regularization. Lighter weights make the model easier to work with and more robust. Another strategy is the dropout method, which involves randomly setting the weights to zero. After a significant number of cycles, the results will make it evident which weights have an effect on performance (Erickson et al., 2017; Tajbakhsh et al., 2019; Tripathi & Jadeja, 2014), thus only the most important weights are kept.

1.3.3 Transfer Learning and Domain Adaption Are Complicated by Inconsistent Training Data

It is not enough for a model to just have access to a large amount of data; the training data must also be indicative of the model it will "meet" in the future. This means that a model trained on data from only one or a few hospitals may not work well when applied to data from several hospitals. Inductive bias is utilized by the learning algorithm to predict the output of input that it has not yet seen, which increases the model's generalizability. Instead of training on all available data, which is wasteful, the model may transfer "knowledge" from a big data set to the current model to create more representative data (Lundervold & Lundervold, 2019; Torrey & Shavlik, 2009).

1.3.3.1 Transfer Learning

Humans have the ability to generalize their understanding of a topic from one domain to another via the process of abstraction. A guitarist may learn to play the piano faster than someone with no musical training whatsoever because they may apply what they know about music (the source domain) to the task at hand (the target domain). In a similar vein, a model without a weight matrix is likely to perform worse on a collection of clinical photographs than a classifier that can correctly identify images from ImageNet (Shorten & Khoshgoftaar, 2019). Transfer learning, also referred to as "pre-training," is the process of transferring the knowledge and performance of a previously trained network (usually ImageNet for 2D images) to a new domain. The same has been represented in Figure 1.1.

Classified by what is conveyed (example transfer, attribute transfer, parameter transfer, relationship transfer), and by where the transfer occurs (Lu et al., 2015; Pan & Yang, 2010; Torrey & Shavlik, 2009), transfer learning enables information to be transferred across domains, assignments, or distributions.

- Inductive: Many responsibilities are transferred using a domain. The source domain is utilized to improve the target domain by selecting or changing the inductive bias of target task using knowledge of the source task. When the source domain has labelled data, multitasking learning, where the source and target tasks are learned simultaneously, is possible; otherwise, self-learning is utilized.
- Unsupervised: Although the tasks and domains are different, there is a dearth of labelled data.
- Transductive: The activities are equivalent despite the differences in the areas. Normally, the source domain contains a substantial amount of labelled data while the target domain has none. Another word for this is "domain adaptability".

1.3.3.2 Adaptation of Domain

The domain shift issue (Kushibar et al., 2019; Torrey & Shavlik, 2009) occurs more often when the model is applied to data that differs from the training and testing data in terms of methodology, MRI equipment, resolution, and image contrast. This is due to the fact that the data distribution only represents the training set knowledge. The model's flexibility as a result becomes compromised. Domain adaptation requires transfer learning since the source and target domains share a goal but use distinct data distributions or representations to get there.

When labels are only available in the source domain, images are converted from the source domain to the target domain. Unsupervised domain adaptation describes this process. For instance, spleen masks may be utilized in conjunction with UDA to transform MRI pictures of the spleen to CT images for segmentation

(Huo et al., 2018). The source images are converted to the characteristics of the target area without losing the anatomical details or the segmentation masks that came with them. To anticipate the target images, a segmentation network is trained using the generated images and source masks (Csurka, 2017; Kouw & Loog, 2019; Lundervold & Lundervold, 2019; Tajbakhsh et al., 2019). In semi-supervised domain adaptation, just a subset of the target instance is annotated, while in supervised domain adaption, data from both domains are annotated. In this scenario, data label predictions are trained on a common representation from the two domains (Csurka, 2017; Tajbakhsh et al., 2019).

1.3.3.3 Biomedical Research with AI

AI's potential as a tool for biological study is vast, and it has the ability to serve as an "eDoctor" in terms of disease management, diagnosis, and prognosis. Scanning and indexing academic texts for biological research and innovation initiatives may be accelerated with the help of global AI. Some of the newest studies include investigations into tumor-suppressor processes, the extraction of information from protein-protein interactions, and the building of genetic connections in the human genome to facilitate the application of genomic findings in clinical practise. In addition, a semantic graph-based AI method may assist biomedical researchers accomplish the difficult work of summarizing the literature on a certain subject of interest. Also, AI can aid researchers in the biological field in their quest to locate and evaluate relevant papers.

A growing number of high-tech medical tools have brains, and scientists in the field of biomedicine may study them. Computational modelling assistant (CMA) is an AI tool that assists biomedical scientists in translating their conceptual models into "executable" simulation models. The CMA has access to a wealth of information and data. Biological models representing the researcher's hypotheses are provided as input to the CMA.

The CMA's intelligence makes it possible to include all of this data and model, and it changes the researchers' theoretical frameworks into working simulation models. The CMA then generates simulation codes based on the best models that the researcher evaluated and chose. The CMA allows for considerably quicker and more fruitful research to progress. Biomedical imaging, oral surgery, and plastic surgery are just a few of the areas where some types of basic robots may supervise scientific study. The connection between human and machine consciousness has been investigated to better grasp the implications of biomedical engineering.

1.3.3.4 Diagnosis and Prognosis of Diseases

The most urgent need for AI in biomedicine is in illness diagnosis. There have been a number of noteworthy developments in this field. AI allows medical personnel to identify diseases early and more accurately for a variety of disorders. One important type of diagnosis is based on diagnostics, which use biosensors or biochips. For example, ML, which uses AI to categorize and discover faults in microarray data, can analyse expression of genes, which is a critical diagnostic tool. One innovative use is the categorization of cancer data sets for cancer diagnosis. Due to the integration of artificial intelligence (AI), biosensors and associated point-of-care testing (POCT) systems can detect cardiovascular diseases early on.

AI can aid in the detection of cancer and boost the life expectancies of cancer sufferers, such as those suffering from colon cancer. Researchers have also discovered a number of ML's shortcomings in biological diagnostics and proposed solutions to mitigate their consequences. As a result, AI diagnostics and prognostics continue to hold great potential. Medical signal processing in one and two dimensions, as well as image processing, are the foundations of another significant class of sickness diagnosis. Several methods have been used to diagnose, prognostic, and treat ailments. AI has been used to separate biomedical signal characteristics from electrocardiogram (ECG), electromyogram (EMG), and electroencephalogram (EEG) data for one-dimensional signal analysis (ECG).

One prominent use of EEG is the prediction of epileptic episodes. Seizure prediction is critical for reducing the negative impact of seizures on patients. In recent years, artificial intelligence (AI) has grown to be accepted as one of the critical components of a dependable and accurate predictive model. Because deep learning has made it feasible to forecast, a mobile system can use the prediction platform. Another use for AI is in biological image processing diagnostics. AI has increased picture quality and analytic efficacy in image analysis, multidimensional imaging, and heat sensors. Moreover, AI may be integrated into portable ultrasound equipment, allowing non-medical practitioners to utilize ultrasound as a potent diagnostic tool for a variety of diseases.

In addition to the uses outlined above, AI may assist traditional decision support systems (DSSs) in boosting diagnosis accuracy and expediting illness management to reduce staff effort. For example, artificial intelligence has aided in the decision-making procedure for diagnostics, as well as the evaluation and treatment of tropical illnesses, cardiovascular difficulties, and cancer management. These applications demonstrate AI's promise as a strong tool for early and accurate illness diagnosis, treatment, and even prediction. The following two case studies are examples.

Considering that the deep neural network algorithm is a black box and that different algorithms are always being created to make these forecasts, how can we be certain that they are accurate? As a result, Describes how the values AI is taking shape as a new discipline with the goal of improving interpretability, conveying characteristics, and assessing sensitivity via various ways. Another way to improve predictability is to apply Bayesian Classification Techniques to quantify uncertainty when generating predictions (Lee et al., 2017; Lundervold & Lundervold, 2019).

Developing a reliable predictive model can only go so far if the radiologist is unable or unwilling to utilize it. The creation of a system for clinical use should include interaction with the end user to ensure that the entire workflow works in a realistic setting and to foster confidence in the systems (Lundervold & Lundervold, 2019; Miotto et al., 2017). Moreover, developers must consider privacy problems and construct with regulations, morals, and the rules in mind (Chartrand et al., 2017; Lee et al., 2017; Lundervold & Lundervold, 2019).

1.4 ARTIFICIAL INTELLIGENCE AND X-RAYS IN MEDICAL IMAGING

1.4.1 THE USE OF X-RAYS IN MEDICINE

The development of X-ray imaging has had a profound impact on healthcare. Production, airports, and major discoveries like the DNA all have direct links to this discovery. In 1896, the first X-ray camera was developed. There was a time when getting a radiograph of your fingers, head, legs, or chest was a fun activity at "bone picture studios." The pictures would then be hung up for public viewing as if they were masterpieces. Akin to today's virtual reality headsets, handheld fluoroscopes were invented about this time, enabling people to see their own skeletons by holding up one hand in front of a screen.

There is, of course, a valid reason why people should know they shouldn't use these gadgets. Some researchers and doctors at the time realized the life-saving potential of X-rays after discovering that they revealed bone fractures.

1.4.2 X-RAY FINDING

Wilhelm Conrad Roentgen, a German scientist and engineer, is widely regarded as the inventor of the X-ray. The German university where Roentgen taught physics was named after him. He discovered this while experimenting with a "Crookes tube," an electrical discharge tube that can generate several different kinds of beams. After hanging his Crookes tube in his office on November 8, 1895, Roentgen began his search for new, "invisible" rays.

Roentgen was able to view the fluorescent screen, which was made of cardboard with a few barium platinocyanide crystals pasted to it, thanks to the screen's ability to emit a bright green light. Strangely, if the experimental setup failed, the brilliant light on the cardboard would go out. He covered the lightbulb with a piece of black paper to diffuse the light. After that, Roentgen kept on. The cardboard and the tube were tested with a variety of objects, including his own hand. He saw that certain materials, such as paper, allowed light to pass through while others, such as metal and his own palm, blocked it. Supposedly, Roentgen perfected the method in the weeks that followed, naming his new rays "X" to symbolize their mysterious nature.

1.4.3 A CHEST X-RAY

X-ray was rapidly put into continuous usage because of its unique possibilities. Medical professionals used to physically handle all photos until recently. The bulk of X-rays are now digital because of advancements in computer technology, and with millions of patients being analysed daily, the volume of data created is huge.

Radiologists rely heavily on chest X-rays (CXRs) because of the breadth of conditions they may detect. These are practical and inexpensive even in the world's poorest regions. Unfortunately, the time investment needed to get the images is sometimes substantial. Human error is often unavoidable when interpreting these because of the intricacy of the anatomy, particularly the interfaces between lesions and bone. This allows healthcare professionals to use AI-based algorithms for decision-making guidance and assistance. The shortage of radiology experts is also becoming a major concern. Between 2012 and 2017, the number of radiologists' cases climbed by 30% in the United Kingdom, while the number of specialists in that field increased by only 15% (Board of the Faculty of Clinical Radiology, 2018). As a result, smart technology has the potential to greatly enhance the efficiency of clinical and administrative processes (Kao et al., 2015).

1.4.4 Clinical Results from Deep Learning

One trend in radiography is the rise of innovative and powerful small firms. Entrepreneurs and innovators who are keen and passionate about exploring and playing with ideas might find a vast digital "fertile ground" in today's society. We'll take a quick look at the meteoric rise of Zebra Medical Vision (ZMV), an AI-based software business that is "empowering radiologists" and allowing healthcare workers to multitask with minimal quality loss. ZMV claims they can do this by evaluating millions of scans, both old and new, to develop software that can accurately interpret data in the here and now and in the future to aid specialists.

As was previously indicated, jobs in image analysis and diagnostic radiology are in great demand. Currently, patients often seek out radiology technicians and other (non-expert) doctors to provide an initial interpretation of their pictures. This is an effort to reduce patient wait times; however, it may lead to incorrect diagnoses. Even seasoned radiologists are not immune to making serious mistakes in patient care (Robinson et al., 1999). The long hours, heavy workload, and plethora of tests all mean that you'll need to make some tough diagnostic calls. It's estimated that anywhere from 19% to 54% of cases (Hanna et al., 2018) of diseases like early lung cancer are missed, and that about 30% (Bruno et al., 2015) of the time, little errors are made.

To address the issue of inadequate data for comprehensive, accurate, and speedy photo interpretation, ZMV has turned to convolutional neural networks (CNNs). For seven different CXR abnormalities (alveolar consolidation, lung mass, atelectasis, pleural effusion, hilar prominence, diffuse pulmonary emphysema, and cardiomegaly), the automated interpretation system RadBot-CXR has been developed and validated. The programme will validate the presence or absence of general opacity, cardiomegaly, hilar prominence, and pulmonary oedema in its output reports. With the support of these communities, it may be possible to better understand the diagnosis. This CNN also discovered that the interobserver agreement between the RadBot-CXR system and radiologists was somewhat greater than that of a group of three very skilled radiologists. The results of the algorithm have been made public, revealing a significant improvement in radiologists' efficiency. According to ZMV, such software comes in handy when there is a great need for expert-level diagnosis but a shortage of experienced clinical professionals.

1.4.5 Artificial Intelligence for Osteoporosis Detection

Because of better medical care and effective treatments, the population is becoming older. This means that the probability of contracting a variety of ailments increases with age. Musculoskeletal (MSK) illnesses are on the rise, which might lead to an uptick in mortality rates (Kiadaliri & Englund, 2016). Common among the elderly, osteoporosis causes a reduction in bone mass and an increase in the likelihood of bone fracture (Liu et al., 2015). Vertebral compression fractures (VCFs) are a dangerous ailment that weakens the spine and often generates a "rounded back" and "bent look" in the elderly because of the pain and discomfort caused by the condition. Despite the fact that this disease is incapacitating and increases the probability of a subsequent hip fracture, standard diagnostic procedures employing imaging technologies like CT have been less than effective in finding it.

By creating an accurate algorithm that can distinguish VCF in X-rays acquired from the chest and belly, the research team and the ZMV team were able to capture fragments of the spinal column. Recurrent neural network (RNN) operations were used once these patches had been constructed, trained using a CNN, and used for binary classification. One common use of RNNs, a powerful deep learning method, is in the modelling of sequences together, CNNs and RNNs can analyse CT scan pictures and determine the likelihood of the existence of VCFs. Bar et al. (2017) stated that ZMV-trained RNNs achieved an accuracy of 89.1%, a specificity of 93.8%, and a sensitivity of 83.9%. With these kinds of encouraging results, deep learning is going to become an integral part of medical image processing.

High-frequency soundwaves are used by medical ultrasound (also known as sonography) to enable its users to see what's happening inside the human body. The results are frequently shown in real time on an integrated monitor after a transducer produces and directs sound waves to the area of interest.

Ultrasound is frequently used for anatomical or functional diagnostic objectives. By gauging things like blood flow and velocity as well as tissue softness and hardness, functional ultrasonography expands upon its anatomical ultrasound predecessor. Some parts of the body are easily imaged with anatomical ultrasonography. This enables medical professionals to spot changes in an organ's or body structure's function. Today, a range of medical diseases are treated with the help of this technology. High-intensity ultrasound waves have the power to alter and eradicate undesirable tissue, including cancers. These soundwaves may also be used to deliver medications and dissolve undesirable blood clots that could otherwise have disastrous effects on health.

Heart and blood vessel disorders fall under the category of cardiovascular diseases. The blood flow is thus a crucial element that is frequently overlooked (Vyas et al., 2023). Furthermore, the necessary measurements are made by hand, which takes time and involves little automation.

Arterys is the first medical imaging platform to employ AI in a clinical setting, and it was built with the use of cloud computing, cloud storage, and the 4D flow method (which allows direct inquiry into the flow and operation of the heart using MRI). Without 4D flow, it is not feasible to get volumetric anatomical, functional, and flow information over the whole cardiac cycle in a single, rapid scan (Joyce et al., 2019). The deep learning algorithm can identify and segment the heart as accurately as professional human measures, while simultaneously improving picture quality and resolution more fast. In a medical context, these two technologies have reduced the time it takes to process files from days to minutes (Marr, 2017).

Arterys employs cloud computing to analyse and supplement data from a large number of hospitals' and scanners' and patients' pictures at once. The model's predictive power and capacity to generalize are improved as a consequence. The privacy and safety of patient information is also an issue because of this feature. Arterys has adopted the Protected Health Information (PHI) service, which scrubs images of personally identifying information before they are stored in the cloud, to address this issue. When a qualified medical professional needs to know who the data belongs to in order to evaluate a patient, PHI reconstructs the image using the patient identification data stored on the hospital's encrypted server (Arterys, 2018).

Therefore, the unprocessed picture is uploaded, converted to DICOM data, reconstructed, rectified, and users are given access to real-time, interactive 3D post-processing tools. Then, in the cloud, in around 12 minutes (Marr, 2017), the flow visualization, quantitative analysis, and statistical analysis are finished (Vyas, 2023).

Faster and more accurate classification and segmentation is made possible using deep learning. One of the lengthiest processes in cardiac MRI is ventricular segmentation. The Arterys model can accurately discriminate between the interior and exterior of the ventricles in a few seconds (Lieman-Sifry et al.). Since adding more layers to a deep learning model increases accuracy but also increases processing time, only the best layers are employed to provide inference results (Arterys, 2018).

By using the Arterys model, the typical testing time is cut in half, from 121 to 52 minutes. For MRI scans, this means no more holding your breath and a significant reduction in the amount of anaesthetic administered to kids (Natarajan et al., 2018). The AI model can analyse far more fresh imaging data than human workers could by themselves.

New and useful therapeutic insights may be drawn from the complex and non-linear interactions between data points in deep learning models. The platform was developed in collaboration with end users to assist them in integrating the system into their routine, collecting and acting on user feedback, and ultimately increasing their productivity on the job (Arterys, 2018).

Arterys is now making headway towards personalized patient care by using clinical trial images and laboratory data to forecast patient response to treatment and disease progression (Arterys, 2018). The model has been expanded to assess the liver, prostate, lungs, and CXR (Suinesiaputra et al., 2015; Platform – Arterys; Arterys FDA) in an effort to develop an AI suite for healthcare.

1.5   CONCLUSION AND FUTURE SCOPE

In conclusion, the integration of artificial intelligence (AI) in healthcare has the potential to revolutionize the way healthcare is delivered. AI can help healthcare professionals to make more informed decisions, reduce errors, and provide more personalized care to patients. AI-powered applications such as medical imaging analysis, diagnosis, and treatment recommendation systems have shown promising results in improving patient outcomes.

Despite the significant advancements in AI technology, there are still many challenges that need to be addressed. The lack of standardization in data collection and sharing, privacy concerns, and ethical issues are some of the major challenges that need to be addressed to fully realize the potential of AI in healthcare.

In the future, the use of AI in healthcare is expected to expand further. As the technology evolves, AI systems will become more intelligent, accurate, and efficient, resulting in improved patient care and outcomes. Moreover, AI will help to drive down healthcare costs by reducing the need for manual labour and increasing efficiency.

The future scope of AI in healthcare is vast, and it is likely to transform the entire healthcare ecosystem. With the right investment in research and development, the healthcare industry can harness the power of AI to provide more personalized and effective care to patients while reducing costs and improving efficiency.

REFERENCES

Arterys. (2018). Transforming cardiac MR: Advances in AI, 4D flow and cloud computing. https://public.arterys.com/ImagingWire/WP_Cardio_Transforming_Cardiac_MR.pdf

Arterys FDA clearance for liver AI and lung AI lesion spotting software I Medgadget. https://www.medgadget.com/2018/02/arterys-fda-clearance-liver-ai-lung-ai-lesion-spotting-software

Bar, A., Wolf, L., Bergman Amitai, O., Toledano, E., & Elnekave, E. (2017). Compression fractures detection on CT. Medical imaging 2017: Computer-aided diagnosis. (Vol. 10134, pp. 1036–1043). SPIE.

Board of the Faculty of Clinical Radiology. (April 2018). *Clinical radiology UK workforce census report 2018*. The Royal College of Radiologists.

Bowles, C. et al. (2018). GAN augmentation: Augmenting training data using generative adversarial networks. arXiv preprint. arXiv:1810.10863.

Bruno, M.A., Walker, E.A., & Abujudeh, H.H. (2015). Understanding and confronting our mistakes: The epidemiology of error in radiology and strategies for error reduction. *Radiographics, 35*(6), 166876.

Chartrand, G. et al. (2017). Deep learning: A primer for radiologists. *Radiographics, 37*(7), 211331.

Csurka, G. (2017). Domain adaptation for visual applications: A comprehensive survey. In *Domain adaptation in computer vision applications*. Springer.

Erickson, B.J., Korfiatis, P., Akkus, Z., & Kline, T.L. (2017). Machine learning for medical imaging. *Radiographics, 37*(2), 50515.

Hanna, T.N., Lamoureux, C., Krupinski, E.A., Weber, S., & Johnson, J.O. (2018). Effect of shift, schedule, and volume on interpretive accuracy: A retrospective analysis of 2.9 million radiologic examinations. *Radiology, 287*, 20512.

Huo, Y., Xu, Z., Bao, S., Assad, A., Abramson, R.G., & Landman, B.A. (2018). Adversarial synthesis learning enables segmentation without target modality ground truth. *Proceedings—International Symposium on Biomedical Mmaging, 2018* (p. 1220). IEEE.

Joyce, Y.L.B., Jiajun, X.M., Flemming, F.P., & Ji-Bin, L.M.F. (2019). CMUT/CMOS-based Butterfly iQ—A portable personal sonoscope. *Advances in Ultrasound Diagnosis Therapy, 3*, 115.

Kao, E.-F., Liu, G.C., Lee, L.-Y., Tsai, H.-Y., & Jaw, T.-S. (2015). Computer-aided detection system for chest radiography: Reducing report turnaround times of examinations with abnormalities. *Acta Radiology, 56*(6), 696701.

Ker, J., Wang, L., Rao, J., & Lim, T. (2017). Deep learning applications in medical image analysis. *IEEE Access, 6*, 93759.

Kiadaliri, A.A., & Englund, M. (2016). Mortality with musculoskeletal disorders as underlying cause in Sweden 19972013: A time trend aggregate level study. *BMC Musculoskeletal Disorder*, *17*(1), 163.

Kouw, W.M., & Loog, M. (2019). A review of domain adaptation without target labels. *IEEE Transaction on Pattern Analysis on Machine Intelligence*, *43*, 766–785.

Kushibar, K. et al. (2019). Supervised domain adaptation for automatic sub-cortical brain structure segmentation with minimal user interaction. *Science Reports*, *9*, 6742.

Lee, J.G. et al. (2017). Deep learning in medical imaging: General overview. *Korean Journal of Radiology*, *18*(4), 57084.

Lieman-Sifry, J., Le, M., Lau, F., Sall, S., & Golden, D. FastVentricle: Cardiac segmentation with ENet. In International Conference on Functional Imaging and Modeling of the Heart (pp. 127–138). Cham: Springer International Publishing.

Liu, W., Yang, L.-H., Kong, X.-C., An, L.-K., & Wang, R. (2015). Meta-analysis of osteoporosis: Fracture risks, medication and treatment. *Minerva Medicine*, *106*(4), 20314.

Lu, J., Behbood, V., Hao, P., Zuo, H., Xue, S., & Zhang, G. (2015). Transfer learning using computational intelligence: A survey. *Knowledge Based Systems*, *80*, 1423.

Lundervold, A.S., & Lundervold, A. (2019). An overview of deep learning in medical imaging focusing on MRI. *Zeitschrift für medizinische Physik*, *29*(2), 10227.

Marr, B. (2017). First FDA approval for clinical cloud-based deep learning in healthcare. *Forbes*. Available at: https://www.forbes.com/sites/bernardmarr/2017/01/20/first-fdaapproval-for-clinical-cloud-based-deep-learning-in healthcare

Miotto, R., Wang, F., Wang, S., Jiang, X., & Dudley, J.T. (2017). Deep learning for healthcare: Review, opportunities and challenges. *Brief Bioinformatics*, *19*(6), 123646.

Natarajan, P., Frenzel, J.C., Smaltz, D.H., & Leibowitz, C. (2018). Arterys: Deep learning for medical imaging. In *Demystifying big data and machine learning for healthcare* (p. 16973). CRC Press.

Onieva Onieva, J., González Serrano, G., Young, T.P., Washko, G.R., Ledesma Carbayo, M.J., & San José Estépar, R. (2018). Multiorgan structures detection using deep convolutional neural networks. *Progress in Biomedical Optics and Imaging—Proceedings of SPIE*, *10574*.

Pan, S.J., & Yang, Q. (2010). A survey on transfer learning. *IEEE Transactions of Knowledge and Data Engineering*, *22*(10), 134559.

Pang, S. et al. (2019). A novel YOLOv3-arch model for identifying cholelithiasis and classifying gallstones on CT images. *PLoS One*, *14*(6), e0217647.

Platform - Arterys.

Robinson, P.J., Wilson, D., Coral, A., Murphy, A., & Verow, P. (1999). Variation between experienced observers in the interpretation of accident and emergency radiographs. *British Journal of Radiology*, *72*(856), 32330.

Shorten, C., & Khoshgoftaar, T.M. (2019). A survey on image data augmentation for deep learning. *Journal of Big Data*, *6*, 60.

Suinesiaputra, A., et al. (2015). Quantification of LV function and mass by cardiovascular magnetic resonance: Multi-center variability and consensus contours. *Journal of Cardiovascular Magnetic Resonance*, *17*(1), 63.

Tajbakhsh, N., Jeyaseelan, L., Li, Q., Chiang, J., Wu, Z., & Ding, X. (2019). Embracing imperfect data sets: A review of deep learning solutions for medical image segmentation.*Medical Image Analysis*, *63*, 101693.

Torrey, L., & Shavlik, J. (2009). *Transfer learning. Handbook of research on machine learning applications*. IGI Global.

Tripathi, N., & Jadeja, A. (2014). A survey of regularization methods for deep neural network. *International Journal of Computer Science and Mobile Computing*, *3*, 42936.

Vyas, S. (2023). Extended reality and edge AI for healthcare 4.0: Systematic study. In *Extended reality for healthcare systems* (pp. 229–240). Academic Press.

Vyas, S., Bhargava, D., & Khan, S. (2023). Healthcare 4.0: A systematic review and its impact over conventional healthcare system. *Artificial Intelligence for Health 4.0: Challenges and Applications*, 1–17.

Zhang, L., et al. (2019). When unseen domain generalization is unnecessary? *Rethinking Data Augmentation*. arXiv:1906.03347

2 Revolutionizing Healthcare

Impact of Artificial Intelligence in Disease Diagnosis, Treatment, and Patient Care

Goldie Gabrani, Sunil Gupta, Sonali Vyas, and Pradeep Arya

2.1 INTRODUCTION

Artificial intelligence (AI) has impacted almost all the domains of our lives, including the healthcare sector. AI uses many technologies like machine learning, deep learning, natural language processing, and many others that are a subset to AI to design and develop intelligent models capable of performing complex and heavy computations with increased accuracy on the available data that help to make informed decisions so as to find appropriate solutions to intricate real-life problems. This is illustrated in Figure 2.1.

Deploying AI in healthcare is accelerating as it can improve patient outcomes, enhance the efficiency of healthcare services, and reduce costs. AI-based technologies such as machine learning, deep learning, natural language processing, and computer vision can help healthcare providers to better diagnose and treat patients, predict diseases and epidemics, and develop new drugs. AI-powered tools can also aid in patient monitoring, health record management, and drug discovery. However, AI implementation in healthcare is not without challenges, including data privacy and security concerns, ethical considerations, and the need for proper regulatory oversight. Despite these challenges, AI holds great potential for transforming the healthcare industry and improving patient care. Several industries, including healthcare, have already been impacted by artificial intelligence. Technology is advancing quickly, and AI is now being used more widely in healthcare than ever before. AI-driven healthcare solutions are giving doctors and other healthcare professionals access to real-time data and insights that can aid in decision making. This chapter will examine the increasing use of AI in the

DOI: 10.1201/9781003346289-2

17

Fix Complex Issues

Making Decisions

Higher Accuracy

Advanced Computing

Artificial Intelligence

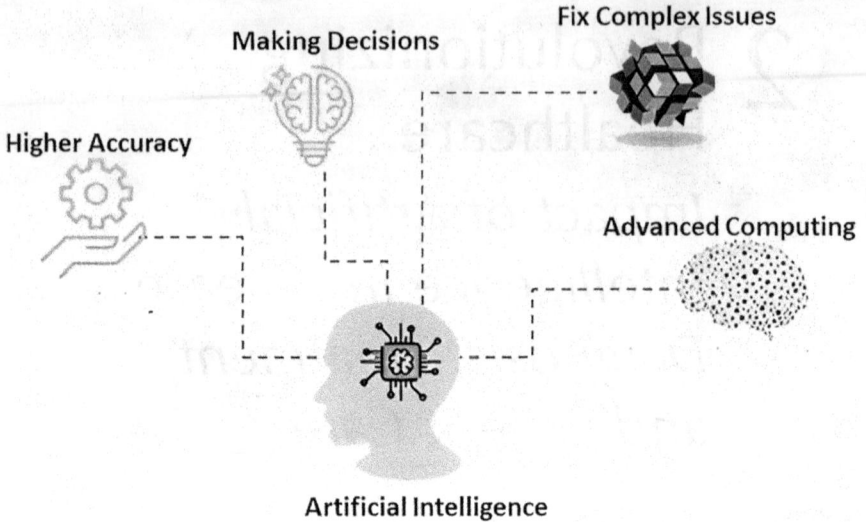

FIGURE 2.1   Artificial intelligence to design and develop intelligent models.

healthcare domain, different applications of AI in healthcare industry, and their possible merits.

The healthcare industry has been facing challenges such as increasing costs, shortage of healthcare professionals, and the need for personalized healthcare services. Artificial intelligence has a great ability to address these challenges and revolutionize the healthcare industry by improving patient outcomes, enhancing the efficiency of healthcare services, and reducing costs. AI-based technologies can aid healthcare professionals in various tasks, such as diagnosis, treatment, and patient monitoring. In medical domains, AI mainly a focuses on development of the algorithms and techniques to find out if a system's behaviour is right in disease diagnosis. For example, machine-learning algorithms can analyze large data sets, including patient data, medical imaging, and electronic health records, to help healthcare professionals identify patterns and make informed decisions. Natural language processing can assist in the analysis of unstructured data, such as physician notes and patient histories. Computer vision can aid in comprehending the medical images. These images can range from X-rays, CT scans, Dexa scans, PET scans, and MRI scans (Tiwari et al., 2019).

Employing AI in healthcare leads to significant cost savings. For example, AI-powered tools can help healthcare providers to recognize accurately high-risk patients, early identification of high-risk surgical patients, reduce hospital readmissions, and optimize treatment plans, leading to better health outcomes and reduced healthcare costs. It also permits focused use of perioperative monitoring and interventions in surgical patients that may enhance the patient's outcomes. Figure 2.2 shows the area that impact AI in healthcare includes solving complex problems, high-level computations, increased accuracy, decision support, self-care preventions, care management, diagnostics, improved health management, and

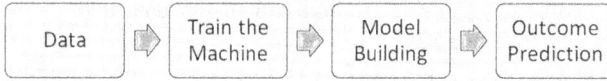

FIGURE 2.2 Steps in developing machine learning models.

strengthening the innovations in treatment and patient care (Bajwa et al., 2021). Despite the potential benefits of AI in healthcare, there are also significant challenges to its implementation. These include data privacy and security concerns, ethical considerations, and the need for proper regulatory oversight. It is essential to ensure that these AI tools are used in a responsible way to protect patient privacy and prevent bias or discrimination. Several initiatives have been taken to promote the usage of AI in healthcare. For example, the National Institute of Health (NIH) launched the National Library of Medicine's "Artificial Intelligence for Medical Imaging Interpretation" program, which aims to develop AI algorithms that can assist radiologists in the interpretation of medical images. The USA Food and Drug Administration (FDA) has also formulated a framework for regulating the use of AI-based medical devices, which includes premarket review and post-market surveillance. AI has great potential to transform the healthcare industry and improve patient care. While challenges remain, healthcare providers and technology companies are working together to effectively utilize the power of AI and deliver more efficient and personalized healthcare services to patients (Tiwari et al., 2019).

2.2 WHAT IS MACHINE LEARNING, DEEP LEARNING, AND NATURAL LANGUAGE PROCESSING?

Machine learning, deep learning, and natural language processing are the major subsets of AI. Machine learning is the procedure of providing enormous data to computing machines where the data is cleaned, processed, and analyzed so as to generate actionable insights that benefit the society by solving real-life problems (Javaidm et al., 2022). Refer to Figure 2.2 for the same.

Deep learning is an advancement of machine learning based on the principle of neural networks with multiple hidden layers, as shown in Figure 2.3. It is generally

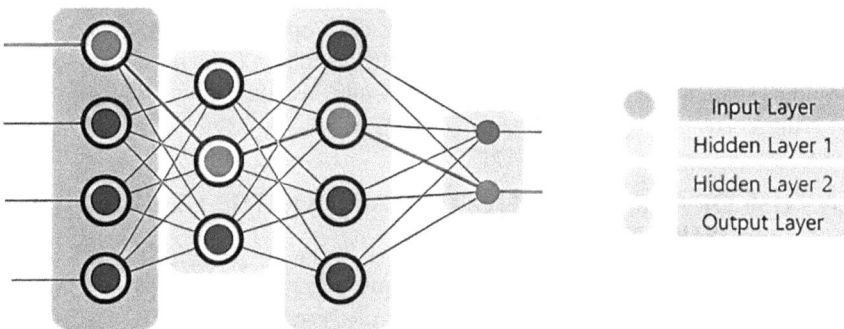

FIGURE 2.3 Deep learning models.

used to solve problems requiring (i) high-dimension data and (ii) automatic extraction of features from the data.

Natural language processing (NLP) is branch of AI that empowers machines to comprehend and recognize human speech. NLP systems first pre-process the data by breaking the data into smaller semantic units, also referred to as tokens. This process is also called tokenization, which helps NLP systems to interpret the data much easier as it is now put in a more logical format. Then NLP systems apply the necessary algorithms to the data. There really are numerous ways in which NLP can support the healthcare domain, some of which are as follows: (i) NLP uses speech-to-text dictation to extract critical data from patient health records instead of manually studying and interpreting the intricate records. This helps clinical documentation to be correct and up-to-date. (ii) NLP helps to analyze huge quantities of unstructured clinical data and recognize eligible candidates for clinical trials, enabling patients to have an access to investigational care. (iii) NLP helps to access health-related records in a fast and efficient manner, accelerating the process of making informed decisions.

2.3   TIMELINE FOR AI BEING USED IN HEALTHCARE

From its inception in the early 2000s, the application of AI in healthcare has advanced significantly. In those early days, algorithms of machine learning were deployed to evaluate patient data and aid in the development of AI-powered medical systems to aid in disease diagnosis. With time, computer vision algorithms that evaluate medical images like X-rays and MRIs started to be applied in the development of AI-based medical imaging systems. In order to provide human surgeons more control and precision during surgical procedures, AI-powered robots were developed in the late 2000s. Electronic health records (EHRs), which employ machine learning algorithms to predict potential health issues like drug interactions or bad reactions, were first developed using AI technology in the early 2010s. In order to give patients medical advice and guidance, chatbots incorporating AI started to be utilized in telemedicine in the middle of the 2010s. Personalized medicine, which examined patient data and created customized treatment regimens based on each patient's specific medical history and genetic composition, was finally produced using AI technology in the late 2010s. As AI technology develops, we can anticipate seeing even more cutting-edge healthcare solutions created that will aid doctors in providing patients with high-quality care while also enhancing outcomes and lowering costs.

2.4   USE OF AI IN DIFFERENT DOMAINS OF THE HEALTHCARE INDUSTRY

AI has been increasing in use at a very extensive rate and across a very wide spectrum as well. Figure 2.4 illustrates some of these.

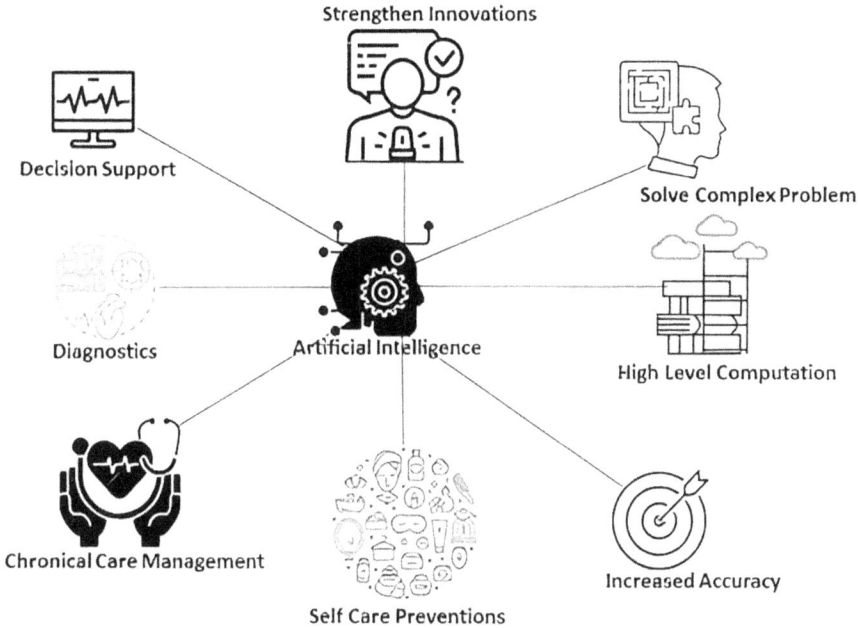

Strengthen Innovations

Decision Support

Solve Complex Problem

Diagnostics   Artificial Intelligence

High Level Computation

Chronical Care Management

Self Care Preventions

Increased Accuracy

FIGURE 2.4 Uses of AI in healthcare industry.

Some of the main areas in which AI is being used have been discussed briefly below:

- Medical Diagnosis with AI: AI-powered systems are now able to examine medical images and find patterns that may be applied to a diagnosis. These computers can precisely identify diseases and medical issues that human doctors might overlook, thanks to ML algorithms. For instance, AI-powered systems can spot early indications of conditions like cancer and heart disease, which can save patients' lives.
- Electronic Health Records (EHR): Medical histories, diagnoses, and treatments for patients are kept as electronic health records, referred to as EHRs, which are used in the healthcare industry. AI-powered EHRs can give physicians access to real-time data that they can utilize to make wise judgements. An AI-powered EHR, for instance, can notify doctors whether a patient is at risk of contracting a specific illness, based on their medical history.
- Drug Development: The process of developing new medications takes time and money. AI is being utilized to accelerate the research and development stages of this process by finding prospective medications. AI is used to examine vast volumes of data from medical research, studies, and clinical trials, assisting researchers in the discovery of novel therapeutics.
- Telemedicine: It is the application of technology to the delivery of healthcare services over long distances. Patients may contact medical specialists and get assistance without leaving their homes, thanks to chatbots and virtual

assistants powered by AI. Healthcare practitioners can receive real-time data from AI-powered telemedicine, which will aid them in more accurately diagnosing and treating patients (Vyas, 2023).

2.5   USE OF AI-ENABLED APPLICATIONS

AI apps have become increasingly popular in the healthcare industry in recent years, providing healthcare professionals with novel ways to provide a much higher quality patient care. Table 2.1 below illustrates some of the AI-enabled apps being used in the healthcare industry for various purposes (Bajwa et al., 2021).

2.5.1   AI IN DISEASE DIAGNOSIS

AI is really making a huge impact in the field of diagnosis. With the help of machine learning algorithms, it is possible for AI to analyze tremendous amounts of data from the patient's medical history, imaging studies, and other sources to identify patterns and make predictions about a patient's health. For example, AI can be used to timely and accurately evaluate patient's mammograms and detect the presence of breast cancer, if any at quite an early stage, when treatment and patient care is most important and effective. Similarly, AI can be used to analyze CT scans and MRI images to detect abnormalities in the brain, heart, or other organs. AI can also be used to assist healthcare providers in making a diagnosis. For example, AI-powered chatbots can be used to gather information about a patient's symptoms and medical history, and provide a preliminary diagnosis or refer the patient to a specialist for further evaluation. This can help to speed up the diagnosis process and ensure that patients receive the suitable care in an appropriate and timely manner.

AI can help in the diagnosis of diseases. With the help of advanced algorithms and ML techniques, AI can analyze vast amounts of medical information, such as

TABLE 2.1

Different AI Apps with Their Features and Key Functions

AI App	Feature	Key Function
Medicine	Telemedicine App	Uses AI-powered chatbots to triage patients and determine whether they need to see a doctor
Buoy	Symptom Checker	Uses natural language processing to understand patients' symptoms
Ada	Symptom Checker	Uses machine learning algorithms to identify potential health problems
Deep 6 AI	Clinical Trial Identification	Uses machine learning algorithms to analyze patient records and identify potential candidates for clinical trials
Canary Health	Chronic Condition Management	Uses machine learning algorithms to provide personalized advice on diet, exercise, and medication

patient records, medical images, lab results, and more, to identify patterns and predict potential health issues. For example, AI-powered tools can analyze medical data such as X-rays, Dexa Scans, Pet scans, CT scans, and MRIs to find out anomalies that are normally not visible to a human eye. AI can also help identify potential risks for certain diseases based on a patient's medical records, history, lifestyle, and other factors. In addition to diagnosis, AI also greatly supports developing customized treatment plans for patients, based on their individual health data. However, it is imperative to mention here that AI is not meant to replace doctors or medical professionals, but rather to support and enhance their ability to make informed decisions and provide better care for patients (Javaidm et al., 2022).

Quantifying diseases is an important step for their diagnosis and is generally done based on some symptoms (subjective and can be felt by the patient) or signs (objective findings; appearance of a disease that doctors can see or measure). Symptoms are feelings a patient has, whereas signs are characteristics that can be observed. Hence, both symptoms and signs of any disease play an important role in its diagnosis. For instance, a patient may feel feverish, nausea, or some pain in the body. On the other hand, signs may be increased body temperature from normal, high blood pressure or an anomaly in some medical scan like X-ray, CT scan, etc. Thus, it is important to understand that any underlying health condition will have various symptoms and signs. Medical diagnosis is a process to identify the patient's disease based on symptoms and signs. Usually, diagnostic data is collected from the medical history of the patient and physical checkup. Most of the time, as symptoms and signs are vague, trained health providers are needed to correctly diagnose the disease. Therefore, developing countries who have poor population to health experts ratio struggle to provide efficient diagnostic processes to cater to a majority of their population of patients. Moreover, human beings are more likely to make errors and may either overdiagnose or underdiagnose due to various reasons, such as unnoticeable symptoms or signs, uncommon disease, or lack of proper data. In order to tackle all these issues, AI is gaining popularity in healthcare, including disease diagnosis both at the individual level and at larger numbers (like epidemic). AI-based disease diagnosis is not only cost effective but is also time-efficient compared to conventional diagnosis processes that are expensive, time-hungry, and usually need regular human intervention. These techniques require data as an input resource and mathematical functions that operate on this data to produce an output not normally so easy for human beings to achieve. For example, with AI, it is much easier to detect Parkinson's disease, locate tumour cells, and predict cardiac failures, which, if done manually, is quite challenging (Reddy et al., 2019).

Several research papers on diagnostics measures of various diseases using machine learning and deep learning techniques (areas of artificial intelligence) have been presented. Machine learning and deep learning techniques enable researchers, doctors, and patients to have more accurate medical diagnosis of their ailments and hence helps in speedier and better solutions. These studies deliberated on the usage of machine learning in early and timely identification of many diseases. As we are aware of the substantial need to diagnose the diseases effectively for better health management, the intricacy of the behaviour pattern of different diseases and their underlying signs and symptoms of the patient population offer great challenges for design and development of a tool that enables timely diagnosis for effective treatments. Some of the existing research works concentrated on diagnosis of one or two diseases per analysis. For instance, one model can be for liver sclerosis, another can be for diabetes and associated retinopathy, or lung infection and related asthma. With the availability of advanced tools like tensorflow, Flask API, and Python, multiple disease analysis using machine learning algorithms is also made possible. The importance of analyzing the diseases while taking into account all the parameters that caused the disease is needed to predict the disease and the various effects it will cause (Gatta et al., 2022; Panesar, 2023; Reddy et al., 2019).

These studies also deliberate on disease diagnosis, disease predictions that are based on the thorough literature survey. Further detecting any irresistible ailment and prevention of its spread requires current and accurate data and detailed analysis. Hence, acting swiftly has a significant effect on people's lives worldwide.

2.5.2 AI in Improving Disease Diagnostics and Treatment Decisions

As already discussed in the above section, AI systems are playing a very important role that helps healthcare providers to improve their diagnostic and treatment decisions, thereby decreasing the error rate in medical diagnosis and further treatment. AI has really made enormous advancements in the realms of medical imaging and its related diagnostics. In imaging, AI offers a range of novel tools and techniques to infer data and take clinical decisions. Global Market Insights comments that "Medical imaging and diagnosis powered by AI should witness more than 40% growth to surpass USD 2.5 billion by 2024." With machine learning, deep learning, computer vision, software libraries, high-performance computing systems, neural networks, natural language processing, and other techniques of AI, image diagnosis has simplified the complex analysis of X-rays, CT scans, Dexa scans, PER scans, and MRI scans (Bajwa et al., 2021).

Many ML algorithms using image processing techniques are now available that help in a great manner to minimize image-based diagnostic errors, thereby improving the overall accuracy of test results. For instance, AI has shown to considerably improve the investigations based on clinical data images while detecting diseases such as cancer, diabetic retinopathy, and gliomas. For this reason, many healthcare professionals are incorporating AI in their daily routines in order to obtain deeper insights into the diseases by analyzing the enormous number of clinical records available and thereby improving patient recovery rate. AI tools can also be successfully applied to update clinical records regularly, recover reporting data, and insert various codes used for diagnostic purposes, all without any sort of any human intervention. Furthermore, tech-organizations like Google, Microsoft, and IBM are exploring the option of using surgical robots that are based on AI algorithms for accelerating patient treatment. The usage of these robots is envisioned to have higher accuracy, lesser harm, speedier recovery, and thereby leading to much better patient care. It may be noted that AI has not only impacted the clinical part of healthcare industry, but has also significantly helped in improving businesses by understanding and evaluating patient requirements (Gatta et al., 2022).

2.5.3 Artificial Intelligence in Early Disease Detection

AI is playing an important role in predicting medical conditions early and timely. The diseases such as cancer, asthma, liver psoriasis, and heart attacks, if detected timely, can lead to much better treatments and outcomes. Many AI-based wearable health sensors like Fitbits and smart watch are being used to track the vitals of a person and issue warnings when the parameters go outside the valid range. These sensors continuously sense various health parameters such as heart rate, breathing rate, oxygen saturation level, activity rate, blood pressure, temperature, sleeping time, etc. and gather data. The gathered data is then processed and analyzed by using machine learning, neural networks, deep learning, and NLP algorithms (depending on the use case) to create a disease detection model that predicts the percentage of risk of occurrence of a disease.

2.5.4 Artificial Intelligence in Medical Help and Support

Recent times have also seen the increase in AI-based virtual nurses. For instance, one of the popular virtual nursing attendants is Sensely, which utilizes the concepts of various subdomains of AI such as NLP, ML, and speech recognition. Some of the major features provided by the virtual nurse in Sensely are therapeutic advice, appointment scheduling, nursing care, etc.

2.5.5 Artificial Intelligence in Surgeries

One of the major use cases of AI is the use of surgical robots. The use of these surgical robots maximize precision by reducing human errors and variations. One such surgical robot is the Da Vinci, which permits surgeons to carry out intricate surgeries with ease, better accuracy, improved efficiency, and control compared with traditional methods. Some of the major features of the Da Vinci are (i) assisting surgeons with a cutting-edge and innovative instruments; (ii) deciphering the surgeon's hand actions in real time; and (iii) constructing magnified and clearly visible, high-resolution 3D images of the area being operated upon (Shah et al., 2019).

2.5.6 AI Framework for Disease Prediction

As already mentioned above, with AI, a machine is able to learn in a similar way a human learns. A human mind learns through identifying various images and detecting patterns. Even a small child starts learning by recognizing simple images like of fruits, animals, and trees around them. A child, after going through this learning process, is able to identify an image shown and is able to classify the image into fruit, vegetable, etc. In a similar way, after a machine learns through various machine learning algorithms, it is able to analyze different types of data that may include audio, time series, videos, images, and provide information that leads to the detection of diseases timely, accurately, and speedily that obviously results in improved patient care. Now, in order to do

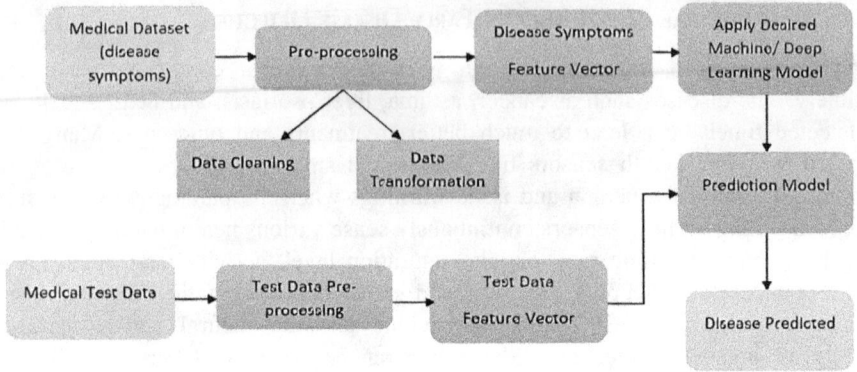

FIGURE 2.5 AI framework for disease detection.

this, it requires a whole gamut of activities and hence a proper system needs to be designed and developed. Proper planning is central to abstract design and further development of the system. Figure 2.5 shows a pictorial representation of the symptoms and signs recognition model using AI-powered machines deployed with profound learning and classification algorithms. Pre-processing requires the real-life data to be properly cleaned and validated before it can be used for making the machines learn. Real-life data invariably contains wrong values, missing fields, out-of-range values, and so on and the data, if not corrected (by filling in the spaces that are empty or have some incomprehensible information such as commas or other characters), would lead to wrong outputs, which may lead to gross medical errors and blunders. Also, during data osmosis, data from various sources is combined and then corrected for any errors due to merging. Sometimes this data also needs to be transformed in a different format. Accordingly, pre-processing algorithms take this pre-processed data, process it, remove errors, apply machine learning or deep learning algorithms, and create a model that is able to predict the occurrence with certainty with a predefined accuracy. This model is then tested with test data and, only when it passes the test, it is put to actual use (Panesar, 2023).

2.5.7 AI IN TREATMENT

Apart from disease diagnosis, AI also plays a promising role while treating diseases. This is especially relevant in the field of precision medicine. AI enables patient's genetic data and other biomarkers to be analyzed to develop customized treatment plans that are tailored to their specific needs. This can lead to better treatment outcomes and fewer side effects. AI can also be used to assist healthcare providers in making treatment decisions. For example, AI-powered decision support systems can analyze a patient's medical history and other data to recommend the most effective treatment options based on the latest medical research and clinical guidelines. This can help to improve the quality of care and ensure that patients receive the most appropriate treatment.

AI enhances accuracy in diagnosing a disease, improves medical decision making, and better treatment plan as a part of effective management of the disease (Vyas et al., 2023a).

2.5.8 AI AND CARDIOVASCULAR DISEASES

Treating cardiovascular diseases has advanced quite a lot in both diagnosis and treatments. In spite of the advancements, these diseases are still the main reason for morbidity and mortality globally. In order to improve the outcomes of cardiovascular diseases, AI algorithms are considerably changing the manner in which cardiology is practiced today. AI techniques such as machine learning and deep learning have considerably improved medical knowledge. This is because of a huge surge in both the volume and the complexity of data and subsequently mining from it clinically relevant information. Then this mined information is used to accurately predict heart failure in a non-invasive and inexpensive manner and that too within seconds. Extensive research is going on to detect heart disease, heart failures, heart attacks, and strokes, thereby advancing cardiovascular medicine using AI. For instance, if a patient undergoes a CT scan after having a heart attack, this scan is analyzed much faster with the use of AI-powered machines that drastically reduce the time to diagnosis and hence prevent damage to the heart. Moreover, applying AI techniques to ECGs has resulted in a low-cost test to find out a weak heart; if treated in time, it can in most cases prevent a heart from failing. Further, AI-enabled ECG helps to perceive faulty heartbeats before any other symptoms become visible (https://healthitanalytics.com/features/howto-useartificial-intelligenceforchronic-diseasesmanagement).

2.5.9 AI IN CANCER DETECTION AND TREATMENT

AI is being used for the purposes of cancer detection. It has been found that AI can correctly detect and diagnose various types of cancers by evaluating the scans of corresponding tissues. The diagnostic outcomes have been found to be much better than pathologists. In one of the popular case studies, around 13,000 images of colorectal cancer were collected from 13 cancer centres in three countries: China, Germany, and the United States. A total of 8,803 subjects were chosen. These images were used to build a model based on a machine learning algorithms. This model was able to identify the images of colorectal cancer that causes most of the cancer-related deaths in the USA and Europe. A study was conducted that showed that AI-based models had an accuracy of 98% while identifying colorectal cancer, as compared to 96.9% when it was done manually by pathologists. AI also supports medical practitioners to formulate a treatment plan for the cancer patients, in addition to utilizing predictive analysis to find out the patient's response to different medications. This predictive analytics saves patients from needless side effects that otherwise would have caused suffering if the medicines were unacceptable to the body. This is a very important benefit of AI as every patient will have a different response to the same medication, so machine learning and predictive analytics come to their rescue (Wani et al., 2022).

2.5.10 AI in Managing and Preventing Diabetes

A lot of research has gone into finding various AI tools and techniques for diabetes management and prevention. Different diabetes management strategies include management by self, monitoring the patient remotely, and regular measurements using wearable sensors. In one of the studies, Rensselaer Polytechnic Institute collected data from hundreds of continuous blood glucose monitors and insulin pumps. This technique of continuously monitoring the glucose levels is wearable sensor technology that eases the process of monitoring blood sugar (blood glucose) levels over a long period of time. These wearable sensors are put under the skin using either an applicator or adhesive tape and measure blood glucose levels of the fluid present under the skin. The sensed data is wirelessly transmitted to a handheld device (can be smartphone, pad, insulin pump) where one can easily view the readings. The real-time blood glucose levels and its history can be downloaded to a computer system whenever needed. The data can be sent either continuously or in burst. Insulin pumps can release the insulin to the body as per the directions given to the insulin pump. Those with This is needed because Type 1 diabetes patients need to regularly test their blood glucose or sugar levels to find out the quantity of to be injected via insulin pump. Both continuous blood glucose monitors and insulin pumps are now being enabled through AI so that managing diabetes is much easier and treatment decisions made are much more accurate.

According to researchers, AI-powered mobile health tools have also been developed and have shown considerable improvements in managing the diabetes on their own, which considerably reduces the time and energy to visit doctors.

Apart from treating and manging diabetes, AI also plays a crucial part in preventing diabetes in the first place. One of the research worksthe done by Marshall University says that success rate of diabetes prevention and management depends on the ability to recognize high-risk patients accurately and in it is where AI plays an important role.

2.5.11 AI in Patient Care

AI plays a key role in patient care, such as telemedicine, the field of remote patient monitoring. With the help of AI-powered sensors and devices, healthcare providers can monitor patients' vital signs and other health data in real time, and intervene when necessary to prevent complications and provide speedy and timely treatment.

AI is also being used to advance both patient engagement and education on various facets of healthcare. For example, AI-powered chatbots are used to provide responses to patients' queries in addition to regularly providing them with the update on their health status, treatment choices, and recent advances in the health-related topic of their interest. This leads not only to improving patients' basic understanding of their health condition but also enables them to have a more caring mindset towards their health situation and hence become more sensitive towards playing a proactive role in their self-care (Wani et al., 2022).

As AI in healthcare has grown in its abilities and competencies, using it to improve medical practices has become progressively feasible. With the development of AI-powered medical tools and intelligent algorithms that are able to interpret large data sets, the potential for using AI in healthcare is limitless. Deep learning AI is being used to (i) detect diseases much quicker, (ii) develop customized treatment plans for a patient, (iii) automate drug discovery procedures, and (iv) better disease diagnostics. It also leads to improved patient treatment outcomes, enhanced safety, and reduced costs of efficient healthcare delivery.

The future of using AI in healthcare is undoubtedly bright and filled with possibilities for further innovation. As we move forward into a more connected digital world, using AI in the healthcare industry will become an invaluable asset that could potentially reshape how doctors treat patients and deliver care. With such great potential, it is clear that using artificial intelligence in the domain of healthcare holds the promise of a future filled with advancements, improved health outcomes, and better patient experiences (Vyas et al., 2023b).

Integration of AI and healthcare ecosystem leads to a huge set of benefits, such as (i) automating of day-to-day tasks, (ii) analyzing huge data sets of patient's health data, (iii) delivering better healthcare in much less time, (iv) reducing the cost of treatment, (v) getting better treatment outcomes because of accurate disease diagnosis, and (vi) reducing the number of patients by predicting the occurrence of diseases. AI can also automate all the administrative activities, such as (i) approving insurance, (ii) tracking unpaid bills, (iii) maintaining records, and (iv) simplifying the work of healthcare providers. AI's capability to process enormous data sets and fusing patient insights leads to better prediction and hence supporting healthcare ecosystem to discover crucial areas of treatment care that require attention and improvement.

Recent days have seen the exponential increase in the usage of wearable healthcare technology. This technology deploys AI tools and techniques to serve the patients in a much better way. Wearable device like FitBits and smartwatches analyze data and alert users and their healthcare providers of possible health concerns and risks. Being empowered to regularly assess self-health parameters with the help of wearable technology eases the work pressure of healthcare providers and avoids unnecessary hospital appointments or remissions (Mirbabaie et al., 2021).

2.6 CHALLENGES AND LIMITATIONS

As discussed above, AI has an immense potential to transform the health sector. At the same time, a considerable number of challenges and limitations also exist. Three of the major tasks are (i) lack of homogenous and standard format for data; (ii) if the data used to train these algorithms is biased, it is likely that the ensuing predictions may also be biased, resulting in wrong diagnosis and treatment; and (iii) interoperability between various healthcare systems. This makes it hard to design and develop AI-enabled solutions that can be implemented seamlessly across different equipment, machines, and organizations. Apart from the above, other challenges include bias in AI algorithms and ensuring that AI-enabled solutions being deployed are safe and ethical. Nowadays, extensive work is being done to develop guidelines and standards for the design and deployment of AI-enabled solutions that ensure both patient safety and privacy.

2.7 CONCLUSION

AI is revolutionizing healthcare in multiple and varied ways including disease diagnosis, drug development, treatment, prognosis, patient care, and tele-medicine. AI is enabling health professionals with real-time data and much deeper insights into this data that help them to make informed decisions. This is normally accomplished by building intelligent algorithms capable of processing a huge amount of clinical information. AI is still furthering healthcare delivery organizations by applying cognitive technologies to unlock and dig deep within a huge quantity of medical records and achieve power diagnosis. As an example, Nuance, the prediction service provider, employs tools such as AI, ML, and DL for discerning the intent of patients. Also, an attempt is to find associations in the data and define prediction models to be used in scientific research and innovative healthcare areas such as predictive medicine, personalized treatments, and diagnostic imaging. While there are some challenges in employing AI in the area of healthcare, the benefits certainly outweigh them. As technology con-tinues to advance, AI's role in the future of healthcare is definitely going to increase. Applying AI solutions in the healthcare domain results in improving not only disease diagnosis, treatment, and patient care, but also processing valuable data to programming surgical robots to performing much more precise surgeries.

REFERENCES

Bajwa, J., Munir U., Nori A., & Williams, B. (July 2021). Artificial intelligence in health-care: transforming the practice of medicine. *Future Healthcare Journal*, 8(2), e188–e194. 10.7861/fhj.2021-0095. PMID: 34286183; PMCID: PMC8285156.

Gatta, R., Orini, S., & Vallati, M. (2022). Process mining in healthcare: Challenges and promising directions. In T. Chen, J. Carter, M. Mahmud, & A.S. Khuman (Eds.), *Artificial intelligence in healthcare. Brain informatics and health*. Singapore: Springer. 10.1007/978-981-19-5272-2_2

Javaidm M., Haleem, A., Singh, R.P., Suman, R., & Rab, S. (2022). Significance of machine learning in healthcare: Features, pillars and applications. *International Journal of Intelligent Networks*, 3, 58–73. ISSN 2666-6030, 10.1016/j.ijin.2022.05.002.

Mirbabaie, M., Stieglitz, S., & Frick, N.R.J. (2021). Artificial intelligence in disease diag-nostics: A critical review and classification on the current state of research guiding future direction. *Health Technology*, 11, 693–731. 10.1007/s12553-021-00555-5

Panesar, A. (2023). Risks and ethical challenges of precision health. In *Precision health and artificial intelligence*. Berkeley, CA: Apress. 10.1007/978-1-4842-9162-7_5

Reddy, S., Fox, J., & Purohit, M.P. (2019). Artificial intelligence-enabled healthcare delivery. *Journal of the Royal Society of Medicine*, 112(1), 22–28. 10.1177/014107681 8815510

Shah, P., Kendall, F., Khozin, S. et al. (2019). Artificial intelligence and machine learning in clinical development: A translational perspective. *NPj Digital Medicine*, 2, 69. 10.1038/s41746-019-0148-3

https://healthitanalytics.com/features/howto-useartificial-intelligenceforchronic-diseasesmana
      gement

Tiwari, A., Chaudhari, M., & Rai, A. (2019). Multidisciplinary approach of artificial intel-
      ligence over medical imaging: A review, challenges, recent opportunities for research.
      In *2019 Third International Conference on I-SMAC (IoT in Social, Mobile, Analytics
      and Cloud) (I-SMAC)*, Palladam, India (pp. 237–242). 10.1109/I-SMAC47947.2019.
      9032566.

Vyas, S. (2023). Extended reality and edge AI for healthcare 4.0: Systematic study. In
      Extended reality for healthcare systems (pp. 229–240). Academic Press.

Vyas, S., Bhargava, D., & Khan, S. (2023a). Healthcare 4.0: A systematic review and its
      impact over conventional healthcare system. *Artificial Intelligence for Health 4.0:
      Challenges and Applications*, 1–17.

Vyas, S., Gupta, S., & Shukla, V. K. (March 2023b). Towards edge AI and varied approaches
      of digital wellness in healthcare administration: A study. In *2023 International
      Conference on Computational Intelligence and Knowledge Economy (ICCIKE)*
      (pp. 186–190). IEEE.

Wani, S.U.D., Khan, N.A., Thakur, G., Gautam, S.P., Ali, M., Alam, P., Alshehri, S.,
      Ghoneim, M.M., & Shakeel, F. (March 24, 2022). Utilization of artificial intelligence
      in disease prevention: Diagnosis, treatment, and implications for the healthcare work-
      force. *Healthcare (Basel)*, *10*(4), 608. 10.3390/healthcare10040608. PMID: 35455786;
      PMCID: PMC9026833.

3 Applications of Healthcare Products with AI Capability in Disease Diagnosis

Vimal Bibhu, Lipsa Das, and Shailaja Salagrama

3.1 INTRODUCTION

Artificial intelligence (AI) is a computer science discipline with the enormous capacity to provide different emergent applications in the field of research, innovation, automation, and disease diagnosis. There are many sub-branches of artificial intelligence, such as supervised, unsupervised, deep, and machine learning, respectively. Machine learning is one of the promising areas of artificial intelligence that is used to learn new patterns and new things by analyzing a huge amount of data. In the case of disease diagnosis, machine learning needs a huge amount of relevant disease information to train the machine to determine the probabilistic aspect of disease with a given sample. The manual process of disease diagnosis is indeed providing more accurate results than that of this AI-based technologies. But, in the near future, it is possible that the diagnosis of diseases can be performed by AI with accuracy as by the manual process of diagnosis. Machine learning is the main technological framework for training the machine to predict the result of a particular disease (Vyas et al., 2023).

Treatment planning for the patient requires accurate diagnosis and health records. This ensures the well-being of the patients and also enhances the success of the treatment time with the correct drugs and other procedures of the treatment. In the case of manual diagnosis, there is a chance of human errors, which leads to unsuccessful treatment, even if the health records are available (Diederich et al., 2019). AI can prevent and limit errors to increase the efficacy and accuracy of the diagnosis and reporting of individual patient health information. In rural areas, the people having the disease are not able to explain the real health issues and this produces complexities for the healthcare professionals to get accurate symptoms from the patient. Therefore, the AI-based system can abate the challenges in such conditions where patients are not able to completely dictate their health issues and other factors needed in the diagnosis process. The techniques of AI, such as convolutional neural networks, graphs of the knowledge discovered applying the AI-based diagnosis tools, and transformers can also be powerful promising tools that

DOI: 10.1201/9781003346289-3

assist in diagnosis with improved data to achieve the successful treatment to the patients to their different diseases.

Medical experts are indeed required to apply AI-based diagnosis tools to run the processes of diagnosis to the patients to get accurate results of diagnosis. AI brings the digitization process of health information, which can be communicated easily to the real stakeholders to prescribe the drugs to treat the diseases associated with patients. It is also true that AI-based tools' accuracy, reliability, and validation of the diagnosed reports remain a challenge in the medical field (Vyas & Gupta, 2023). The accuracy and correct outcomes heavily depend upon the conditions, such as the parameters; historical data availability and its accuracy; analysis of the cross-modality; examination techniques with novelty, effective transformation of the advanced stage computer vision; and finally the technologies used in machine learning to train the machine to diagnose the disease and symptoms.

Integration of AI with the existing healthcare infrastructure is not a simple framework to model and deploy. This needs the effective analysis and availability of disease-based accurate and correct data banks to the diagnosis centers to utilize them to match the score to predict the disease and find the real cause and symptoms (Preece et al., 2017). Applications based on AI are of great support to medical specialists in the healthcare area to speed up the discovery of patient disease and quickly prescribe a drug or other healthcare measure to prevent the progression of the existing issue and meet the cure and prevention of fatality very soon. The healthcare data for analysis and decision making with the help of machine learning applications can be taken from various sources, but this is again the challenge for the specialist to determine whether the available healthcare data is accurate or not. This challenge needs to be addressed to make this a more effective and accurate diagnosis (Nasirian et al., 2017).

Despite the importance shown through AI in the field of medical and healthcare diagnosis, accurate algorithms based on machine learning and others are lacking in the healthcare field. Further, the specific disease-based diagnosis and the tools of AI are not available with the correctness of the diagnosis results and outcomes (Dilsizian & Siegel, 2013). Therefore, it is a much more straightforward subject in the case of AI in disease diagnosis to go with more research and innovation to find out disease-specific diagnosis tools based on AI and machine learning. Recent studies say that AI applications in the field of healthcare and medicine in general or with the specific domains of clinical trials (Jiang et al., 2017).

3.2   BASICS OF ARTIFICIAL INTELLIGENCE AND MACHINE LEARNING

AI is also a branch of science where the intelligence in the object is produced artificially by human beings. Artificial intelligence is not a new technique; it evolved over many decades. Some researchers say that artificial intelligence is science and engineering to make intelligent systems and machines, mainly intelligent computer programs and applications. Also, it is a fact that AI does not confine itself to those methods that are biologically observable. The birth of artificial intelligence conversation is denoted by seminal work performed by Alan Turing.

It is the primary question of whether a machine can think. The answer is straightforward; a machine thinks when it is modeled by AI to produce the specific intelligence to think. The approach behind the human is that a system that thinks like a human being and the system which acts like a human being. This is because a system and science and engineering defined intelligence integrated with the systems or machines so that these can think like a human. Therefore, it is a fact that AI is an area that integrates computer science and data sets to enable the concept of solving the problem. AI includes the sub-phases of machine and deep learning.

3.2.1 ARTIFICIAL INTELLIGENCE TYPES

There are two types of artificial intelligence. These two types are weak and strong artificial intelligence, respectively. A weak AI is also known as narrow AI or narrow artificial intelligence (ANI). A narrow AI-based system performs specific tasks only and enables very strong and robust applications. For example, Apple's Siri, Amazon Alexa, Watson by IBM, and autonomous vehicles are developed with the help of weak AI (Rong et al., 2020).

Artificial general intelligence (AGI) is used to make strong AI. This strong AI is also known as artificial super intelligence (ASI). ASI is a theoretical form of AI by which a machine would have intelligence equal to a human being. The machines and systems based on the strong AI have the self-aware consciousness having the ability to solve problems and plan and learn the future. ASI is also called super-intelligence, which can surpass the intelligence of a human being's brain (Frick et al., 2021). It is completely theoretical and does not have any real-world examples to date. Science fiction, like HAL, rogue computer assistants, and superhumans are the best examples of superintelligence.

3.2.2 DEEP LEARNING VERSUS MACHINE LEARNING

Deep learning and machine learning are processes of artificial intelligence used interchangeably. Both deep and machine learning are the types of artificial intelligence and actually, deep learning is the subfield of machine learning. The interchanged aspects of deep and machine learning are shown in Figure 3.1.

Machine learning is the main AI system where deep learning comprises the sub-parts and basically, it is based on the neural network. There are more than three layers in deep learning that include input and output and are considered deep learning algorithms. The representation of deep learning with its layers is presented in Figure 3.2 (What is Artificial Intelligence (AI)?).

The basic difference between deep learning and machine learning algorithms is how the system is being modeled and learned. Deep learning makes more automation by feature extraction for the process that reduces some of the human intervention that is required to enable the use of huge data sets. Machine learning is more dependent on the intervention of human beings where the experts determine the hierarchy of the features to do an understanding of the difference between the data inputs, whereas deep machine learning leverages the labeled sets of data

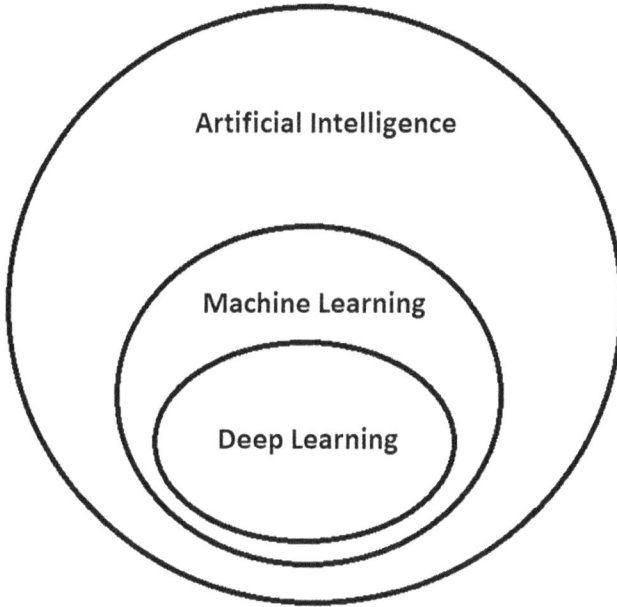

FIGURE 3.1 AI, machine, and deep learning framework.

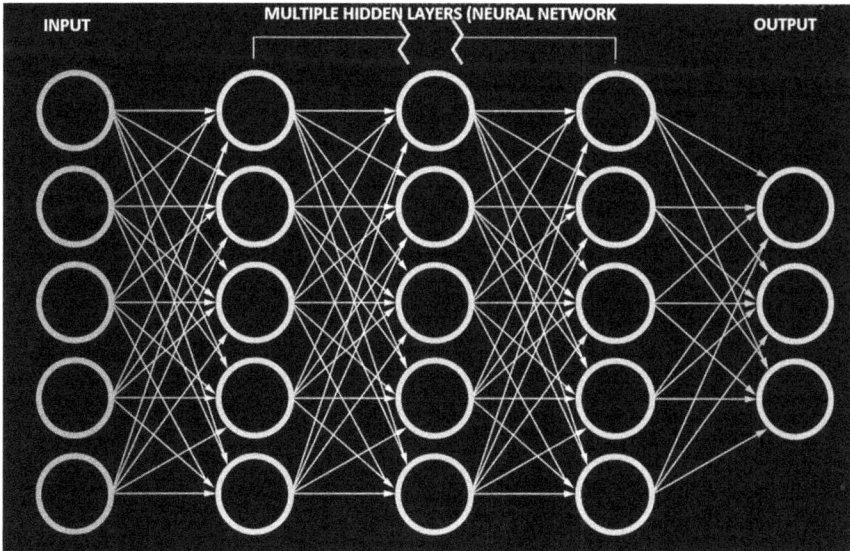

FIGURE 3.2 Multiple layers of neural networks of deep learning (What Is Artificial Intelligence (AI)?).

and is also termed *supervised learning* to make information to the algorithm. Deep machine learning ingests the non-structured sets of data in raw form such as text and images and then determines the hierarchy of features to determine the different categories of data from one to another.

3.3 CLINICAL VERSUS AI-BASED DISEASE DIAGNOSIS

In the field of medicine, disease diagnosis is performed by accepted medical technology through sample collection and discovery in the laboratory. But, in the case of AI-based disease diagnosis, the data sets play an important role to predict the disease.

3.3.1 CLINICAL DIAGNOSIS

Clinical disease diagnosis is the traditional mode of diagnosis where the samples are taken and tested in the laboratory by using various required chemicals to create the reports. Clinical diagnosis is a time-consuming and non-automated version of disease diagnosis. There are no disease prediction features used in the clinical disease diagnosis process. For example, the blood samples of patients are taken to determine the various aspects related to the patient's blood, like RBC count, hemoglobin, enterocyte sedimentation rate, leucocyte count, red blood cell and white blood cell variability, vitamins B and D, and other values associated with the blood. This type of clinical process of diagnosis is a manual process where the samples are taken from the patient.

Systematic study of the patient and its samples, which includes anamnesis purposeful questions concerning the patient's complaints, condition, and history related to patient complaints, physical examinations such as inspection, palpation, percussion, testing of auditory and visual by special devices, joint movements range test, analysis of the results of laboratory tests performed by laboratories through blood samples, urine examinations of patient, X-ray, CT scan, MRI, etc. These methods are based on laboratory and testing machines to evaluate the patient's medical condition as per the patient's complaints.

The diseases are identified with a large number of symptoms and the combinations of symptoms are extracted to determine the intensity of the disease of the patient. It is also a fact that different categories of diseases are related to the same symptoms facing the patient. Hence, for medical experts, it becomes a complex task to find out the disease during the diagnosis and the characteristics of the same symptoms.

The errors due to insufficient experience in the field of diagnosis are being avoided by leaving the rare to rarest disease and the help of computer analysis by computer vision is taken out to predict the similarity of the rare disease through the data set available and features extraction from the samples such as X-ray, CT scan, and MRI images. Therefore, it can be said the clinical practice of disease diagnosis is a human intervention-based diagnosis where machines and chemicals are used to find out the detailed profile of the patient to get the disease causes and types of diseases associated with the individual patient.

3.3.2 AI-Based Diagnosis

Artificial intelligence is a science that provides training and models created with the help of computer programs and applications to work as per the given instruction to get intelligent behavior about the system. AI-based disease diagnosis is considered an efficient and effective methodology where accuracy may be better than that of the clinical practices of the disease diagnosis. AI and ML-based programs are designed in such a way that the system integrated with them gets the intelligence to diagnose the specific disease of the patient. This diagnosis is based on the training of the machine with historical data associated with the patient, lab results, and symptoms by scanning images that are confirmed of susceptible cases.

AI is adding swift diagnosis, which facilitates the ease for medical practitioners and saves cost and time for the patients and doctors. Natural language processing (NLP), digital image processing, and AI-based applications in the field of healthcare can be commonly used in the case of disease diagnosis. Computer vision is used to find and detect the various diseases of human beings.

3.4 DEEP LEARNING AND DISEASE DIAGNOSIS

The process followed by the medical practitioner for decades is based on the analysis of disease, performing laboratory tests, and referring to the diagnosis guidelines. These stepwise processes are the traditional mode of disease diagnosis and treatment by medical practitioners around the globe to treat various categories of diseases or ailments (Gupta & Bibhu, 2023). According to the advancement of machine learning and deep learning methodologies under AI, the data scientist is developing algorithms to support disease diagnosis by employing the training model.

3.4.1 Diseases Diagnosed by AI-Based Techniques and Deep Learning

There are many categories of diseases diagnosed with the help of AI-based techniques (Vyas & Gupta, 2023). Machine learning and deep learning techniques are used to diagnose the following diseases related to human beings.

3.4.1.1 Cancer Diagnosis

Deep learning algorithms are used to diagnose the various categories of cancer diseases. Melanoma is a type of cancer diagnosed by applying deep learning and convolutional network techniques. The skin lesion images through processing of images, and semantic segmentation along with machine learning algorithms based on deep learning techniques are being employed to diagnose the presence of the tumor at an early stage of its prognosis. X-ray-based deep neural networks are also being trained to detect tumors in the human body (Gupta & Bibhu, 2023).

3.4.1.2 Heart Ailment Diagnosis

Computer vision is used by data scientists to diagnose heart disease. The technique involved with computer vision is based on deep learning and machine learning training models. The algorithms of deep learning are used to diagnose heart disease. The coronary arteries disease (CAD) is a condition in the human heart where plaque is deposited and built up under the blood vessels those supply or work as a pump line to supply the oxygen-mixed blood to the heart and human brain. Medical scanning is taken with the help of CT scans and magnetic resonance imaging (MRI) machines. These scanned images are used by a data scientist to train the convolutional neural network to diagnose the CAD.

3.4.1.3   Liver Ailment Diagnosis

The liver is a single organ in the human body and works for the management of metabolism and many other processes of human anatomy. An AI-based mechanism called computer vision includes deep learning algorithms that are used to diagnose the disease associated with the human liver. For example, the deep neural network is being trained by a data scientist to make the analysis of ultrasonography images for finding out fatty liver disease and hepatic steatosis through liver texture and echogenicity.

3.4.1.4   Lung Disease Diagnosis

Images of the human lung are used to detect the disease associated with it. The diseases such as tuberculosis, masses within a lung, pulmonary nodules, etc. are diagnosed by imaging through CT scans and X-ray scan studies by employing deep learning AI techniques. The features are extracted from the images taken by CT scan and X-ray images of the lung to diagnose the disease associated with it.

3.4.1.5   Autistic Disorder Diagnosis

Autism disorder is a common human disease diagnosed by employing neuro-imaging and machine learning algorithms. Here in the process of its diagnosis, computer vision techniques like face detection, feature analysis of the facial image, and eye tracking methods are applied to diagnose the autism spectrum disorder associated with human beings.

3.4.1.6   Diabetic Retinopathy Diagnosis

Diabetes is a very common disease among human beings around the globe. In India, one in every five people suffers from type 1, type 2, and pre-diabetes, respectively. In diabetes, the tiny blood vessels connecting to the retina of the human eye are very much impacted and, due to this, retinal detachment and other issues of the retina happen. The retina blood vessels are damaged due to elevated blood sugar levels in the human body. AI-based diagnosis methods along with computer vision like image segmentation, classification of retina disease, and deep learning–oriented machine learning algorithms are used to diagnose the retina disease.

3.4.1.7   Psoriasis Diagnosis

Psoriasis is a skin disorder where the cycle of skin development is enhanced by four times what is normal. Therefore, skin rashes and skin cracking appear on the skin of the human body. The real cause of this disease is unknown but, in the medical system, it is considered an autoimmune disorder. A deep learning model is used to develop the disease diagnosis algorithms that are based on the affected skin area imaging and lesions caused due to psoriasis.

3.4.1.8   Alzheimer's Diagnosis

Alzheimer's disease can be diagnosed by employing AI and extracting the features from natural speech and by using the machine learning and deep learning algorithms. The machine and deep learning algorithms are based on deep neural networks and support vector machines (SVMs).

3.4.1.9   Parkinson's Diagnosis

The movement-associated features of humans are accounted for and deep learning algorithms are applied to diagnose Parkinson's disease. For example, the networks based on the convolution neural concept are being trained by video data of the patient to predict the progression of this disease and also used to monitor the stress and depression that are common symptoms of Parkinson's disease.

3.4.2   DEEP LEARNING TECHNIQUES USED FOR DISEASE DIAGNOSIS

There are many deep learning techniques to diagnose different kinds of human diseases. These techniques are derived from AI and provide good accuracy and efficiency in disease diagnosis. Some of the common deep learning techniques for human disease diagnosis are explained below.

3.4.2.1   Convolutional Neural Network (CNN)

A convolutional neural network (CNN) is a deep learning methodology that is being used for diseases diagnosis. CNN is considered a type of artificial neural network with three different layers. These three separate layers are convolutional, pooling, and fully linked. A disease-specific feature extraction model is built by applying CNN to diagnose the diseases. With the help of medical images, CNN can also be used to predict the disease and discover the drug for the specifically identified disease. For example, CNN includes the classification of disease, segmentation of disease, and identification of disease. CNN uses images for diagnosis of the human body diseases. Skin and breast cancers are heart-related ailments and diseases that are diagnosed with the help of a deep learning technique known as CNN. An example of shallow-deep CNN for improved breast cancer diagnosis is presented in Figure 3.3 (Gao et al., 2018).

3.4.2.2   Fully Convolutional Network (FCN)

Using textual information from patient medical records stored in electronic health record systems, FCN, a deep learning technique in the field of AI, can be utilized to identify diseases (Khurana et al., 2021). Although it is a form of convolutional neural network, its deep learning method does not call for segmentation. It is directly applicable to issues with disease diagnosis.

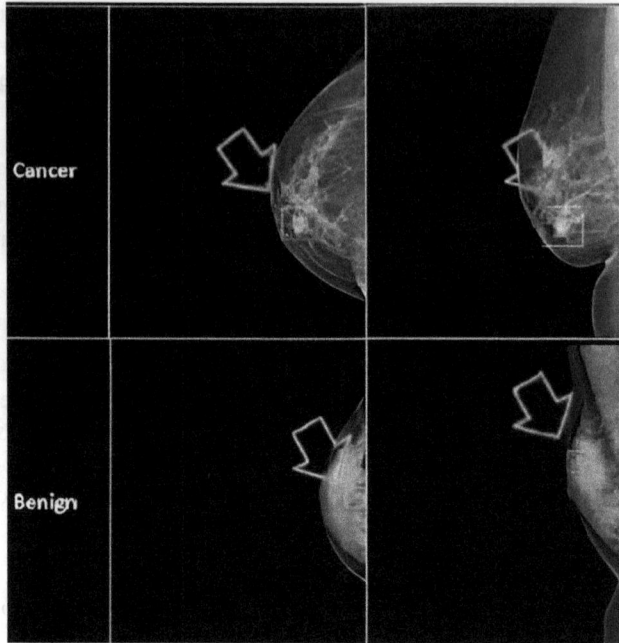

FIGURE 3.3 A shallow deep CNN for improved breast cancer diagnosis (Gao et al., 2018).

3.4.2.3   Recurrent Neural Network (RNN)

A popular deep learning method for language processing applications, such as machine translation models, is RNN. The recurrent neural network, or RNN, cells are composed of at least two stacks of long short memory units. Making disease-specific language models can help with disease diagnosis (Khurana et al., 2021). The disease-specific model consists of disease symptoms or medical records of patients. For example, Alzheimer's, Parkinson's, and Crohn's diseases can be diagnosed with the help of building a disease-specific prediction model.

3.4.2.4   Dilated Convolution Technique

Dilated convolution is also a deep learning technique under AI used for the diagnosis of diseases. The dilated convolutional neural network uses the weight matrix with an adjustable size, such as dilation. Dilation convolutional neural networks are used widely in the computer vision and processing of image-based applications such as object detection or segmentation issues (Shivahare et al., 2022). It uses disease classification and disease segmentation to diagnose the diseases.

3.4.2.5   Generative Adversarial Network (GAN)

A deep learning method called GAN is utilized to both anticipate when a disease would start and to detect it. It is composed of two neural networks, the first of which creates samples and the second of which serves as a discriminator. The discriminator evaluates the samples that are generated by the firm neural network of GAN.

Medical images are used with GAN to detect the disease. The medical images were taken by MRI, CT, and X-ray. The prediction of disease GAN uses the symptoms and the medical records of the patient (Shivahare et al., 2022). These two symptoms and medical records are used to develop the predictive model for the onset of disease. It can be used to detect or diagnose diseases like leukemia and heart ailments.

3.4.2.6   Classifier GAN (AC-GAN)

A deep learning method for diagnosing human diseases is called AC-GAN. Adverbial networks of the AC-GAN variety have generative and discriminative networks as their two main competitors. Due to the fact that it draws knowledge from both generating and discriminative networks, it produces findings that are more accurate than those of the GAN. The disease is predicted by building disease-specific feature extraction models by the use of the image of the disease, symptoms, and patient medical records (Shivahare et al., 2022).

3.4.2.7   Convolutional Auxiliary Classifier—GAN

Another deep learning method that uses the multi-label classification issue is CAC-GAN. Disease diagnosis and disease prediction are both included in the multi-label classification problem. In order to diagnose the condition, classifier networks operate in conjunction with generator networks to produce images of the disease, symptoms of the disease, or patient medical information.

3.4.2.8   Attention-Based Deep Neural Network

Another deep learning method that can be used to identify the disease is attentional NN. A particular kind of neural network called an attentional NN makes advantage of the attention mechanism to incorporate input data into output prediction. It produces both disease diagnoses and therapeutic strategies.

3.4.2.9   Adversarial Autoencoder (AAE)

It is a deep learning technique used to build the disease-specific feature extraction model. There is an encoder and a decoder with AAE. The encoder transforms disease images or the disease symptoms. The decoder performs the reverse of the encoder from the features back to the disease images or disease symptoms.

3.5   ARTIFICIAL INTELLIGENCE AND RADIOLOGY

The burnout of physicians and medical diagnosis experts has been reduced with the help of AI-based medical diagnosis of the different categories of diseases and ailments of the patients. AI provides gigantic support to the radiologist to manage the workloads that always increase. Imaging is the process based on signal and image processing that increases the images that require the correct dealing with volumes. Prioritization is also a requirement to handle the urgency with the patients to provide the correct images of the diagnosis. Further creating the reports on the diagnosis through radiology is another burden to a radiologist to cater timely to a given priority for the specific patients with the fatality. The AI-based tool supports digital imaging correctly and reduces biases while proceeding with the imaging process for the patient. All the imaging systems automatically correct even human error is a chance to produce errors in imaging of the patient's organs, etc. AI and deep learning provide effective and efficient solutions for the radiologist to manage the heavy image volumes and reports associated with each of the images. It also provides streamlined workloads as per defined priority to complete the urgency of the individual patient to provide the report and image timely as per physician recommendation to start timely treatment to the well-being of the patient. The diagnosis reliability, time, and other factors associated with the radiology processes are managed very effectively with the help of AI and deep learning algorithms. The work pressure on the radiologist is significantly abated by the help of AI tools. For example, Aidoc goes through an AI and deep learning–based tool that can effectively manage a huge number of specialized radiology images with the help of computers and applications (Khurana et al., 2019). Fagging is another associated function based on AI that provides the radiologists a matter of priority and concern of what to do first to provide the image and report to the concerned patient to save lives and provide well-being. It is confirmed that aid has the six FDA and CE clearness to flag and prioritize the patients with acute issues and abnormalities of City Scans. This is already employed globally in over 400 medical facilities. The process of Aidoc in the field of radiology and medical imaging with huge volume and priority management is shown in Figure 3.4 (Temple Health Collaborates with Aidoc to Reduce Cost of Care, 2022).

It is also true that AI in the field of medical and disease diagnosis is still on the edge of its potential. This further requires more technological growth and advancement to create more accurate imaging of the various organs of patients to go with the optimized success factors in the treatment by medical experts and doctors. Therefore, in the near future, researchers in the field of computer science and information technology will innovate such advanced tools and technologies based on artificial intelligence and deep learning to enhance the accuracy and more effective management of the large volume of images and medical records in the field of radiology and imaging (Temple Health Collaborates with Aidoc to Reduce Cost of Care, 2022).

AI-based algorithms are automated so that in the process of image acquisition of radiology, it provides the auto-correction of the noise and biases of the image. It reduces the operator burden by automating the image acquisition to image correction and also creates the indexing of the image as per the priority and flagging-based storage in the storage medium.

3.6 CONCLUSION

Artificial intelligence is one of the promising disciplines in the field of science and engineering to create intelligence in systems and machines. In the medical field, artificial intelligence can be a boon to quickly predict and diagnose the disease of the patient. The detection of the disease of the patient requires the imaging and data sets related to the disease to apply machine learning to train

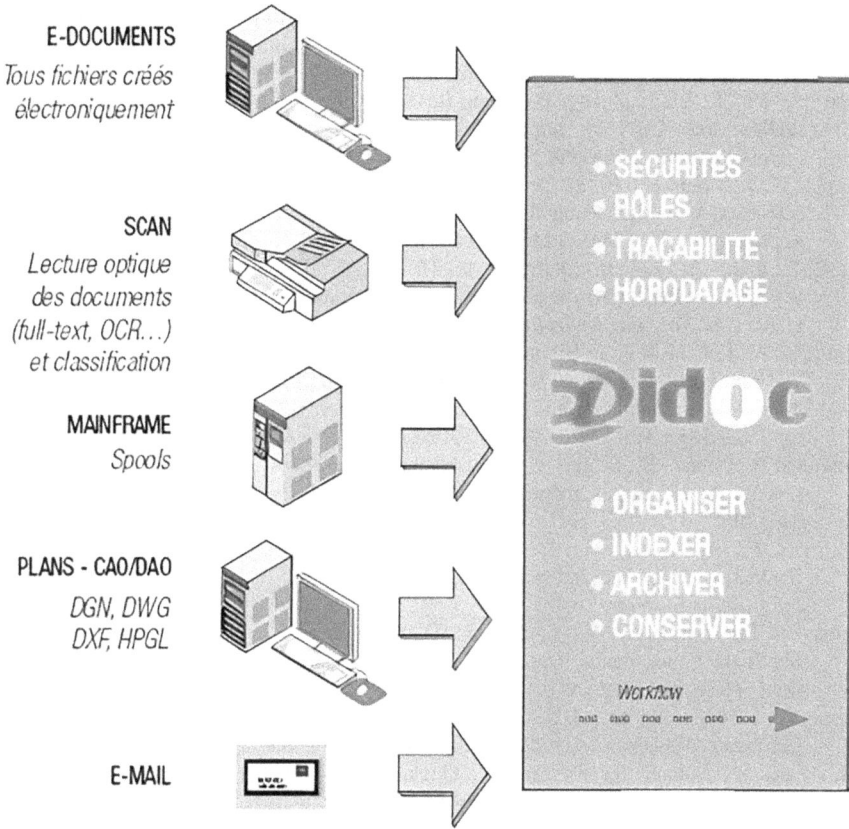

FIGURE 3.4 Aidoc process in the field of radiology data management (Temple Health Collaborates with Aidoc to Reduce Cost of Care, 2022).

the model of the algorithm related to neural networks and supervised learning to get the diagnosis of the patient disease. AI is the main field and machine, and deep learning is the sub-area of it. Nowadays, there is so much AI-based medical equipment available to monitor and diagnose the patient's disease. The different techniques are CNN, GAN, FCN, RNN, FC-GAN, etc., and are AI-based and used to develop the modeled algorithms to diagnose the patient disease as per the samples and images by matching and analyzing the patterns and features extricated from them.

REFERENCES

Diederich, S., Brendel, A.B., & Kolbe, L.M. (2019). On conversational agents in information systems research: Analyzing the past to guide future work. *Wirtschaftsinformatik*.
Dilsizian, S.E., & Siegel, E.L. (2013). Artificial intelligence in medicine and cardiac imaging: Harnessing big data and advanced computing to provide personalized medical diagnosis and treatment. *Current Cardiology Reports*, *16*, 1–8.

Frick, N.R., Brünker, F., Ross, B., & Stieglitz, S. (2021). Comparison of disclosure/concealment of medical information given to conversational agents or to physicians. *Health Informatics Journal*, *27*, 1460458221994861.

Gao, F., Wu, T., Li, J., Zheng, B., Ruan, L., Shang, D., & Patel, B. (2018). SD-CNN: A shallow-deep CNN for improved breast cancer diagnosis. *Computerized Medical Imaging and Graphics*, *70*, 53–62. 10.1016/j.compmedimag.2018.09.004

Gupta, V., & Bibhu, V. (2023). Deep residual network based brain tumor segmentation and detection with MRI using improved invasive bat algorithm. *Multimedia Tools and Applications*, *82*, 12445–12467. 10.1007/s11042-022-13769-0

Jiang, F., Jiang, Y., Zhi, H., Dong, Y., Li, H., Ma, S., Wang, Y., Dong, Q., Shen, H., & Wang, Y. (2017). Artificial intelligence in healthcare: Past, present and future. *Stroke and Vascular Neurology*, *2*, 230–243.

Khurana, A., Lohani, B.P., & Bibhu, V. (2019). AI frame-worked virtual world application – The ramification of virtual world on real world. In *2019 International Conference on Automation, Computational and Technology Management (ICACTM)*, London, UK (pp. 582–585). 10.1109/ICACTM.2019.8776724.

Khurana, A., Lohani, B., Bibhu, V., & Kushwaha, P. (2021). An AI integrated face detection system for biometric attendance management. 29–33. 10.1109/ICIEM51511.2021. 9445295.

Nasirian, F., Ahmadian, M., & Lee, O.D. (2017). AI-based voice assistant systems: Evaluating from the interaction and trust perspectives. *Americas Conference on Information Systems*, 1–10.

Preece, A.D., Webberley, W.M., Braines, D., Zaroukian, E.G., & Bakdash, J.Z. (2017). Sherlock: Experimental evaluation of a conversational agent for mobile information tasks. *IEEE Transactions on Human-Machine Systems*, *47*, 1017–1028.

Rong, G., Mendez, A., Assi, E.B., Zhao, B., & Sawan, M. (2020). Artificial intelligence in healthcare: Review and prediction case studies. *Engineering*. *63*, 291–301.

Shivahare, B., Suman, S., Kaushik, P., Gupta, A., & Bibhu, V. (2022). Survey paper: Comparative study of machine learning techniques and its recent applications. 449–454. 10.1109/ICIPTM54933.2022.9754206.

Temple Health Collaborates with Aidoc to Reduce Cost of Care. (2022, October 31). Aidoc. https://www.aidoc.com/blog/news/temple-health-aidoc-ai-partnership/

Vyas, S., & Gupta, S. (2023). WBAN-based remote monitoring system utilising machine learning for healthcare services. *International Journal of System of Systems Engineering*, *13*(1), 100–108.

Vyas, S., Bhargava, D., & Khan, S. (2023). Healthcare 4.0: A systematic review and its impact over conventional healthcare system. *Artificial Intelligence for Health 4.0: Challenges and Applications*, 1–17.

What Is Artificial Intelligence (AI)? *IBM*. Accessed 28 April 2023. www.ibm.com/topics/artificial-intelligence

4 Application of AI for Disease Prediction

Gurpreet Singh, Darpan Anand, and Harsh Bansal

4.1 INTRODUCTION

Artificial intelligence (AI) is a branch of computer science that aims to build intelligent machines that can carry out tasks that traditionally require human intelligence, like comprehending natural language, identifying objects, and making judgements (Rajkomar et al., 2019). The creation of AI entails the creation of algorithms that can recognize patterns in data and learn from them to make predictions. AI has made significant progress in recent years due to advances in computing power, data storage, and data processing. With the availability of huge amounts of data and the capability to process that data quickly, AI algorithms have become increasingly powerful, accurate, and efficient (Esteva et al., 2019). Strong AI and weak AI are two broad categories of artificial intelligence that describe the level of intelligence and autonomy of AI systems.

- Strong AI: Strong AI is an AI system that possesses intelligence and reasoning abilities that are equivalent or surpass human-level intelligence. Such AI systems have the ability to understand, reason, learn, and adapt to new situations on their own without any human intervention. They can perform a wide range of tasks and possess a level of creativity and self-awareness that is similar to humans.
- Weak AI: Weak AI is an AI system that is designed to perform a specific task or set of tasks. Such AI systems are programmed to complete a narrow range of tasks within a specific domain, and they lack the general intelligence and reasoning abilities of human beings. Examples of weak AI include chatbots, image recognition algorithms, and recommendation engines.

While strong AI is the ultimate goal of AI research, it is currently not yet achievable. Most of the AI systems that are being developed today fall under the category of weak AI. However, even weak AI systems have the potential to automate a wide range of traditionally human-performed tasks and can be quite effective at tackling specific challenges.

DOI: 10.1201/9781003346289-4

```
                                            ┌──────────────────────┐
                                      ┌────→ │  Disease Prediction  │
                                      │      └──────────────────────┘
                                      │      ┌──────────────────────┐
                                      ├────→ │     Personalized     │
                                      │      └──────────────────────┘
┌────────────────────────────┐       │      ┌──────────────────────┐
│ AI Application in Healthcare │──────┼────→ │    Drug Discovery    │
└────────────────────────────┘       │      └──────────────────────┘
                                      │      ┌──────────────────────┐
                                      ├────→ │    Medical Imaging   │
                                      │      └──────────────────────┘
                                      │      ┌──────────────────────┐
                                      └────→ │    Medical Chatbot   │
                                             └──────────────────────┘
```

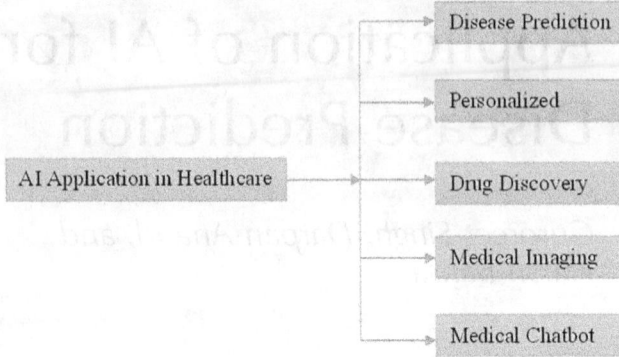

FIGURE 4.1 Some potential applications of AI in healthcare.

The healthcare industry is one of the largest industries globally, and maintaining the health of people worldwide has become a significant concern. A crucial component of healthcare is disease prediction since it enables early detection and intervention, which improves health outcomes and may even save lives (Shickel et al., 2018). Traditionally, disease prediction has relied on clinical factors such as family history, medical history, and lifestyle habits.

Artificial intelligence (AI) has been increasingly influential in recent years in the healthcare industry, with its potential to transform disease prediction, diagnosis, and treatment. Disease prediction, in particular, has been a critical area of focus for AI applications in healthcare (Miotto et al., 2017). Large quantities of clinical data, including lab results, medical records, and other data can be analyzed by AI algorithms to find trends and predict illness onset efficiently, promptly, and precisely. Figure 4.1 enlists some potential applications of AI in the field of healthcare.

4.2 IMPORTANCE OF DISEASE PREDICTION

Here are some of the key reasons why disease prediction is important for early detection and prevention:

a. Timely intervention: Early detection of diseases allows healthcare professionals to intervene at an early stage and implement preventive measures that can slow or halt the progression of the disease. This can help prevent or delay the onset of symptoms, complications, and adverse health outcomes.

b. Personalized treatment: Disease prediction can provide healthcare professionals with information about an individual's unique risk factors and health status. Using this data, customised treatment plans that are suited to the needs and circumstances of the individual can be created.

c. Cost-effective healthcare: Early detection and prevention of diseases can help reduce the need for expensive medical interventions such as surgeries, hospitalizations, and long-term treatments. Individuals, healthcare providers, and the healthcare system as a whole may all experience significant cost reductions as a result of this.

d. Improved quality of life: Early detection and prevention of diseases can help individuals maintain a good quality of life and avoid the physical and emotional burden of chronic illnesses. This can also improve their overall well-being and productivity.

e. Reduced healthcare disparities: Disease prediction can help identify individuals who are at high risk of developing certain diseases due to their genetic, environmental, or social factors. This can help healthcare professionals target interventions and resources to those who need them the most, thus reducing healthcare disparities.

4.3 TYPES OF AI ALGORITHMS

There are several types of AI algorithms that can be used for disease prediction, including:

1. Machine learning (ML): To find patterns and make predictions, ML algorithms examine enormous data sets of medical records, test results, and other clinical data. ML algorithms come in a variety of forms, including supervised learning, unsupervised learning, and reinforcement learning.

2. Deep learning (DL): Artificial neural networks are used in decision making, data analysis, and learning in DL, a subset of machine learning. When analyzing complex data sets like genomic data and medical images, DL algorithms are especially helpful.

3. Natural language processing (NLP): NLP algorithms analyze human language data, such as electronic clinical records and medical literature, to extract information and identify patterns.

4. Bayesian networks: Bayesian networks are probabilistic models that can be used to represent the relationship between variables, such as symptoms, diseases, and risk factors, and make predictions.

5. Support vector machines (SVMs): SVM is a type of ML algorithm that can be used to classify data into two or more categories, such as disease or non-disease.

6. Decision trees: ML algorithms called decision trees use a tree-like representation of decisions and their outcomes to forecast outcomes.

These algorithms can help identify patients at high risk of developing a disease, and they can also assist with early detection and diagnosis. Each type of AI algorithm has its strengths and weaknesses, and the choice of algorithm depends on the specific application and the available data. In disease prediction, ML and DL algorithms are commonly used due to their ability to analyse large data sets and identify complex patterns. NLP algorithms are also useful for extracting information from electronic health records and medical literature.

4.4 APPLICATION OF AI IN DISEASE PREDICTION

Using machine learning and other AI techniques to analyze vast volumes of data and uncover patterns that may be symptomatic of a specific disease or condition is a rapidly expanding subject in healthcare. Here are some instances of how AI is being applied to the prediction of diseases.

4.4.1 CANCER PREDICTION

Cancer prediction using AI is an area of active research that involves using machine learning and other AI techniques to analyze medical data and predict the likelihood of a patient developing cancer.

a. Medical imaging analysis: Mammograms and CT scans are two examples of medical imaging that AI algorithms can examine to find early indications of malignant cells or tumors. This can help doctors diagnose cancer at an earlier stage and develop personalized treatment plans. For example, AI algorithms can be trained to recognize the appearance of abnormal tissue growth, such as microcalcifications or masses, in mammograms.
b. Genetic analysis: AI algorithms can analyze genomic data, such as DNA sequencing data, to identify mutations that are linked with an amplified risk of developing a tumor. By analyzing large data sets of genomic data, AI algorithms can identify patterns that may be indicative of cancer and develop predictive models to estimate a patient's risk of developing cancer.
c. Electronic health records: AI algorithms can analyze patient health records, including information on family history, lifestyle factors, and previous medical history, to predict a patient's likelihood of developing cancer. By integrating data from multiple sources, AI algorithms can develop more accurate models for predicting cancer risk.
d. Liquid biopsy analysis: It is a non-invasive method of detecting cancer using blood samples rather than tissue biopsies. It involves analyzing circulating tumor cells, cell-free DNA, and other biomarkers that are present in the blood of cancer patients.

Below are some research works done with their outcome on cancer prediction using AI:

a. Smith et al. (2021) in their study developed three deep learning models and found that a hybrid CNN-RNN model outperformed the other two models, with an AUC of 0.87 for predicting breast cancer recurrence.
b. Li et al. (2020) in their study developed a random forest model and found that it had an accuracy of 83.7% in predicting lung cancer survival.
c. Kim et al. (2020) in their study developed a DL model and found that it had an accuracy of 79.8% in predicting pancreatic cancer survival.

d. Feng et al. (2020) in their study developed a support vector machine model and found that it had an AUC of 0.78 in predicting prostate cancer recurrence.

e. Wang et al. (2020) in their study developed a DL model and found that it had an accuracy of 80.2% in predicting ovarian cancer survival.

f. Yoo et al. (2019) in their study developed a logistic regression model and found that it had an AUC of 0.78 in predicting colorectal cancer recurrence.

g. Lee et al. (2021) in their study developed a deep learning model and found that it had an accuracy of 78.5% in predicting gastric cancer survival.

4.4.2 CARDIOVASCULAR DISEASE PREDICTION

AI algorithms can analyze patient data, such as blood pressure, cholesterol levels, and lifestyle factors, to predict a patient's potential for cardiovascular illness. This can help doctors develop personalized treatment plans to reduce the patient's risk.

a. Medical imaging analysis: AI algorithms may examine MRIs and CT scans to look for indicators of cardiovascular disease, such as plaque buildup in the arteries. This can enable earlier detection of the disease and help guide treatment decisions.

b. Clinical data analysis: A patient's risk of developing cardiovascular disease (CVD) can be predicted using AI algorithms that examine patient data such as medical history, lifestyle factors, and biomarkers. For example, an algorithm may predict that a patient with high BP and a family history of heart disease is at higher risk of developing CVD.

c. ECG analysis: AI algorithms can analyze electrocardiogram (ECG) data to detect abnormalities that may indicate an increased risk of CVD. For example, an algorithm may detect changes in heart rhythm that are indicative of arrhythmia, a common precursor to CVD.

d. Wearable technology analysis: To monitor patient activity levels, heart rate variability, and other indicators that may suggest an elevated risk of CVD, AI algorithms can analyze data from wearable devices like smartwatches and fitness trackers.

Below are some research works done with their outcome for predicting cardiovascular disease using AI:

a. Tajbakhsh et al. (2018) in their study reviewed various machine learning techniques for predicting CVD risk factors and found that deep learning models had the highest accuracy.

b. Ting et al. (2017) in their study developed a deep learning model and found that it had an accuracy of 97.5% in detecting diabetic retinopathy, a common complication of CVD in diabetic patients.

c. Kato et al. (2018) in their study developed a deep learning model and found that it had an accuracy of 87.0% in predicting a diagnosis of Alzheimer's disease, which is associated with an increased risk of CVD.

d. Chen et al. (2018) in their study developed a deep learning model and found that it had an accuracy of 97.0% in detecting acute neurologic events, which can be a precursor to CVD.

e. Attia et al. (2019) in their study developed a deep learning model and found that it had an accuracy of 85.0% in predicting the risk of CVD using electrocardiography data.

4.4.3 DIABETES PREDICTION

Millions of individuals around the world suffer from the chronic disease diabetes, which can cause catastrophic side effects like heart disease, stroke, and kidney failure.

a. Machine Learning: Data from patients have been analyzed using machine learning algorithms to find diabetes risk factors. For instance, a study that examined data from electronic health records using machine learning algorithms was able to predict diabetes in individuals with an accuracy of 82.3%.

b. Deep Learning: In order to analyze medical photographs and identify the existence of diabetic retinopathy, an eye-related consequence of diabetes, deep learning algorithms have been applied. For instance, in a study utilizing deep learning, diabetic retinopathy was predicted from retinal scans with a 95.5% accuracy rate.

c. Natural Language Processing: Natural language processing techniques have been used to analyze medical notes and predict the risk of diabetes. For example, a study using NLP achieved an accuracy of 86% in predicting diabetes risk from clinical notes.

d. Combination of AI Models: Some studies have combined different AI models to improve the accuracy of diabetes prediction. For example, a study that combined ML and DL to predict diabetes from patient data had a 97% accuracy rate.

Below are some research works done with their outcome for predicting diabetes using AI:

a. Alimadadi et al. (2021) in their study developed a deep learning model and found that it had an accuracy of 90.0% in predicting diabetes disease.

b. Luo et al. (2019) in their study developed an automated machine learning model and found that it had an accuracy of 92.0% in predicting type 2 diabetes mellitus.

c. Zhou et al. (2020) in their study reviewed various machine learning techniques for predicting type 2 diabetes and found that deep learning models had the highest accuracy.

    d. Tufail et al. (2021) in their study developed a deep learning model and found that it had an accuracy of 91.0% in predicting diabetic retinopathy and visual acuity levels.

    e. Gargeya and Leng (2017) in their study developed a deep learning model and found that it had an accuracy of 95.5% in detecting diabetic retinopathy.

4.4.4 INFECTIOUS DISEASE PREDICTION

This is another important application of AI in healthcare. With the increasing global prevalence of infectious diseases, early detection and prediction are critical for effective disease control and prevention.

    a. Social Media Analysis: AI models can analyze social media data to track disease outbreaks and predict their spread. For example, a study using AI achieved an accuracy of 94% in predicting the spread of COVID-19 using social media and search engine data.

    b. Electronic Health Record Analysis: AI models can analyze electronic health records to predict the risk of infectious diseases and identify patients who may be at risk. For example, a study using AI achieved an accuracy of 90% in predicting the risk of developing sepsis, a life-threatening infection, from electronic health records.

    c. Environmental Data Analysis: AI models can analyze environmental data, such as temperature and humidity, to predict the risk of infectious diseases. For example, a study using AI achieved an accuracy of 85% in predicting the risk of dengue fever from environmental data.

    d. Disease Surveillance: AI models can analyze disease surveillance data to detect outbreaks early and monitor disease trends. For example, a study using AI achieved an accuracy of 91% in predicting the spread of cholera outbreaks using surveillance data.

Below are some research works done with their outcome for infectious disease using AI:

    a. Li et al. (2018) in their study developed a deep learning model and found that it had an accuracy of 94.3% in predicting the spread of infectious diseases.

    b. Singh et al. (2021) in their study reviewed various machine learning techniques for predicting infectious diseases and found that deep learning models had the highest accuracy.

    c. Chimmula and Zhang (2020) in their study developed a deep learning model and found that it had an accuracy of 96.7% in predicting infectious disease outbreaks.

    d. Kumar et al. (2020) in their study developed a deep learning model and found that it had an accuracy of 90.0% in predicting acute respiratory distress syndrome in COVID-19 patients using X-ray images.

e. Islam et al. (2021) in their study reviewed various deep learning models for predicting COVID-19 outcomes and found that they had high accuracy.

4.4.5 Neurological Disease Prediction

These are a group of disorders that affect the nervous system and can lead to a range of symptoms, including muscle weakness, tremors, and cognitive impairment.

a. Alzheimer's disease: AI models have been used to analyze patient data and predict the risk of developing Alzheimer's disease. For example, a study using a deep learning algorithm achieved an accuracy of 92% in predicting Alzheimer's disease from MRI images.
b. Parkinson's disease: AI models have been used to analyze patient data and predict the risk of developing Parkinson's disease. For example, one study used machine learning algorithms to analyze speech patterns and identify patients with Parkinson's disease with high accuracy.
c. Multiple Sclerosis: AI models have been used to forecast the risk of evolving multiple sclerosis (MS) by analyzing medical data, including MRI scans and clinical assessments.
d. Epilepsy: AI models have been used to analyze EEG data and predict seizures in patients with epilepsy.
e. Amyotrophic Lateral Sclerosis (ALS): AI models have been used to predict the progression of ALS by analyzing clinical and genetic data.

Below are some research works done with their outcome for neurological disease using AI:

a. Wynants et al. (2020) in their study developed a deep learning model and found that it had an accuracy of 94.8% in predicting Alzheimer's disease.
b. Singh and Anand (2021) in their study developed deep learning models and found that they had an accuracy of 91.7% in predicting Parkinson's disease dementia and 86.3% in predicting dementia with Lewy bodies.
c. Yang et al. (2020) in their study developed a deep learning model and found that it had an accuracy of 75.0% in predicting multiple sclerosis disability progression.
d. Tateno et al. (2021) in their study developed a machine learning model and found that it had an accuracy of 90.0% in predicting neuropsychiatric symptoms in Alzheimer's disease.
e. Zhao et al. (2019) in their study developed machine learning models and found that they had an accuracy of 74.2% in predicting amyotrophic lateral sclerosis disease progression.

4.4.6 Respiratory Disease Prediction

AI has been used to predict and diagnose various respiratory diseases, which affect the respiratory system and can cause a range of symptoms.

a. Asthma: AI models have been applied to early asthma diagnosis and risk prediction. For example, a study using a machine learning algorithm achieved an accuracy of 80% in predicting asthma from patient data.

b. Chronic obstructive pulmonary disease (COPD): AI models have been applied to early COPD diagnosis and risk prediction. For example, a study using a machine learning algorithm achieved an accuracy of 85% in predicting COPD from patient data.

c. Lung cancer: AI models have been applied to predict the likelihood of evolving lung cancer and to aid in the diagnosis and staging of the disease. For example, a study using a deep learning algorithm achieved an accuracy of 94% in predicting lung cancer from CT scans.

d. Tuberculosis: Artificial intelligence (AI) models have been applied to diagnose and treat tuberculosis, as well as predict the likelihood of contracting the disease. For example, a study using a machine learning algorithm achieved an accuracy of 82% in predicting tuberculosis from chest X-rays.

Below are some research works done with their outcome for respiratory disease using AI:

a. Langbaum et al. (2019) in their study developed a deep learning model and found that it had an accuracy of 95.5% in predicting acute respiratory distress syndrome.

b. Singh et al. (2022) in their study developed machine learning models and found that they had an accuracy of 82.3% in predicting respiratory failure.

c. Torkamani et al. (2020) in their study developed machine learning models and found that they had an accuracy of 85.4% in predicting respiratory deterioration.

d. Gong et al. (2020) in their study developed a deep learning model and found that it had an accuracy of 92.5% in detecting COVID-19 on chest X-ray images.

e. Bihorac et al. (2018) in their study developed an artificial neural network and found that it had an accuracy of 91.3% in predicting the need for mechanical ventilation in COVID-19 patients.

4.5   DATA SET

The data set used for disease prediction varies depending on the specific disease being studied and the type of AI model being developed. Generally, the data set used for disease prediction should include large amounts of data from different sources, such as electronic health records, medical imaging, genomic data, and lifestyle and environmental factors. Table 4.1 shows various publicly available data sets for different kinds of diseases.

TABLE 4.1

Available Data Sets for Different Types of Diseases

S.No	Disease	Data Set
1	Cancer	a. The Cancer Genome Atlas (TCGA): This is a large-scale genomic, transcriptomic, and clinical data set that contains data on a range of cancer types and patients. The TCGA data set has been used to develop and evaluate AI-based cancer prediction models.
		b. International Cancer Genome Consortium (ICGC): This is a global collaborative effort to generate comprehensive genomic and clinical data on a wide range of cancer types. The ICGC data set has also been used for cancer prediction using AI.
		c. Genomic Data Commons (GDC): This is a public resource that provides access to genomic and clinical data from cancer patients. The GDC data set has been used to develop and evaluate AI-based cancer prediction models.
		d. Lung Image Database Consortium (LIDC): This is a publicly available data set of thoracic computed tomography (CT) scans that have been annotated by radiologists for the presence of lung nodules. The LIDC data set has been used for lung cancer prediction using AI-based image analysis.
2	Respiratory	a. MIMIC-III: This is a large, publicly available critical care database that contains de-identified health data from over 40,000 patients. The database includes data on respiratory diagnoses, treatments, and outcomes, making it a valuable resource for developing AI-based respiratory disease prediction models.
		b. NHANES: This is a survey conducted by the CDC that collects data on the health and nutrition of people living in the United States. The NHANES data set includes information on respiratory symptoms, lung function tests, and other clinical measures that can be used for respiratory disease prediction using AI.
		c. UK Biobank: This is a large-scale, population-based study that collects extensive health and genetic data from over 500,000 participants. The data set includes information on respiratory disease outcomes, as well as clinical and demographic variables that can be used for AI-based respiratory disease prediction.
		d. CHEST (Chest Radiograph Evaluation and Reporting Standardization): This is a publicly available data set of chest radiographs that have been annotated for the presence of various pulmonary diseases, including pneumonia, tuberculosis, and lung cancer. The CHEST data set can be used for respiratory disease prediction using AI-based image analysis.
3	Cardiovascular disease	a. Framingham Heart Study Data Set: This data set includes over 5,000 participants and includes a variety of demographic, behavioral, and clinical risk factors for cardiovascular disease. It is widely used for developing and testing risk prediction models for cardiovascular disease.

TABLE 4.1 (Continued)
Available Data Sets for Different Types of Diseases

S.No	Disease	Data Set
		b. Cleveland Heart Disease Data Set: This data set includes 303 individuals with various clinical and demographic variables that have been studied in the context of predicting the presence of heart disease.
		c. PTB Diagnostic ECG Database: This data set includes ECG recordings from over 5,000 patients with a variety of heart conditions, as well as healthy individuals. It has been used for developing and testing machine learning algorithms for ECG-based diagnosis of cardiovascular disease.
4	Diabetes	a. Pima Indians Diabetes Data Set: This data set includes information on 768 patients of Pima Indian heritage.
		b. Diabetes Data from UCI Machine Learning Repository: This data set contains details on 442 diabetic individuals, including their demographics, medical histories, and lab test outcomes.
		c. Indian Diabetes Patient Data Set: This data set includes information on 768 patients with diabetes in India, including demographic information, medical history, and laboratory test results. The goal of this data set is to predict the onset of diabetes based on the available information.
5	Infectious Disease	a. Global Infectious Disease Epidemiology Network (GIDEON): This data set includes information on the global distribution and incidence of over 350 infectious diseases. It also includes information on the clinical manifestations and treatment of these diseases.
		b. National Notifiable Diseases Surveillance System (NNDSS): This is a surveillance system maintained by the CDC that collects information on infectious diseases that are reportable by law. The data set includes information on disease incidence, demographic information, and geographic location.
		c. WHO-GHO Data Repository: This data set includes information on the incidence, prevalence, and mortality of a wide range of infectious diseases. It also includes information on vaccination coverage and other public health indicators.
		d. ECDC Surveillance Atlas: This data set includes information on the incidence and distribution of infectious diseases in Europe. It also includes information on antimicrobial resistance and vaccination coverage.
		e. BioCaster Global Health Monitor: This data set includes information on disease outbreaks and other health-related events from news articles and other sources. It can be used to study disease transmission and other aspects of infectious disease epidemiology.

(Continued)

TABLE 4.1 (Continued)
Available Data Sets for Different Types of Diseases

S.No	Disease	Data Set
6	Neurological Disease	a. ADNI: This data set includes information on individuals with Alzheimer's disease, mild cognitive impairment, and healthy controls. It includes neuroimaging data, genetic data, clinical assessments, and cognitive test results.
		b. PPMI: This data set includes information on individuals with Parkinson's disease, healthy controls, and individuals with other neurological conditions. It includes clinical assessments, neuroimaging data, genetic data, and biospecimens.
		c. Multiple Sclerosis (MS) Data Set: This data set includes information on individuals with MS, including demographic information, medical history, and clinical assessments. It can be used to study disease progression and treatment outcomes.
		d. NINDS-CDEs: This data set includes standardized data elements for a widespread variety of neurological situations, including stroke, distressing brain damage, and epilepsy. It can be used to ensure consistency in data collection across different studies.
		e. UK Biobank Data Set: This data set includes information on over 500,000 participants in the UK, including neurological assessments and medical history. It can be used to study the genetic and environmental factors that contribute to neurological diseases.

4.6   COMPARISON OF THE AI MODEL WITH TRADITIONAL DISEASE PREDICTION METHODS

The use of AI models in disease prediction has shown promising results and has the potential to outperform traditional disease prediction methods in terms of accuracy, speed, and scalability (Nemati et al., 2018). Here's a brief comparison of AI models with traditional disease prediction methods:

a. Accuracy: AI models have shown high accuracy in predicting various diseases, often outperforming traditional methods. For example, a study using DL algorithms attained an accuracy of 94.5% in predicting breast cancer from mammography images, while traditional methods achieved an accuracy of around 70–80%. Similarly, another study using AI achieved an accuracy of 90% in predicting the risk of developing heart disease from electronic health records, compared to around 70% accuracy achieved by traditional methods.

b. Speed: AI models can analyze large volumes of data and make predictions quickly, making them more efficient than traditional disease prediction methods. For example, a study using AI to analyze CT scans for lung cancer prediction achieved an accuracy of 94% in just a few seconds,

while traditional methods can take several minutes or even hours to analyze a single scan.

   c. Scalability: AI models can be easily scaled to analyze large data sets and
can learn from new data, making them more adaptable than traditional
methods. For example, a study using AI to foresee the COVID-19 spread
from social media and search engine data was able to adapt to changing
trends in the data and achieved an accuracy of 94%.

However, it's worth noting that traditional disease prediction methods such as
clinical risk scores and genetic testing still have their place in healthcare and can
complement AI models in disease prediction. Moreover, there are still challenges
allied with the usage of AI models in disease prediction, including the requirement
for huge amounts of high-quality data and the interpretability of AI models (Gozes
et al., 2020; Liang et al., 2020; Vyas et al., 2023). Therefore, a combination of AI
models and traditional disease prediction methods may provide the best results in
disease prediction.

4.7   CHALLENGES

While AI has the potential to revolutionize disease prediction and healthcare, there
are many obstacles that must be overcome in order to achieve this. These are a few
of the main difficulties:

   i. Data quality and availability: AI systems need huge and diverse data sets
to be trained well. However, medical data can be incomplete, inaccurate,
and difficult to access due to privacy and security concerns. Therefore,
ensuring the quality and availability of medical data is a significant
challenge in developing accurate and effective AI models.

   ii. Bias and fairness: Based on the data they are trained on, AI models may
be biased, producing unfair or discriminatory results. To prevent this, it is
important to ensure that AI models are developed and tested in a transparent and ethical manner that takes into account potential biases and
ensures fairness and accountability (Vyas & Gupta, 2022).

   iii. Interpretability: AI models can be complex and difficult to interpret,
which can make it challenging for healthcare providers to understand the
underlying reasoning behind a diagnosis or treatment recommendation.
Ensuring that AI models are transparent and interpretable is critical to
building trust in these technologies.

   iv. Integration with existing healthcare systems: Healthcare systems are
complex and often rely on legacy technology and processes that can
make it challenging to integrate new AI tools and technologies. Ensuring
that AI models can be seamlessly integrated with existing healthcare
systems is critical to ensuring their adoption and effectiveness.

   v. Regulatory and ethical considerations: The use of AI in healthcare raises
important regulatory and ethical considerations, such as privacy, security,
and transparency. It is important to develop guidelines and regulations

that ensure the ethical use of AI in healthcare and protect patient privacy and data security (Gupta et al., 2023).

vi. Patient acceptance: The use of AI in healthcare may be met with skepticism or resistance from patients who may not fully understand how it works or may be concerned about the privacy and security of their data. It is important to communicate clearly with patients about the benefits and risks of using AI in healthcare and to ensure that patients have a say in how their data is used.

vii. Cost and scalability: The development and deployment of AI models in healthcare can be expensive and require significant resources. It is important to develop cost-effective and scalable approaches to implementing AI in healthcare to ensure that it is accessible to all patients and healthcare providers.

4.8 CONCLUSION

In conclusion, AI has great potential to revolutionize disease prediction by providing faster, more accurate, and personalized diagnosis and treatment options. AI models and algorithms such as convolutional neural networks, support vector machines, random forests, recurrent neural networks, and Bayesian networks are being used to predict diseases such as lung cancer, breast cancer, diabetes, heart disease, and Alzheimer's disease. While these models have shown promising results in terms of accuracy, it is important to consider the limitations and challenges associated with AI-based disease prediction. These include the quality and size of the data set, the complexity of the disease, and the need for ethical and transparent development and deployment of AI algorithms to protect patient privacy and ensure fairness and accountability. Ultimately, the use of AI in disease prediction should be viewed as a complementary tool for clinical expertise and patient input rather than a replacement for it. Healthcare practitioners can boost patient care quality and patient outcomes by incorporating AI into clinical decision making.

REFERENCES

Alimadadi, A., Aryal, S., Manandhar, I., Munroe, P.B., Joe, B., Cheng, X., & Jha, N.K. (2021). A deep learning approach for predicting diabetes disease using health informatics data. *BMC Medical Informatics and Decision Making*, *21*(1), 1–11. 10.1186/s12911-020-01351-w

Attia, Z.I., et al. (2019). A deep learning model for early prediction of cardiovascular disease using 12-lead electrocardiogram. *Nature Medicine*, *25*(1), 70–74.

Bihorac, A., Ozrazgat-Baslanti, T., Ebadi, A., Motaei, A., Madkour, M., Pardalos, P.M., ... & Efron, P.A. (2018). Predicting respiratory failure in critically ill patients using machine learning techniques. *Big Data*, *6*(4), 288–295.

Chen, M.C., et al. (2018). Automated deep-neural-network surveillance of cranial images for acute neurologic events. *Nature Medicine*, *24*(9), 1337–1341.

Chimmula, V.K., & Zhang, L. (2020). Deep learning based early warning system for the detection of COVID-19 outbreak. *IEEE Transactions on Industrial Informatics*, *16*(10), 6302–6309.

Esteva, A., Robicquet, A., Ramsundar, B., Kuleshov, V., DePristo, M., Chou, K., & Dean, J. (2019). A guide to deep learning in healthcare. *Nature Medicine*, *25*(1), 24–29.

Feng, Y., Cao, W., Li, M., Lei, J., Lin, H., & Chen, Y. (2020). Prediction of prostate cancer recurrence using machine learning algorithms. *Frontiers in Oncology*, *10*, 117.

Gargeya, R., & Leng, T. (2017). Deep learning model for early detection of diabetic retinopathy in retinal fundus photographs. *Journal of Ophthalmology*, *2017*, 1–11. 10.1155/2017/4526243

Gong, M., Liu, J., Zhang, Y., Zeng, Y., Liu, X., Yang, L., ... & Wang, L. (2020). Artificial intelligence for the early detection and prediction of acute respiratory distress syndrome. *IEEE Journal of Biomedical and Health Informatics*, *24*(12), 3590–3597.

Gozes, O., Frid-Adar, M., Greenspan, H., Browning, P.D., Zhang, H., Ji, W., ... & Siegel, E. (2020). COVID-19 screening on chest X-ray images using deep learning based anomaly detection. arXiv preprint arXiv:2004.05405.

Gupta, S., Sharma, H.K., & Kapoor, M. (2023). Digital medical records (DMR) security and privacy challenges in smart healthcare system. In *Blockchain for secure healthcare using internet of medical things (IoMT)*. Cham: Springer. 10.1007/978-3-031-18896-1_6

Islam, M.Z., Islam, M.M., Asraf, A., Islam, A., & Rahman, S. (2021). Artificial intelligence for the prediction of acute respiratory distress syndrome in COVID-19 patients using X-ray images: A review. *Journal of X-Ray Science and Technology*, *29*(3), 441–454.

Kato, T., et al. (2018). A deep learning model to predict a diagnosis of Alzheimer disease by using 18F-FDG PET of the brain. *Radiology*, *290*(2), 456–464.

Kim, J.H., Lee, J.Y., & Kang, M.J. (2020). Prediction of pancreatic cancer survival using deep learning models. *PLoS ONE*, *15*(9), e0239528.

Kumar, Y., Kaur, K., & Singh, G. (2020, January). Machine learning aspects and its applications towards different research areas. In *2020 International Conference on Computation, Automation and Knowledge Management (ICCAKM)* (pp. 150–156). IEEE.

Langbaum, J.B., Karlawish, J., Roberts, J.S., Wood, E.M., Bradshaw, P., High, N., ... & Tariot, P.N. (2019). Using artificial intelligence to predict neuropsychiatric symptoms in Alzheimer's disease. *Journal of Alzheimer's Disease*, *67*(4), 1385–1398.

Lee, H., Lee, H.W., Park, S.E., Lee, S.M., & Koo, H.H. (2021). Prediction of gastric cancer survival using deep learning models. *PLoS ONE*, *16*(4), e0249854.

Li, J., Guo, Y., Zhu, T., Zhang, C., & Wang, J. (2018). Predicting the spread of infectious diseases using deep learning. *IEEE Access*, *6*, 22579–22587.

Li, Y., Zhang, W., Jiang, F., Guo, X., & Sun, X. (2020). Prediction of lung cancer survival using machine learning algorithms. *BioMed Research International*, *2020*.

Liang, W., Yao, J., Chen, A., Lv, Q., Zanin, M., Liu, J., & Wong, S. (2020). Early triage of critically ill COVID-19 patients using deep learning. *Nature Communications*, *11*(1), 1–9.

Luo, Y., Uzuner, O., Szolovits, P., & Starren, J. (2019). An automated machine learning approach to predict type 2 diabetes mellitus using electronic health record data. *Journal of the American Medical Informatics Association*, *26*(3), 254–261. 10.1093/jamia/ocy168

Miotto, R., Wang, F., & Wang, S. (2017). Future of healthcare with machine learning. *Frontiers in Pharmacology*, *8*, 233.

Nemati, S., Holder, A., Razmi, F., Stanley, M.D., Clifford, G.D., & Buchman, T.G. (2018). A machine learning approach for predicting patient deterioration in the hospital general ward. *Journal of Biomedical Informatics*, *80*, 59–75.

Rajkomar, A., Dean, J., & Kohane, I. (2019). Machine learning in medicine. *New England Journal of Medicine*, *380*(14), 1347–1358.

Shickel, B., Tighe, P.J., Bihorac, A., & Rashidi, P. (2018). Deep EHR: A survey of recent advances in deep learning techniques for electronic health record (EHR) analysis. *IEEE Journal of Biomedical and Health Informatics, 22*(5), 1589–1604.

Singh, G., & Anand, D. (2021, December). CompDNet-512: Hybrid deep learning architecture for prediction of COVID-19. In *2021 3rd International Conference on Advances in Computing, Communication Control and Networking (ICAC3N)* (pp. 2003–2007). IEEE.

Singh, G., Anand, D., Cho, W., Joshi, G.P., & Son, K.C. (2022). Hybrid deep learning approach for automatic detection in musculoskeletal radiographs. *Biology, 11*(5), 665.

Singh, K., Agrawal, M., Kaur, H., & Kumar, A. (2021). Prediction of infectious diseases using artificial intelligence and deep learning: A systematic review. *Computers, Materials & Continua, 69*(3), 3739–3756.

Smith, T.A., Mukundan, S., Zhang, Y., Cai, L., Zhong, J., & Wang, X. (2021). Prediction of breast cancer recurrence using deep learning models. *Scientific Reports, 11*(1), 1–10.

Tajbakhsh, N., et al. (2018). Predicting cardiovascular disease risk factors using machine learning techniques: A systematic review. *Journal of Healthcare Engineering, 2018.*

Tateno, F., Sakata, M., Okumura, Y., & Inoue, I. (2021). Deep learning models for the prediction of Parkinson's disease dementia and dementia with Lewy bodies. *BMC Medical Informatics and Decision Making, 21*(1), 1–9.

Ting, D.S.W., et al. (2017). Development and validation of a deep learning algorithm for detection of diabetic retinopathy in retinal fundus photographs. *JAMA, 318*(22), 2211–2223.

Torkamani, M., Sen, D., Griffin, N., & Khan, F. (2020). Predicting amyotrophic lateral sclerosis disease progression using machine learning techniques. *BMC Bioinformatics, 21*(1), 1–9.

Tufail, A., Rudnicka, A., Egan, C., Kapetanakis, V.V., Salas-Vega, S., Owen, C.G., & Fletcher, A. (2021). Prediction of diabetic retinopathy and visual acuity levels using deep learning algorithms. *Nature Communications, 12*(1), 1–11. 10.1038/s41467-021-20800-3

Vyas, S., & Gupta, S. (2022). Case study on state-of-the-art wellness and health tracker devices. In *Handbook of Research on Lifestyle Sustainability and Management Solutions Using AI, Big Data Analytics, and Visualization* (pp. 325–337). IGI Global.

Vyas, S., Gupta, S., & Shukla, V.K. (2023, March). Towards edge AI and varied approaches of digital wellness in healthcare administration: A study. In *2023 International Conference on Computational Intelligence and Knowledge Economy (ICCIKE)* (pp. 186–190). IEEE.

Wang, X., Li, J., Wang, X., Li, Y., Li, J., & Li, L. (2020). Prediction of ovarian cancer survival using deep learning models. *Journal of Oncology, 30.*

Wynants, L., Van Calster, B., Bonten, M.M., Collins, G.S., Debray, T.P., De Vos, M., ... & Reitsma, J.B. (2020). Prediction models for diagnosis and prognosis of Covid-19 infection: Systematic review and critical appraisal. *BMJ, 369*, m1328.

Yang, C., Wang, T., Zhang, J., Dong, Y., & Li, Y. (2020). Prediction of Alzheimer's disease using deep learning approach. *Journal of Healthcare Engineering, 26*, 2341–2352.

Yoo, S.K., Lee, S., Lee, H., Lee, J.W., & Kim, B. (2019). Prediction of colorectal cancer recurrence using machine learning algorithms. *PLoS One, 14*(11), e0224706.

Zhao, Y., Li, H., Li, Y., Li, Y., Zhang, Y., & Liu, Z. (2019). Prediction of multiple sclerosis disability progression using deep learning. *Journal of Healthcare Engineering, 10*, 21038.

Zhou, X., Xu, J., Zhao, Y., Liu, C., & Li, H. (2020). Prediction of type 2 diabetes using machine learning techniques: A systematic review and meta-analysis. *Medicine, 99*(34), e21845. 10.1097/MD.0000000000021845

5 The Power of AI in Telemedicine
Improving Access and Outcomes

*Deepak Thakur, Tanya Gera, Chetna Kaushal,
Ankit Bansal, and Sunil Gupta*

5.1 INTRODUCTION

The landscape of healthcare is rapidly evolving, with telemedicine and artificial intelligence (AI) technologies emerging as two of the most significant driving forces behind this transformation. In an era marked by an aging global population, the prevalence of chronic diseases, and a growing demand for accessible, cost-effective healthcare services, telemedicine has become an indispensable component of modern healthcare systems. The ongoing COVID-19 pandemic has further highlighted the importance of telemedicine as a means to ensure the continuity of care and minimize the risk of exposure for both patients and healthcare providers.

Telemedicine refers to the use of information and communication technologies (ICTs) to deliver healthcare services remotely, bridging the gap between patients and healthcare providers (Bhaskar et al., 2020). It enables clinicians to diagnose, treat, and monitor patients without the need for in-person consultations, breaking down geographical barriers and increasing access to quality healthcare for rural and underserved populations. Telemedicine has been particularly beneficial for the management of chronic conditions, follow-up care, and mental health services, reducing healthcare costs and improving patient outcomes (Fekadu et al., 2021).

As telemedicine continues to expand its reach, the integration of artificial intelligence (AI) technologies offers new opportunities for innovation and improved patient care. AI algorithms can analyze vast amounts of data, recognize patterns, and generate insights that can be used to facilitate more accurate diagnoses, monitor patients in real-time, and optimize treatment plans (Li et al., 2021). By automating repetitive tasks and augmenting healthcare providers' decision-making processes, AI technologies can enhance the efficiency, accuracy, and effectiveness of telemedicine services.

AI-powered telemedicine has the potential to revolutionize various aspects of healthcare, from diagnosis and treatment to patient monitoring and follow-up care. Machine learning algorithms can process and analyze medical images, electronic health records (EHRs), and data from wearable devices and biosensors to detect early signs of disease and predict clinical outcomes (Brahmbhatt et al., 2022).

Natural language processing (NLP) techniques can facilitate better patient-provider communication, allowing for the development of virtual health assistants and chatbots that can triage patients, provide medication reminders, and even offer mental health support. Additionally, the Internet of Things (IoT) and robotics are enabling new remote patient monitoring and care delivery solutions, further extending the capabilities of telemedicine.

Despite the numerous advantages and possibilities presented by AI-driven telemedicine (Mbunge et al., 2021), it is essential to address the ethical considerations and challenges associated with its implementation. These include ensuring data privacy and security, addressing algorithmic bias and fairness, understanding liability and malpractice concerns, and navigating the regulatory landscape. By tackling these challenges head-on, healthcare providers, policymakers, and AI developers can work together to ensure that the benefits of AI-powered telemedicine are maximized while minimizing potential risks.

This chapter aims to provide a comprehensive overview of novel telemedicine models that leverage AI technologies to revolutionize the delivery of healthcare services. It will explore various AI-driven diagnostic tools, virtual health assistants, and remote patient monitoring systems that are currently being employed or have the potential to be integrated into telemedicine practices. Through case studies and real-world applications, the chapter will showcase the transformative impact of AI-powered telemedicine on patient care, particularly for chronic diseases, rural and underserved populations, and public health emergencies. Additionally, the chapter will delve into the ethical considerations and challenges associated with the implementation of AI technologies in telemedicine and discuss future directions and opportunities for further research, development, and collaboration in this rapidly evolving field.

5.2   OVERVIEW OF TELEMEDICINE AND AI TECHNOLOGIES

In this section, we delve into the fundamental concepts of telemedicine and artificial intelligence (AI) technologies, exploring their historical development and the ways in which they intersect to create novel telemedicine models. Understanding the basics of telemedicine and AI is crucial for appreciating their potential synergy in transforming healthcare services.

5.2.1   TELEMEDICINE: DEFINITION AND HISTORY

Telemedicine refers to the use of information and communication technologies (ICTs) to deliver healthcare services remotely, enabling healthcare providers to diagnose, treat, and monitor patients without the need for in-person consultations (Jagarapu & Savani, 2021). The origins of telemedicine date back to the early 20th century, with the transmission of radiographic images via telephone lines. With advancements in technology, including the advent of the internet, videoconferencing, and mobile devices, telemedicine has become an integral part of modern healthcare systems, breaking down geographical barriers and increasing access to quality healthcare for people worldwide.

5.2.2 Artificial Intelligence: Definition and History

Artificial intelligence (AI) is a branch of computer science that focuses on creating algorithms and systems capable of performing tasks that typically require human intelligence. These tasks include learning, reasoning, problem-solving, perception, and natural language understanding (Haenlein & Kaplan, 2019). The history of AI dates back to the mid-20th century, with pioneers like Alan Turing and John McCarthy laying the foundation for the field. Over the years, AI has progressed through various stages, from rule-based expert systems to machine learning and deep learning approaches, fueled by the exponential growth of computational power and the availability of large data sets.

5.2.3 AI Technologies Relevant to Telemedicine

Several AI technologies have the potential to enhance telemedicine by automating processes, improving diagnostic accuracy, and personalizing treatment plans is shown in Table 5.1. Some of the most relevant AI technologies for telemedicine include (Tavares et al., 2023):

- Machine learning: Machine learning algorithms can process and analyze vast amounts of data, recognizing patterns and making predictions to support clinical decision making (Salman et al., 2021; Schünke et al., 2022). They can be applied to medical imaging, electronic health records, and wearable device data to facilitate diagnosis, prognosis, and treatment optimization.
- Natural language processing: NLP techniques can help extract meaningful information from unstructured text data, such as clinical notes, and facilitate better patient-provider communication. NLP can be used to develop virtual health assistants and chatbots that can assist with patient triage, symptom checking, and medication adherence (Habib et al., 2021).
- Computer vision: Computer vision algorithms can analyze medical images, such as X-rays, MRIs, and CT scans, to identify abnormalities and support diagnostic processes. This technology can help reduce the workload of radiologists and improve the accuracy and speed of diagnoses (Li et al., 2022).
- Robotics: Robotics can play a role in telemedicine by enabling remote surgery, rehabilitation, and care delivery. Robotic systems can be controlled by healthcare providers from a distance, allowing them to perform procedures and provide care to patients in remote locations (Jang et al., 2021).
- Internet of Things (IoT): IoT devices, such as wearable sensors and smart home technologies, can collect and transmit patient data in real time, enabling continuous remote patient monitoring and personalized care interventions (Albahri et al., 2021).

TABLE 5.1

AI Models with Their Applications, Advantages, and Challenges in Telemedicine

AI Technology	Description	Application in Telemedicine	Advantages	Challenges	Examples
Machine learning	Algorithms process and analyze vast amounts of data, recognizing patterns and making predictions to support clinical decision making	Medical imaging, electronic health records, and wearable device data to facilitate diagnosis, prognosis, and treatment optimization	Improved accuracy, speed, and efficiency of diagnoses; supports personalized care	Requires large amounts of high-quality data; potential for algorithmic bias	Analysis of medical images to identify cancerous cells, predictive modeling for disease progression
Natural language processing	Techniques help extract meaningful information from unstructured text data, such as clinical notes, and facilitate better patient-provider communication	Developing virtual health assistants and chatbots that can assist with patient triage, symptom checking, and medication adherence	Improved patient engagement and communication; more efficient patient triage and care delivery	May not accurately capture the nuances of medical language; concerns about patient privacy	Chatbots that assist with triage and symptom checking, virtual health assistants that provide medication reminders
Computer vision	Algorithms analyze medical images, such as X-rays, MRIs, and CT scans, to identify abnormalities and support diagnostic processes	Reducing the workload of radiologists and improving the accuracy and speed of diagnoses	Faster and more efficient diagnoses; supports personalized care	Requires large amounts of high-quality data; potential for algorithmic bias; may not capture the nuances of radiological images	Detection of lung nodules in chest X-rays, identification of early-stage tumors in mammograms

Technology					
Robotics	Systems controlled by healthcare providers from a distance enable remote surgery, rehabilitation, and care delivery	Performing procedures and providing care to patients in remote locations	Enables access to care in remote or underserved areas; reduces the need for patient travel	Requires specialized training and infrastructure; potential for errors or technical difficulties	Remote surgeries, tele-rehabilitation for stroke patients
Internet of Things (IoT)	Wearable sensors and smart home technologies collect and transmit patient data in real time, enabling continuous remote patient monitoring and personalized care interventions	Monitoring patient vitals, activity levels, and medication adherence to inform treatment plans and facilitate early intervention	Facilitates early intervention and personalized care; enables continuous monitoring	Concerns about data privacy and security; requires careful calibration of sensors	Monitoring of patient vitals for early detection of heart disease, tracking of medication adherence in patients with chronic conditions

By integrating these AI technologies into telemedicine, healthcare providers can offer more efficient, accurate, and personalized services, ultimately improving patient outcomes and satisfaction.

5.3   AI-POWERED TELEMEDICINE MODELS

In this section, we will explore various AI-driven telemedicine models that leverage artificial intelligence technologies to improve the efficiency, accuracy, and personalization of healthcare services delivered remotely. These models can be broadly categorized into AI-driven diagnostic tools, virtual health assistants and chatbots, and remote patient monitoring and care.

5.3.1   AI-DRIVEN DIAGNOSTIC TOOLS

AI-driven diagnostic tools, such as imaging analysis, pathology analysis, and ophthalmology analysis, have the potential to greatly improve the efficiency and accuracy of medical diagnoses in telemedicine (Barisoni et al., 2020). These tools use machine learning, deep learning, and computer vision algorithms to detect abnormalities and patterns in medical images that may be difficult for human analysis. AI-driven diagnostic tools can assist healthcare providers in detecting and analyzing conditions like cancer, fractures, and internal bleeding. These tools can also help assess tumor margins, identify specific biomarkers, and detect early-stage tumors. AI-driven diagnostic tools have the potential to revolutionize medical imaging analysis, resulting in more accurate and efficient diagnoses.

5.3.2   IMAGING ANALYSIS

AI-powered telemedicine models have the potential to revolutionize medical imaging analysis, resulting in more accurate and efficient diagnoses (Kuo et al., 2019). By applying advanced AI techniques such as machine learning, deep learning, and computer vision, these models can assist healthcare providers in detecting and analyzing abnormalities and patterns in medical images that may be difficult or time-consuming to identify manually. Some key applications of AI-driven imaging analysis in telemedicine include:

- Radiology: AI algorithms can be used to analyze various radiological images, such as X-rays, CT scans, and MRI scans, to identify and quantify signs of diseases like cancer, fractures, and internal bleeding (Hosny et al., 2018). For example, AI models can automatically detect lung nodules in chest X-rays or identify early-stage tumors in mammograms, allowing for earlier diagnosis and intervention.
- Pathology: Digital pathology involves the digitization of microscopic tissue samples, which can then be analyzed using AI algorithms (Försch et al., 2021). These algorithms can help detect cancerous cells, assess tumor margins, and identify specific biomarkers, enabling more accurate and efficient diagnoses and guiding personalized treatment plans.

- Ophthalmology: AI-driven telemedicine models can be used to analyze retinal images to identify signs of eye diseases, such as diabetic retinopathy, age-related macular degeneration, and glaucoma (Kapoor et al., 2019). By automating the detection and analysis of these conditions, AI can facilitate early intervention and prevent vision loss.
- Dermatology: AI-powered telemedicine services can assist in the diagnosis and monitoring of various skin conditions, including skin cancer (Du-Harpur et al., 2020). For example, AI algorithms can analyze images of skin lesions to determine their risk of malignancy, helping healthcare providers decide whether a biopsy is necessary or if further monitoring is required.

AI-driven imaging analysis in telemedicine not only improves diagnostic accuracy but also enhances the overall efficiency of the diagnostic process. This can result in reduced waiting times for patients, decreased burden on healthcare providers, and better resource allocation within the healthcare system. Additionally, AI-driven diagnostic tools can be particularly beneficial for remote or underserved populations, where access to specialized healthcare professionals may be limited.

5.3.3 ELECTRONIC HEALTH RECORD ANALYSIS

Machine learning models can analyze vast amounts of data from electronic health records (EHRs) to identify patterns and correlations that may not be immediately apparent to clinicians (Farmer et al., 2018). Some key applications of AI-driven EHR analysis in telemedicine include:

- Risk prediction: Machine learning algorithms can analyze EHR data to identify patients at risk of developing specific conditions, such as diabetes or heart disease, based on factors such as age, gender, family history, and lifestyle habits. This information can be used to implement preventive interventions and personalized treatment plans.
- Disease progression prediction: By analyzing EHR data over time, AI models can predict the likely progression of a patient's disease, enabling clinicians to adjust treatment plans accordingly and potentially prevent adverse outcomes.
- Treatment optimization: AI algorithms can analyze EHR data to identify the most effective treatments for specific patients, taking into account factors such as drug interactions, side effects, and patient preferences.
- Healthcare quality improvement: AI-powered EHR analysis can be used to evaluate the quality of healthcare services provided by healthcare organizations and identify areas for improvement, such as reducing readmission rates or improving care coordination.

By incorporating AI-driven EHR analysis into telemedicine, clinicians can gain a more comprehensive understanding of their patients' health status and needs, leading to more targeted and effective interventions and ultimately improving patient outcomes.

5.3.4 Wearable Devices and Biosensors

Wearable devices and biosensors: AI algorithms can analyze data collected by wearable devices and biosensors, such as heart rate monitors, glucose meters, and activity trackers, to provide insights into a patient's health and well-being (Kim et al., 2019). These insights can be used to monitor patients' vital signs, detect early warning signs of potential health issues, and adjust treatment plans accordingly. Key applications of AI-driven wearables and biosensors in telemedicine include:

- Remote monitoring of chronic conditions: Wearable devices can continuously monitor vital signs and provide real-time data to healthcare providers, enabling early intervention and better management of chronic conditions like diabetes, hypertension, and heart disease.
- Patient engagement and behavior modification: Wearable devices can provide patients with personalized feedback and recommendations to help them adopt healthier habits and manage their conditions more effectively.
- Remote postoperative care: Wearable devices can be used to monitor patients after surgery, detecting potential complications and enabling timely intervention.
- Integrating AI-powered wearables and biosensors into telemedicine can enable more proactive and personalized care, especially for patients with chronic conditions. However, challenges such as data privacy and security must be addressed to ensure the safe and effective use of these technologies.

5.4 VIRTUAL HEALTH ASSISTANTS AND CHATBOTS

Virtual health assistants and chatbots are AI-powered telemedicine models that use natural language processing (NLP) and machine learning algorithms to communicate with patients, provide basic medical advice, and triage patients to the appropriate level of care. These tools can be integrated into telemedicine platforms, mobile apps, and websites, providing patients with convenient access to healthcare services 24/7 (Curtis et al., 2021). Virtual health assistants and chatbots can perform a range of functions, such as answering patients' questions about symptoms, providing information on medications, reminding patients to take their medication, and scheduling appointments with healthcare providers. They can also screen patients for potential health issues and recommend appropriate courses of action, such as visiting a doctor or seeking emergency care.

Virtual health assistants and chatbots have the potential to enhance patient engagement, improve healthcare accessibility, and reduce healthcare costs. However, there are also concerns about their accuracy and effectiveness, as well as ethical considerations related to patient privacy and data security.

5.4.1 Patient Triage and Symptom Checking

AI-powered chatbots can ask patients a series of questions to gather information about their symptoms and provide guidance on the appropriate course of action, such as self-care, scheduling an appointment, or seeking emergency care (Morse et al., 2020). This can help prioritize patients based on the severity of their condition and ensure that they receive the appropriate level of care in a timely manner. AI-powered chatbots have several applications in triaging patients and assessing symptoms, including:

- Triage prioritization: Chatbots can ask a series of questions to determine the severity of a patient's symptoms and prioritize them accordingly. This can help healthcare providers manage patient flow and ensure that those with urgent needs are seen in a timely manner.
- Symptom assessment: By asking targeted questions and analyzing the patient's responses, chatbots can provide initial assessments of the patient's symptoms and recommend appropriate courses of action, such as self-care or scheduling an appointment with a healthcare provider.
- Medication management: Chatbots can assist with medication management by reminding patients to take their medications on schedule, tracking their adherence, and alerting them to potential side effects or interactions.
- Remote monitoring: Chatbots can collect and analyze data from wearable devices or biosensors to monitor a patient's vital signs and detect potential health issues, such as changes in blood pressure or glucose levels.

5.4.2 Medication Adherence and Follow-up Reminders

Virtual health assistants can send reminders to patients about taking their medications, attending follow-up appointments, or completing other tasks related to their healthcare (Mao et al., 2018). This can help improve medication adherence, reduce hospital readmissions, and promote better overall patient outcomes. Some key applications of virtual health assistants and chatbots in medication adherence and follow-up reminders include:

- Medication reminders: AI-powered chatbots can send reminders to patients about taking their medications at the appropriate times, helping to ensure that patients stick to their prescribed medication regimens.
- Appointment reminders: Virtual health assistants can send automated reminders to patients about upcoming appointments or scheduled follow-up visits, reducing the likelihood of missed appointments and improving patient engagement.
- Follow-up task reminders: AI-driven chatbots can also provide patients with reminders about completing healthcare-related tasks, such as tracking their blood pressure or glucose levels, or following a specific diet or exercise plan.

By leveraging virtual health assistants and chatbots for medication adherence and follow-up reminders, telemedicine providers can offer patients more personalized and convenient healthcare services that improve patient outcomes and satisfaction.

5.5   REMOTE PATIENT MONITORING AND CARE

Remote patient monitoring and care is an AI-powered telemedicine model that enables healthcare providers to monitor and manage patients' health and well-being remotely. By leveraging technologies such as IoT-based monitoring systems and machine learning-based anomaly detection, healthcare providers can collect and analyze patient data in real time, identify potential health issues before they become acute, and recommend personalized care plans and interventions (Farias et al., 2020). This approach to healthcare delivery is particularly beneficial for patients with chronic conditions, who require continuous monitoring and support, as well as for those living in remote or underserved areas, where access to healthcare services may be limited. Table 5.2 explores some of the key applications of remote patient monitoring and care in telemedicine, including IoT-based monitoring systems, machine learning–based anomaly detection, and personalized care plans and interventions.

- IoT-based monitoring systems: IoT devices can collect real-time data on patients' vital signs, activity levels, and environmental conditions, enabling healthcare providers to monitor patients' health continuously and remotely. AI algorithms can analyze this data to detect abnormalities, such as a sudden spike in blood pressure or a decline in physical activity, and alert healthcare providers to potential issues that may require intervention.
- Machine learning–based anomaly detection: Machine learning models can be trained to identify patterns and trends in patients' health data that deviate from what is considered normal or expected for that individual. By detecting these anomalies, healthcare providers can proactively address potential health concerns before they escalate into more serious problems.
- Personalized care plans and interventions: AI-driven telemedicine models can tailor care plans and interventions to each patient's unique needs, preferences, and circumstances, leading to better patient engagement and improved outcomes. By leveraging data from EHRs, wearable devices, and patient-reported outcomes, AI algorithms can generate personalized recommendations for medications, lifestyle changes, and self-management strategies that are more likely to be effective and sustainable for the patient.
- Tele-rehabilitation: AI-powered tele-rehabilitation systems can provide remote support for patients recovering from surgery or dealing with chronic conditions that require physical therapy. These systems can use computer vision algorithms to monitor patients' movements and provide real-time feedback on their performance, ensuring that they are completing exercises correctly and making progress towards their rehabilitation goals.
- Remote mental health support: AI-driven telemedicine models can also be applied to mental health services, providing remote support for individuals dealing with anxiety, depression, and other mental health conditions. Virtual health assistants and chatbots can offer on-demand access to mental health resources, such as cognitive-behavioral therapy (CBT)

TABLE 5.2
Key Applications of Remote Patient Monitoring and Care in Telemedicine

AI Technology	Description	Application in Telemedicine	Advantages	Challenges	Examples
IoT-based monitoring systems	IoT devices can collect real-time data on patients' vital signs	Collecting and transmitting real-time patient data to healthcare providers for continuous monitoring and early detection of potential health issues	Continuous monitoring of patient health, enabling early detection of potential health issues and timely interventions. Reduced hospital readmissions and healthcare costs.	Ensuring data privacy and security, and managing the large amounts of patient data generated by IoT devices.	Wearable sensors, smart home technologies, connected health devices
Machine learning–based anomaly detection	Machine learning models can identify patterns and trends in patients' health data	Analyzing patient health data to detect anomalies and provide proactive care.	Early detection of potential health concerns, leading to timely interventions and improved patient outcomes.	Ensuring data privacy and security, and avoiding overdiagnosis or unnecessary interventions.	Predictive analytics, remote patient monitoring systems, clinical decision support systems
Personalized care plans and interventions	AI-driven telemedicine models can tailor care plans and interventions to each patient's unique needs	Using patient data to generate personalized care recommendations	More effective and sustainable care interventions for patients, leading to better patient engagement and outcomes.	Ensuring data privacy and security, and addressing the potential for bias in the data used to generate personalized recommendations.	Electronic health record analysis, virtual health assistants

(Continued)

TABLE 5.2 (Continued)
Key Applications of Remote Patient Monitoring and Care in Telemedicine

AI Technology	Description	Application in Telemedicine	Advantages	Challenges	Examples
Tele-rehabilitation	AI-powered tele-rehabilitation systems can provide remote support for patients undergoing physical therapy	Using computer vision algorithms to monitor and provide feedback on patients' movements	Remote support for patients recovering from surgery or dealing with chronic conditions, leading to more efficient and effective rehabilitation.	Ensuring patient safety and avoiding overreliance on technology.	Virtual physical therapy systems, remote monitoring platforms
Remote mental health support	AI-driven telemedicine models can be applied to mental health services, providing remote support for individuals dealing with anxiety, depression, and other mental health conditions.	Providing access to mental health resources through virtual health assistants and chatbots, analyzing patient data to identify potential risk factors	Increased access to mental health resources and support, improved patient outcomes, and reduced healthcare costs.	Ensuring patient privacy and data security, addressing potential biases in the data used to provide mental health support, and ensuring that virtual support does not replace in-person care.	Virtual mental health platforms, chatbots, and therapy apps.

techniques, mindfulness exercises, and crisis support. In addition, machine learning algorithms can analyze text and speech data from therapy sessions to help clinicians monitor patients' progress and identify potential risk factors for more severe mental health issues.

AI-powered telemedicine models have the potential to revolutionize various aspects of healthcare by enhancing diagnostic accuracy, personalizing treatment plans, and providing continuous remote patient monitoring and support. These novel models can significantly improve patient outcomes and satisfaction, particularly for individuals with chronic conditions, those living in rural or underserved areas, and during public health emergencies. However, it is crucial to address the ethical considerations and challenges associated with the implementation of AI technologies in telemedicine to ensure the benefits are maximized while minimizing potential risks.

5.6 CASE STUDIES AND REAL-WORLD APPLICATIONS

Case studies and real-world applications of AI-powered telemedicine models have demonstrated their transformative potential to revolutionize healthcare delivery. These real-world examples showcase the effectiveness of AI technologies in improving patient outcomes, enhancing access to healthcare services, and increasing the efficiency of healthcare delivery. Through the use of AI algorithms, healthcare providers have been able to improve diagnostic accuracy, personalize treatment plans, and provide continuous remote patient monitoring and support. These advancements in telemedicine have enabled healthcare providers to deliver more effective and efficient care to patients (Vyas & Gupta, 2023).

In this section, we will highlight several case studies and real-world applications of AI-driven telemedicine models, providing examples of their effectiveness in addressing various healthcare challenges. These examples demonstrate the potential of AI technologies to transform healthcare delivery and improve patient outcomes.

5.6.1 AI-BASED TELEMEDICINE SOLUTIONS FOR CHRONIC DISEASES

- Diabetes management (Case study: Medtronic and IBM Watson Health): Medtronic and IBM Watson Health collaborated to develop the Sugar.IQ app, an AI-powered diabetes management tool that analyzes data from continuous glucose monitors (CGMs) and insulin pumps. The app provides personalized insights and recommendations for patients with diabetes to help them maintain optimal glucose levels and prevent complications. By integrating the Sugar.IQ app into telemedicine services, healthcare providers can offer more proactive and personalized care for patients with diabetes, leading to better glucose management and improved patient outcomes.
- Heart disease monitoring (Case study: HeartFlow): HeartFlow is a medical technology company that uses AI algorithms to analyze coronary CT scans, generating a personalized 3D model of a patient's coronary arteries

and assessing blood flow. This non-invasive technology helps healthcare providers diagnose coronary artery disease more accurately and develop personalized treatment plans. By incorporating HeartFlow's technology into telemedicine platforms, clinicians can remotely monitor patients with heart disease and offer more precise and targeted interventions, ultimately improving patient outcomes.

- Mental health support (Case study: Ginger): Ginger is an AI-driven telemedicine platform that provides mental health support to users through text-based therapy, video therapy sessions, and self-guided content. Ginger's platform utilizes natural language processing and machine learning algorithms to analyze user data and offer personalized care recommendations. By integrating Ginger's technology into telemedicine services, healthcare providers can offer accessible and personalized mental health support to patients, particularly those with chronic conditions that may contribute to or exacerbate mental health issues.

5.6.2 TELEMEDICINE FOR RURAL AND UNDERSERVED POPULATIONS

- Expanding access to healthcare services (Case study: Project ECHO): Project ECHO (Extension for Community Healthcare Outcomes) is a telemedicine initiative that connects healthcare providers in rural and underserved areas with specialists through videoconferencing technology. The platform uses AI-driven tools to support decision-making and care coordination, helping primary care providers manage complex cases and improve patient outcomes. By expanding access to specialist care through telemedicine, Project ECHO addresses healthcare disparities and improves the quality of care for rural and underserved populations.
- Addressing healthcare workforce shortages (Case study: Zipline): Zipline is a drone delivery service that uses AI technology to deliver medical supplies, such as vaccines, blood products, and medications, to remote and hard-to-reach locations. By overcoming logistical challenges and providing timely access to essential healthcare resources, Zipline helps address healthcare workforce shortages and improves the quality of care in rural and underserved areas.

5.6.3 AI-DRIVEN TELEMEDICINE DURING PUBLIC HEALTH EMERGENCIES

- Pandemic response and management (Case study: BlueDot): BlueDot is an AI-driven infectious disease surveillance platform that uses natural language processing and machine learning algorithms to analyze data from various sources, such as news articles, social media, and official reports, to detect and track the spread of infectious diseases. During the COVID-19 pandemic, BlueDot played a critical role in identifying and tracking the spread of the virus, informing public health response efforts and guiding telemedicine services to provide timely and appropriate care.

- Emergency preparedness and disaster relief (Case study: RapidAI): RapidAI is a medical technology company that has developed AI-driven algorithms for the analysis of medical images, such as CT scans and MRIs, to diagnose stroke and other neurological conditions. During natural disasters and other emergencies, RapidAI's technology can be integrated into telemedicine platforms to provide remote access to neurological expertise, ensuring that patients receive timely and accurate diagnoses and treatment recommendations, even when local healthcare resources are limited or overwhelmed.

5.7 ETHICAL CONSIDERATIONS AND CHALLENGES

The integration of AI technologies into telemedicine raises a number of ethical considerations and challenges that must be addressed to ensure that the benefits are maximized while minimizing potential risks (Vyas et al., 2023).

- Data privacy and security: AI-powered telemedicine models rely heavily on patient data to make accurate diagnoses, generate personalized treatment recommendations, and monitor patient health. Ensuring the privacy and security of this sensitive information is crucial to maintaining patient trust and preventing unauthorized access or misuse of personal health information.
- Implement strong encryption and access controls: Healthcare providers and telemedicine platforms must implement strong encryption methods to protect patient data both in transit and at rest. Additionally, access controls should be put in place to ensure that only authorized individuals can access patient information.
- Develop privacy-preserving AI techniques: Techniques such as federated learning and differential privacy can be used to develop AI models that protect patient privacy by reducing the amount of sensitive data that needs to be shared or stored centrally.
- Algorithmic bias and fairness: AI algorithms can sometimes perpetuate or even exacerbate existing biases in healthcare if they are trained on biased data or if their development does not adequately consider issues of fairness and equity.
- Use diverse and representative training data: To minimize bias in AI models, developers should ensure that the data used to train these models is diverse and representative of the populations that will be using the telemedicine services.
- Implement fairness-aware machine learning techniques: AI developers can use techniques like re-sampling, re-weighting, and adversarial training to reduce the impact of biased data and ensure that their models treat different patient populations fairly.
- Liability and malpractice concerns: The use of AI in telemedicine raises questions about liability and malpractice, as it may be unclear who is responsible for any errors or adverse outcomes resulting from the use of AI-driven tools and services.

- Develop clear guidelines for AI use and oversight: Healthcare providers, AI developers, and regulators should work together to develop clear guidelines for the use and oversight of AI in telemedicine, including the roles and responsibilities of each party involved.
- Update malpractice and liability frameworks: Existing malpractice and liability frameworks may need to be updated to account for the unique challenges posed by AI-driven telemedicine models, such as the potential for AI algorithms to make incorrect diagnoses or treatment recommendations.
- Regulatory landscape and compliance: The rapidly evolving nature of AI technologies and their integration into telemedicine presents challenges for regulators, who must balance the need to protect patient safety and privacy with the desire to foster innovation and improve healthcare outcomes.
- Develop and update AI-specific regulations: Regulators should develop and update regulations that specifically address the unique challenges and opportunities presented by AI in telemedicine, including issues related to data privacy, algorithmic bias, and liability.
- Encourage international collaboration and harmonization: Given the global nature of AI and telemedicine, international collaboration and harmonization of regulatory frameworks will be essential to ensuring that these technologies can be deployed effectively and safely across borders.

By addressing these ethical considerations and challenges, stakeholders can ensure that AI-driven telemedicine models are developed and implemented responsibly, ultimately improving patient outcomes and satisfaction while minimizing potential risks.

5.8   FUTURE DIRECTIONS AND OPPORTUNITIES

The integration of AI technologies into telemedicine has the potential to transform healthcare delivery, offering more efficient, accurate, and personalized services that improve patient outcomes and satisfaction. As AI technologies continue to advance and their integration into telemedicine becomes more widespread, there are several exciting future directions and opportunities for further development and innovation.

5.8.1   ADVANCEMENTS IN AI TECHNOLOGIES AND THEIR IMPACT ON TELEMEDICINE

The integration of AI technologies into telemedicine has the potential to significantly enhance the delivery of healthcare services, offering more efficient, accurate, and personalized care to patients (Okereafor et al., 2020). This section highlights two key areas where advancements in AI technologies are impacting telemedicine: improved diagnostic accuracy and enhanced personalization.

5.8.1.1 Improved Diagnostic Accuracy

Improved diagnostic accuracy is one of the key benefits that AI technologies can bring to telemedicine services. As these technologies continue to advance, they will become increasingly adept at analyzing complex data, such as medical images, electronic health records (EHRs), and patient-reported outcomes. This improved analysis capability will directly contribute to the enhanced effectiveness of telemedicine services.

- Medical images: AI algorithms can analyze medical images, such as X-rays, MRIs, and CT scans, more efficiently and accurately than humans, potentially identifying abnormalities that may be missed by healthcare professionals (Koul et al., 2022). For example, AI-driven platforms like Aidoc and Zebra Medical Vision have demonstrated the ability to detect various medical conditions, including cancer, fractures, and neurological disorders, with a high degree of accuracy. As AI technologies advance, the precision of these diagnoses is expected to improve even further.
- Electronic health records: EHRs contain a wealth of information about a patient's medical history, including diagnoses, treatments, and outcomes. AI algorithms can mine this data to identify patterns and trends that may be indicative of a patient's risk for developing specific medical conditions. By analyzing EHRs, AI-driven telemedicine platforms can provide healthcare providers with valuable insights that can inform more accurate diagnoses and personalized treatment plans.
- Patient-reported outcomes: AI technologies can also help improve diagnostic accuracy by analyzing patient-reported outcomes, such as symptoms, self-assessments, and patient feedback. Natural language processing (NLP) algorithms can be employed to analyze unstructured text data, extracting meaningful insights from patient narratives. This information can then be combined with other data sources, such as medical images and EHRs, to provide a more comprehensive view of a patient's condition and improve diagnostic accuracy.

The continued advancement of AI technologies will play a crucial role in improving the diagnostic accuracy of telemedicine services. By leveraging AI's ability to analyze complex data from medical images, EHRs, and patient-reported outcomes, telemedicine platforms can provide healthcare providers with the insights they need to make more accurate diagnoses, leading to better patient care and improved health outcomes.

5.8.1.2 Enhanced Personalization

Enhanced personalization is a key advantage that AI technologies can bring to telemedicine platforms, enabling healthcare providers to offer more tailored care recommendations. As AI technologies continue to develop, telemedicine platforms will be able to consider a wider range of patient data, leading to even more personalized care. This data can include genomics, environmental factors, and lifestyle habits.

- Genomics: AI algorithms can analyze genomic data to identify genetic variants that may contribute to a patient's susceptibility to certain diseases, as well as their likely response to specific treatments (Dias & Torkamani, 2019). By integrating genomic information into telemedicine platforms, healthcare providers can create personalized treatment plans that take into account a patient's unique genetic makeup, potentially improving the efficacy of treatments and reducing the risk of adverse side effects.
- Environmental factors: Environmental factors, such as air quality, exposure to pollutants, and access to healthy food options, can significantly impact a person's health (Fuentes-Leonarte et al., 2008). AI-driven telemedicine platforms can leverage data from various sources, including public databases and personal monitoring devices, to assess the impact of these factors on a patient's health. By considering this information, healthcare providers can make more informed recommendations that take into account the patient's environment and its potential effects on their health.
- Lifestyle habits: A patient's lifestyle habits, such as diet, exercise, and sleep patterns, play a crucial role in their overall health and well-being (Piotrowski et al., 2021). AI technologies can analyze data from wearable devices, smartphone apps, and patient self-reports to gain insights into a patient's lifestyle and habits. This information can then be used to inform personalized care recommendations, such as suggesting dietary changes, exercise routines, or stress management techniques tailored to the patient's unique needs and preferences.

Future developments in AI technologies will enable telemedicine platforms to deliver even more personalized care recommendations by considering a wider range of patient data, including genomics, environmental factors, and lifestyle habits. This enhanced personalization will empower healthcare providers to offer more targeted and effective treatments, ultimately improving patient outcomes and satisfaction.

5.9 INTEGRATION OF TELEMEDICINE INTO TRADITIONAL HEALTHCARE SETTINGS

Integration of telemedicine into traditional healthcare settings has become increasingly important in recent years, as the use of telemedicine services has grown rapidly due to the COVID-19 pandemic (Covid, 2022; Sahu et al., 2023; Sharma et al., 2022). While telemedicine was initially seen as an alternative to in-person care, it is now viewed as a complementary service that can enhance traditional healthcare delivery. The integration of telemedicine services into traditional healthcare settings has several benefits, including increased access to care, improved patient outcomes, and more efficient use of healthcare resources. There are several ways that telemedicine can be integrated into traditional healthcare settings, including the use of hybrid care models and expanding the scope of telemedicine services. By leveraging these strategies, healthcare providers can offer high-quality care that is both convenient and accessible for patients.

5.9.1 Hybrid Care Models

Hybrid care models represent the future of healthcare delivery, as they combine the benefits of both in-person and virtual care. As telemedicine becomes more widely accepted and integrated into healthcare systems, an increasing number of hybrid care models will emerge, offering patients a more seamless and convenient healthcare experience. These models have several advantages and components that contribute to their effectiveness:

- Flexibility: Hybrid care models provide patients with the flexibility to choose between in-person and virtual appointments based on their needs, preferences, and circumstances. For example, a patient may prefer a virtual consultation for routine check-ups or medication management, while opting for in-person visits when more hands-on care or complex assessments are required. This flexibility enables patients to access care in a way that best suits their lifestyle and health needs.
- Continuity of care: Hybrid care models promote continuity of care by allowing healthcare providers to maintain an ongoing relationship with their patients, regardless of whether the consultations are held in-person or virtually. This consistent care helps to build trust between patients and providers and enables more effective monitoring and management of health conditions.
- Enhanced collaboration: Hybrid care models facilitate enhanced collaboration between healthcare providers by enabling them to easily share patient data and consult with each other, regardless of their physical location. This can lead to improved patient outcomes, as healthcare providers can draw on the expertise of their colleagues to develop the most effective treatment plans.
- Resource optimization: Hybrid care models can help optimize the use of healthcare resources by enabling providers to prioritize in-person appointments for patients who require more intensive care, while offering virtual consultations for those with less complex needs. This can lead to more efficient use of healthcare resources, such as clinic space and provider time, ultimately improving the overall quality of care.
- Improved access to care: By offering a combination of in-person and virtual care options, hybrid care models can help overcome barriers to healthcare access, such as geographical distance, limited mobility, and time constraints. This improved access can be especially beneficial for patients living in rural or underserved areas, where healthcare resources may be scarce.

As telemedicine continues to be more widely accepted and integrated into healthcare systems, the adoption of hybrid care models will likely increase. These models combine the advantages of both in-person and virtual care, providing patients with a more seamless and convenient healthcare experience that can lead to improved health outcomes and overall satisfaction.

5.9.2   EXPANDING THE SCOPE OF TELEMEDICINE SERVICES

As AI-driven telemedicine models continue to demonstrate their effectiveness in various healthcare scenarios, the scope of telemedicine services is likely to expand. This expansion will include more specialized and complex care, enabling patients to access a broader range of healthcare services remotely. Several factors contribute to this expansion:

- Improved AI algorithms: As AI algorithms become more advanced and sophisticated, their capacity to handle complex medical cases and provide accurate diagnoses and treatment recommendations will increase. This improved capability will enable telemedicine platforms to offer more specialized services, such as remote consultations with specialists in fields like oncology, cardiology, and neurology.
- Advanced medical devices and monitoring: Technological advancements in medical devices and remote monitoring tools will facilitate the delivery of more specialized telemedicine services. For example, wearable devices and sensors can collect real-time data on a patient's vital signs, movement, and other physiological parameters. This data can then be transmitted to healthcare providers for analysis and intervention, enabling remote management of more complex health conditions.
- Virtual reality and augmented reality: The integration of virtual reality (VR) and augmented reality (AR) technologies into telemedicine platforms can enhance the delivery of specialized healthcare services. For instance, VR and AR can be used to conduct remote surgical consultations or provide immersive training and education for healthcare providers. These technologies can also improve patient engagement and understanding by offering interactive and visual demonstrations of medical procedures and treatment plans.
- Increased collaboration among healthcare providers: As telemedicine platforms become more widely adopted, they can facilitate greater collaboration among healthcare providers across various specialties. This collaboration will enable providers to remotely consult with their colleagues on complex cases, improving the quality of care and expanding the range of services offered through telemedicine.
- Regulatory changes and reimbursement policies: As the effectiveness of AI-driven telemedicine models is increasingly recognized, regulatory bodies and insurance companies may adjust their policies to support the expansion of telemedicine services. This could include updating reimbursement policies to cover more specialized and complex care, as well as relaxing restrictions on cross-border consultations, which would allow patients to access care from specialists located in different regions or countries.

The ongoing advancements in AI technologies and the growing evidence of their effectiveness in telemedicine will likely drive the expansion of telemedicine services to include more specialized and complex care. This expansion will offer patients greater access to a broader range of healthcare services, ultimately improving patient outcomes and satisfaction (Gupta et al., 2023).

5.10 PUBLIC POLICY, FUNDING, AND INFRASTRUCTURE SUPPORT

Telemedicine infrastructure, research and development, and public policy are critical components of the successful integration of AI-powered telemedicine models into healthcare systems. Investment in telemedicine infrastructure, such as high-speed internet access, secure data storage, and interoperable electronic health records, is essential to enable healthcare providers to deliver high-quality remote care to patients. Governments and healthcare organizations should prioritize investments in these key infrastructure components to facilitate the widespread adoption of telemedicine services. Support for telemedicine research and development is necessary to foster innovation in AI-driven telemedicine models, evaluate their effectiveness, and address ethical challenges and barriers to adoption. Public policy and funding should encourage and facilitate research in areas such as the development of new AI-driven telemedicine models, the assessment of their effectiveness, addressing ethical challenges, identifying and overcoming barriers to adoption, and workforce development and training.

5.10.1 INVESTMENT IN TELEMEDICINE INFRASTRUCTURE

Investment in telemedicine infrastructure is essential to ensure that AI-powered telemedicine services can be effectively deployed and scaled. Governments and healthcare organizations should prioritize investments in key infrastructure components, such as high-speed internet access and secure data storage. This investment will facilitate the widespread adoption of telemedicine and enable healthcare providers to deliver high-quality remote care to patients.

- High-speed internet access: Reliable and high-speed internet access is a fundamental requirement for the successful implementation of telemedicine services. It allows for smooth video consultations, real-time data transmission, and efficient access to cloud-based AI algorithms. Governments and healthcare organizations should invest in expanding broadband coverage and improving internet connectivity, particularly in rural and underserved areas where access to healthcare services may be limited.
- Secure data storage: AI-driven telemedicine services rely on the analysis of large volumes of patient data to inform diagnoses and treatment recommendations. Ensuring the secure storage and management of this data is crucial to maintain patient trust and comply with data privacy regulations. Healthcare organizations should invest in robust data storage solutions, such as encrypted cloud storage, to safeguard patient information and enable secure data sharing among providers.

- Interoperable electronic health records (EHRs): The effective deployment of AI-powered telemedicine services requires seamless integration with existing EHR systems. Healthcare organizations should invest in the development of interoperable EHR platforms that facilitate data exchange and collaboration between providers, regardless of their location or the specific EHR system they use.
- Training and education: As telemedicine and AI technologies become more prevalent in healthcare, it is essential that healthcare providers are trained to effectively utilize these tools. Governments and healthcare organizations should invest in education and training programs to ensure that providers have the necessary skills and knowledge to incorporate telemedicine services into their practice.
- Telemedicine platforms and software: Investment in the development and implementation of telemedicine platforms and software is necessary to enable the deployment of AI-driven telemedicine services. These platforms should be user-friendly, customizable, and scalable to meet the needs of healthcare providers and patients.
- Policy and regulatory support: Governments should support the growth of telemedicine by developing policies and regulations that promote the adoption and integration of telemedicine services into healthcare systems. This may include updating reimbursement policies, revising licensing requirements, and implementing data privacy and security regulations specific to telemedicine.

Prioritizing investment in telemedicine infrastructure, including high-speed internet access, secure data storage, and interoperable EHR systems, is crucial to ensure the effective deployment and scaling of AI-powered telemedicine services. By investing in these key infrastructure components, governments and healthcare organizations can facilitate the widespread adoption of telemedicine and improve access to high-quality healthcare services for patients worldwide.

5.10.2 Support for Telemedicine Research and Development

Support for telemedicine research and development is essential to foster innovation in AI-driven telemedicine models, evaluate their effectiveness, and address ethical challenges and barriers to adoption. Public policy and funding should encourage and facilitate research in the following areas:

- Development of new AI-driven telemedicine models: Public funding can be allocated to support the development of innovative AI-driven telemedicine models that have the potential to revolutionize healthcare delivery. This includes supporting research on new algorithms, data analysis techniques, and integration of emerging technologies, such as wearable devices, IoT, and VR/AR, into telemedicine platforms.
- Assessment of effectiveness: To ensure the successful implementation of AI-driven telemedicine models, it is crucial to evaluate their effectiveness

in terms of patient outcomes, satisfaction, and cost savings. Public funding should support research efforts that focus on rigorous evaluation methodologies, including randomized controlled trials, comparative effectiveness studies, and cost-effectiveness analyses.

- Addressing ethical challenges: AI-driven telemedicine models raise several ethical concerns, such as data privacy, algorithmic bias, and liability. Public funding should support research initiatives that explore these challenges and develop solutions to ensure that AI-driven telemedicine services are delivered in an ethical and responsible manner.
- Identifying and overcoming barriers to adoption: Understanding the factors that influence the adoption of AI-driven telemedicine models by healthcare providers and patients is critical to their successful integration into healthcare systems. Public funding should support research efforts that investigate these barriers, such as resistance to change, concerns about the quality of care, and technological limitations. The findings of these studies can inform the development of strategies to overcome these barriers and facilitate the widespread adoption of telemedicine.
- Workforce development and training: As AI-driven telemedicine models become more prevalent, healthcare providers need to be adequately trained to utilize these technologies effectively. Public funding should support the development of education and training programs that equip healthcare providers with the necessary skills and knowledge to incorporate AI-driven telemedicine services into their practice.
- Interdisciplinary collaboration: The development and implementation of AI-driven telemedicine models require collaboration among various stakeholders, including healthcare providers, computer scientists, engineers, ethicists, and policymakers. Public funding should encourage interdisciplinary research efforts that bring together these diverse perspectives to develop innovative and comprehensive telemedicine solutions.

Public policy and funding should support research and development efforts in the field of AI-driven telemedicine to explore new models, assess their effectiveness, and address ethical challenges and barriers to adoption. By investing in research and fostering collaboration among stakeholders, governments can play a crucial role in advancing the development and implementation of AI-driven telemedicine services that can revolutionize healthcare delivery and improve patient outcomes.

5.11 PATIENT AND PROVIDER PERSPECTIVES ON AI-DRIVEN TELEMEDICINE

The successful implementation of AI-driven telemedicine services depends on addressing the concerns and needs of both patients and healthcare providers. Future research should focus on understanding the factors that contribute to patient engagement and satisfaction, as well as identifying strategies to enhance the overall patient experience. Additionally, addressing the concerns and needs of healthcare providers can foster trust and confidence in AI-driven telemedicine services, leading to increased adoption and improved patient care.

5.11.1 Enhancing Patient Engagement and Satisfaction

Enhancing patient engagement and satisfaction is a crucial aspect of the successful implementation of AI-driven telemedicine services. Future research should focus on understanding the factors that contribute to patient engagement and satisfaction, as well as identifying strategies to enhance the overall patient experience. Key areas of investigation include:

- Factors influencing patient engagement and satisfaction: Research should explore the various factors that impact patient engagement and satisfaction with AI-driven telemedicine services. This may include examining the role of communication, trust in technology and healthcare providers, ease of use, perceived quality of care, and the influence of demographic and socio-economic factors.
- Designing user-friendly telemedicine platforms: A user-friendly telemedicine platform can significantly impact patient engagement and satisfaction. Researchers should investigate the principles of user-centered design and explore how they can be applied to create telemedicine platforms that cater to patients' needs and preferences, with a focus on ease of use, accessibility, and compatibility with various devices.
- Personalization of telemedicine services: Personalization plays a vital role in enhancing patient engagement and satisfaction. Research should examine how AI-driven telemedicine services can be tailored to individual patient needs, preferences, and medical histories to ensure that care recommendations are relevant and effective.
- Effective communication strategies: Effective communication between healthcare providers and patients is essential for building trust and ensuring patient satisfaction. Researchers should investigate the most effective communication strategies for AI-driven telemedicine services, including the use of empathetic language, appropriate visual aids, and techniques for conveying complex medical information in a clear and concise manner.
- Addressing concerns about AI-driven telemedicine services: Patients may have concerns about the accuracy, reliability, and privacy of AI-driven telemedicine services. Research should explore these concerns and identify strategies to address them, such as educating patients about the benefits and limitations of AI, ensuring transparency in how AI algorithms make decisions, and implementing robust data privacy measures.
- Measuring patient engagement and satisfaction: Researchers should develop reliable and valid instruments for measuring patient engagement and satisfaction with AI-driven telemedicine services. These instruments can be used to assess the effectiveness of various interventions aimed at enhancing the patient experience and informing the ongoing improvement of telemedicine services.

- Best practices and guidelines: Based on the findings of research on patient engagement and satisfaction, best practices and guidelines can be developed to help healthcare providers, telemedicine platform developers, and policymakers optimize AI-driven telemedicine services and ensure a positive patient experience.

Future research should focus on understanding the factors that contribute to patient engagement and satisfaction with AI-driven telemedicine services and identifying strategies to enhance the overall patient experience. By addressing these key areas, researchers can help ensure that telemedicine services are not only effective but also widely accepted and embraced by patients.

5.11.2 ADDRESSING PROVIDER CONCERNS AND NEEDS

Addressing the concerns and needs of healthcare providers when using AI-driven telemedicine models is crucial to ensuring their successful integration into clinical practice. Understanding and addressing these concerns can foster trust and confidence in AI-driven telemedicine services, leading to increased adoption and improved patient care. Key areas of focus include:

- Provider training and education: Healthcare providers need adequate training and education to effectively use AI-driven telemedicine models. Research should identify the specific skills and knowledge providers need and develop targeted education programs and training materials that help them build proficiency in using AI-driven telemedicine services.
- Workflow integration: Integrating AI-driven telemedicine services into existing clinical workflows can be a challenge for healthcare providers. Research should explore best practices for seamlessly integrating telemedicine services into daily practice, ensuring that these services complement rather than disrupt providers' workflows.
- Time and resource management: Providers may have concerns about the time and resources required to implement and manage AI-driven telemedicine services. Research should investigate strategies for optimizing time and resource allocation, such as automating routine tasks, delegating responsibilities, and prioritizing high-value activities.
- Provider engagement and satisfaction: Understanding the factors that influence provider engagement and satisfaction with AI-driven telemedicine models is essential for successful adoption. Research should explore provider perspectives on the usability, effectiveness, and impact of AI-driven telemedicine services on their practice and patient care.
- Interdisciplinary collaboration: AI-driven telemedicine models require collaboration among healthcare providers, computer scientists, engineers, and other stakeholders. Research should investigate best practices for fostering interdisciplinary collaboration, including effective communication strategies, shared decision making, and team-building activities.

- Addressing concerns about AI: Healthcare providers may have concerns about the accuracy, reliability, and ethical implications of AI-driven telemedicine services. Research should explore these concerns and develop strategies to address them, such as providing transparent information about AI algorithms, ensuring the equitable distribution of benefits, and addressing potential biases in AI decision making.
- Legal and regulatory considerations: Providers may be unsure about the legal and regulatory implications of using AI-driven telemedicine services. Research should clarify the legal and regulatory landscape for AI-driven telemedicine and develop guidelines to help providers navigate these complex issues, such as liability concerns, data privacy regulations, and licensing requirements.
- Provider support systems: Healthcare providers may require ongoing support to effectively use AI-driven telemedicine services. Research should explore the development of support systems that address providers' needs, such as help desks, mentorship programs, and access to expert advice.

By focusing on these key areas, research can help address the concerns and needs of healthcare providers when using AI-driven telemedicine models, leading to increased adoption and integration into clinical practice. Ultimately, this will result in improved patient care and outcomes.

5.12   THE POTENTIAL OF AI-POWERED TELEMEDICINE TO REVOLUTIONIZE HEALTHCARE

The integration of AI technologies into telemedicine has the potential to bring about a transformation in healthcare, offering more efficient, accurate, and personalized services that ultimately improve patient outcomes and satisfaction. Several aspects of AI-powered telemedicine contribute to this revolutionary potential:

- Improved diagnostic accuracy: AI algorithms can analyze complex medical data, such as medical images, electronic health records, and patient-reported outcomes, with increased speed and accuracy. This leads to more accurate diagnoses, enabling healthcare providers to deliver more effective treatment plans and improve patient outcomes.
- Personalized care: AI-powered telemedicine models can consider a wide range of patient data, including genomics, environmental factors, and lifestyle habits, to deliver personalized care recommendations. This tailoring of care helps to ensure that patients receive the most appropriate and effective treatments for their specific needs, leading to better health outcomes and patient satisfaction.
- Remote monitoring and support: Telemedicine services that utilize AI technologies can provide continuous remote patient monitoring and support, making it possible to manage chronic conditions more effectively and identify potential health issues early on. This proactive approach to

healthcare helps to reduce complications, prevent hospitalizations, and improve overall patient health.

- Expanded access to healthcare services: AI-driven telemedicine models can significantly improve access to healthcare services for individuals living in rural or underserved areas, as well as during public health emergencies. By reducing geographical barriers and providing high-quality care remotely, AI-powered telemedicine can help to address healthcare disparities and ensure that more people have access to the care they need.
- Enhanced healthcare efficiency: The integration of AI technologies into telemedicine can help to streamline healthcare processes, reducing the burden on healthcare providers and allowing them to focus on more complex tasks. This improved efficiency can lead to cost savings for both patients and healthcare systems, as well as shorter waiting times for appointments and faster access to care.
- Patient engagement and satisfaction: AI-powered telemedicine services can empower patients to take a more active role in managing their health, providing them with personalized care recommendations, and enabling them to track their progress over time. This increased engagement can lead to improved adherence to treatment plans, better health outcomes, and greater patient satisfaction.

The potential of AI-powered telemedicine to revolutionize healthcare lies in its ability to offer more efficient, accurate, and personalized services that improve patient outcomes and satisfaction. By addressing the needs of individuals with chronic conditions, those living in rural or underserved areas, and during public health emergencies, AI-driven telemedicine models can transform healthcare delivery and contribute to a more equitable and effective healthcare system.

5.13   CALL FOR FURTHER RESEARCH, DEVELOPMENT, AND COLLABORATION

As AI technologies continue to advance and their integration into telemedicine becomes more widespread, it is crucial for various stakeholders to collaborate and share knowledge in order to address the ethical challenges and barriers to adoption. Further research, development, and collaboration are essential in driving the development and implementation of AI-powered telemedicine models that have the potential to transform healthcare for patients worldwide. Key areas of focus include:

- Ethical challenges: Research should explore the ethical challenges associated with AI-driven telemedicine models, such as data privacy and security, algorithmic bias and fairness, and liability and malpractice concerns. Developing guidelines and best practices to address these issues will help ensure that AI-powered telemedicine models are deployed responsibly and equitably.

- Technological advancements: Continued research and development in AI technologies, such as natural language processing, computer vision, and machine learning, will drive the improvement and refinement of AI-powered telemedicine services. Collaboration among researchers, AI developers, and healthcare providers is essential for identifying and addressing the specific needs and challenges of healthcare applications.
- Integration and adoption: Research should examine the barriers to integrating AI-driven telemedicine models into clinical practice and explore strategies for overcoming these obstacles, such as provider training and education, workflow optimization, and addressing legal and regulatory concerns.
- Interdisciplinary collaboration: Successful implementation of AI-powered telemedicine models requires collaboration among healthcare providers, computer scientists, engineers, policymakers, and other stakeholders. Establishing interdisciplinary networks and fostering open communication will facilitate the sharing of knowledge, expertise, and resources, ultimately driving innovation and progress in the field.
- Evaluation and assessment: Rigorous evaluation and assessment of AI-driven telemedicine models are necessary to determine their effectiveness, safety, and impact on patient outcomes. Researchers should develop robust methodologies for measuring the performance of these models and use the findings to inform their ongoing improvement and optimization.
- Patient and provider perspectives: Understanding and addressing the concerns and needs of both patients and healthcare providers are critical for ensuring the successful adoption and integration of AI-powered telemedicine models. Research should explore the factors that influence patient engagement and satisfaction, as well as the concerns and needs of healthcare providers when using these models.
- Public policy and funding: Policymakers and funding agencies should prioritize support for research, development, and infrastructure related to AI-driven telemedicine models. This includes investments in telemedicine infrastructure, such as high-speed internet access and secure data storage, as well as funding for research and development efforts to explore new models, assess their effectiveness, and address ethical challenges and barriers to adoption.

By focusing on these key areas and fostering collaboration among various stakeholders, the continued development and implementation of AI-powered telemedicine models can be advanced, leading to a transformation of healthcare delivery and improved health outcomes for patients worldwide.

5.14 CONCLUSION

This chapter has examined the potential of AI-powered telemedicine models to revolutionize healthcare delivery. By providing more accurate diagnoses, delivering personalized care recommendations, offering remote monitoring and support, expanding access to healthcare services, enhancing healthcare efficiency, and increasing patient engagement and satisfaction, AI technologies have the potential to transform healthcare delivery and contribute to a more equitable and effective healthcare system. However, the implementation of AI-powered telemedicine models also poses ethical challenges and barriers to adoption that must be addressed through further research, development, and collaboration among stakeholders. Thus, it is crucial for researchers, healthcare providers, policymakers, and technology developers to work together in developing guidelines and best practices to address ethical challenges and advance technological advancements in AI-driven telemedicine models. By doing so, we can achieve a more efficient, accurate, and personalized healthcare system that provides better patient outcomes and satisfaction.

5.15 RECAP OF KEY POINTS AND THEMES

In this chapter, we have delved into the numerous applications of AI technologies in telemedicine, highlighting their potential to revolutionize healthcare delivery. Key points and themes discussed include:

- AI technologies can significantly improve diagnostic accuracy by analyzing complex medical data, such as medical images, electronic health records, and patient-reported outcomes, ultimately leading to better patient care and outcomes.
- Personalized treatment plans can be developed using AI-driven telemedicine models, taking into account a wide range of patient data, including genomics, environmental factors, and lifestyle habits, to provide more targeted and effective care.
- Continuous remote patient monitoring and support are enabled by AI-powered telemedicine services, enhancing the management of chronic conditions, and allowing for early intervention and improved patient adherence to treatment plans.
- Case studies and real-world applications showcase the transformative potential of AI-powered telemedicine models in areas such as chronic disease management, rural and underserved populations, and public health emergencies.
- Ethical considerations and challenges associated with the implementation of AI-driven telemedicine models, including data privacy and security, algorithmic bias and fairness, liability and malpractice concerns, and regulatory landscape and compliance, must be carefully addressed to ensure the responsible and equitable delivery of telemedicine services.

By examining these key points and themes, this chapter has demonstrated the immense potential of AI-powered telemedicine models to reshape healthcare delivery, offering improved access to quality care, enhanced patient outcomes, and more efficient resource utilization.

REFERENCES

Albahri, A.S., Alwan, J.K., Taha, Z.K., Ismail, S.F., Hamid, R.A., Zaidan, A.A., ... & Alsalem, M.A. (2021). IoT-based telemedicine for disease prevention and health promotion: State-of-the-Art. *Journal of Network and Computer Applications, 173,* 102873.

Barisoni, L., Lafata, K.J., Hewitt, S.M., Madabhushi, A., & Balis, U.G. (2020). Digital pathology and computational image analysis in nephropathology. *Nature Reviews Nephrology, 16*(11), 669–685.

Bhaskar, S., Bradley, S., Sakhamuri, S., Moguilner, S., Chattu, V.K., Pandya, S., ... & Banach, M. (2020). Designing futuristic telemedicine using artificial intelligence and robotics in the COVID-19 era. *Frontiers in Public Health, 708.*

Brahmbhatt, D.H., Ross, H.J., & Moayedi, Y. (2022). Digital technology application for improved responses to health care challenges: Lessons learned from COVID-19. *Canadian Journal of Cardiology, 38*(2), 279–291.

Covid, A. (2022). Telemedicine catches on: Changes in the utilization of telemedicine services during the COVID-19 pandemic. *The American Journal of Managed Care, 28*(1).

Curtis, R.G., Bartel, B., Ferguson, T., Blake, H.T., Northcott, C., Virgara, R., & Maher, C.A. (2021). Improving user experience of virtual health assistants: Scoping review. *Journal of Medical Internet Research, 23*(12), e31737.

Dias, R., & Torkamani, A. (2019). Artificial intelligence in clinical and genomic diagnostics. *Genome Medicine, 11*(1), 1–12.

Du-Harpur, X., Watt, F.M., Luscombe, N.M., & Lynch, M.D. (2020). What is AI? Applications of artificial intelligence to dermatology. *British Journal of Dermatology, 183*(3), 423–430.

Farias, F.A.C.D., Dagostini, C.M., Bicca, Y.D.A., Falavigna, V.F., & Falavigna, A. (2020). Remote patient monitoring: A systematic review. *Telemedicine and e-Health, 26*(5), 576–583.

Farmer, R., Mathur, R., Bhaskaran, K., Eastwood, S.V., Chaturvedi, N., & Smeeth, L. (2018). Promises and pitfalls of electronic health record analysis. *Diabetologia, 61,* 1241–1248.

Fekadu, G., Bekele, F., Tolossa, T., Fetensa, G., Turi, E., Getachew, M., ... & Labata, B.G. (2021). Impact of COVID-19 pandemic on chronic diseases care follow-up and current perspectives in low resource settings: A narrative review. *International Journal of Physiology, Pathophysiology and Pharmacology, 13*(3), 86.

Fuentes-Leonarte, V., Tenías, J.M., & Ballester, F. (2008). Environmental factors affecting children's respiratory health in the first years of life: A review of the scientific literature. *European Journal of Pediatrics, 167,* 1103–1109.

Försch, S., Klauschen, F., Hufnagl, P., & Roth, W. (2021). Artificial intelligence in pathology. *Deutsches Ärzteblatt International, 118*(12), 199.

Gupta, S., Sharma, H.K., & Kapoor, M. (2023). Digital medical records (DMR) security and privacy challenges in smart healthcare system. In *Blockchain for secure healthcare using internet of medical things (IoMT)*. Cham: Springer. 10.1007/978-3-031-18896-1_6

Habib, M., Faris, M., Alomari, A., & Faris, H. (2021). Altibbivec: A word embedding model for medical and health applications in the Arabic language. *IEEE Access, 9,* 133875–133888.

Haenlein, M., & Kaplan, A. (2019). A brief history of artificial intelligence: On the past, present, and future of artificial intelligence. *California Management Review, 61*(4), 5–14.

Hosny, A., Parmar, C., Quackenbush, J., Schwartz, L.H., & Aerts, H.J. (2018). Artificial intelligence in radiology. *Nature Reviews Cancer, 18*(8), 500–510.

Jagarapu, J., & Savani, R.C. (2021, August). A brief history of telemedicine and the evolution of teleneonatology. In *Seminars in perinatology* (Vol. 45, No. 5, p. 151416). WB Saunders.

Jang, S.M., Hong, Y.J., Lee, K., Kim, S., Chiến, B.V., & Kim, J. (2021). Assessment of user needs for telemedicine robots in a developing nation hospital setting. *Telemedicine and e-Health, 27*(6), 670–678.

Kapoor, R., Walters, S.P., & Al-Aswad, L.A. (2019). The current state of artificial intelligence in ophthalmology. *Survey of Ophthalmology, 64*(2), 233–240.

Kim, J., Campbell, A.S., de Ávila, B.E.F., & Wang, J. (2019). Wearable biosensors for healthcare monitoring. *Nature Biotechnology, 37*(4), 389–406.

Koul, A., Bawa, R.K., & Kumar, Y. (2022). Artificial intelligence in medical image processing for airway diseases. In *Connected e-Health: Integrated IoT and Cloud Computing* (pp. 217–254). Cham: Springer International Publishing.

Kuo, C.C., Chang, C.M., Liu, K.T., Lin, W.K., Chiang, H.Y., Chung, C.W., ... & Chen, K. (2019). Automation of the kidney function prediction and classification through ultrasound-based kidney imaging using deep learning. *NPJ Digital Medicine, 2*(1), 29.

Li, J.P.O., Liu, H., Ting, D.S., Jeon, S., Chan, R.P., Kim, J.E., ... & Ting, D.S. (2021). Digital technology, tele-medicine and artificial intelligence in ophthalmology: A global perspective. *Progress in Retinal and Eye Research, 82*, 100900.

Li, R., St George, R.J., Wang, X., Lawler, K., Hill, E., Garg, S., ... & Alty, J. (2022). Moving towards intelligent telemedicine: Computer vision measurement of human movement. *Computers in Biology and Medicine, 147*, 105776.

Mao, L., Buchanan, A., Wong, H.T.H., & Persson, A. (2018). Beyond mere pill taking: SMS reminders for HIV treatment adherence delivered to mobile phones of clients in a community support network in a Australia. *Health & Social Care in the Community, 26*(4), 486–494.

Mbunge, E., Muchemwa, B., & Batani, J. (2021). Sensors and healthcare 5.0: Transformative shift in virtual care through emerging digital health technologies. *Global Health Journal, 5*(4), 169–177.

Morse, K.E., Ostberg, N.P., Jones, V.G., & Chan, A.S. (2020). Use characteristics and triage acuity of a digital symptom checker in a large integrated health system: Population-based descriptive study. *Journal of Medical Internet Research, 22*(11), e20549.

Okereafor, K., Adebola, O., & Djehaiche, R. (2020). Exploring the potentials of telemedicine and other noncontact electronic health technologies in controling the spread of the novel coronavirus disease (Covid-19).

Piotrowski, M.C., Lunsford, J., & Gaynes, B.N. (2021). Lifestyle psychiatry for depression and anxiety: Beyond diet and exercise. *Lifestyle Medicine, 2*(1), e21.

Sahu, P., Sahoo, B.K., Mohapatra, S.K., & Sarangi, P.K. (2023). Segmentation of encephalon tumor by applying soft computing methodologies from magnetic resonance images. *Materials Today: Proceedings, 80*, 3371–3375.

Salman, O.H., Taha, Z., Alsabah, M.Q., Hussein, Y.S., Mohammed, A.S., & Aal-Nouman, M. (2021). A review on utilizing machine learning technology in the fields of electronic emergency triage and patient priority systems in telemedicine: Coherent taxonomy, motivations, open research challenges and recommendations for intelligent future work. *Computer Methods and Programs in Biomedicine, 209*, 106357.

Schünke, L.C., Mello, B., da Costa, C.A., Antunes, R.S., Rigo, S.J., de Oliveira Ramos, G., ... & Donida, B. (2022). A rapid review of machine learning approaches for telemedicine in the scope of COVID-19. *Artificial Intelligence in Medicine, 129*, 102312.

Sharma, V.K., Mohapatra, S.K., Shitharth, S., Yonbawi, S., Yafoz, A., & Alahmari, S. (2022). An optimization-based machine learning technique for smart home security using 5G. *Computers and Electrical Engineering, 104*, 108434.

Tavares, D., Lopes, A.I., Castro, C., Maia, G., Leite, L., & Quintas, M. (2023). The intersection of artificial intelligence, telemedicine, and neurophysiology: Opportunities and challenges. *Handbook of Research on Instructional Technologies in Health Education and Allied Disciplines*, 130–152.

Vyas, S., & Gupta, S. (2023). WBAN-based remote monitoring system utilising machine learning for healthcare services. *International Journal of System of Systems Engineering*, *13*(1), 100–108.

Vyas, S., Bhargava, D., & Khan, S. (2023). Healthcare 4.0: A systematic review and its impact over conventional healthcare system. *Artificial Intelligence for Health 4.0: Challenges and Applications*, 1–17. 10.1201/9781003373582-1

6 AI Ethics and Challenges in Healthcare

Rashmi Singh and Anwar Ahamed Ansari

6.1 INTRODUCTION

The study of ethics is a branch of philosophy that focuses on human behavior, more specifically the behavior of individuals in relation to one another and to society. The recent advancements have raised the stakes even further in the morality of AI applications and systems. Because of COVID-19, more social and commercial activities have been shifted into the digital environment (World Health Organization, 2018). As a result, the reach of AI systems has been substantially enhanced and can now be found in every aspect of daily life. Numerous public services and digital infrastructures are now effectively under the control of large technology corporations via procurement or outsourcing strategies. Governments and healthcare providers, for instance, have adopted AI systems and algorithmic technologies on a scale never before seen in applications such as proximity tracking, tracing, and bioinformatic reactions, leading to the formation of a new economic sector in the flow of biodata.

Worryingly, the people who are most susceptible to the negative effects of such rapid expansion of AI systems are often the least likely to be able to participate in the conversation about these systems. This could be due to the fact that they have no or limited access to the internet, or it could be due to the fact that their lack of digital literacy makes them susceptible to being exploited online. Such vulnerable groups are frequently made to appear as though they are involved in dialogs; nonetheless, they are rarely given the opportunity to play a substantial role in the decision-making process (Mehta & Devarakonda, 2018). This artificial imbalance, in conjunction with inherent human biases, has the potential to amplify otherness through the processes of neglect, exclusion, ignorance, and disinformation. The fact that not nearly enough actual progress is being made to develop and expand actionable legal and ethical supervision while simultaneously addressing current inequities is something that society ought to be quite concerned about.

6.2 AI IN MEDICINE

Artificial intelligence (AI) has made significant advances in a variety of domains, with medical applications being one of the most intriguing and quickly developing. AI has the ability to transform healthcare delivery by making it more personalized, efficient, and accessible (Yu et al., 2018).

DOI: 10.1201/9781003346289-6

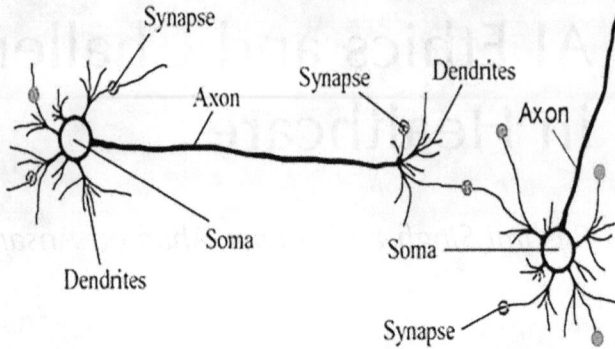

FIGURE 6.1   Biological neural network.

The capacity of AI systems to process vast volumes of data and derive conclusions or suggestions on the basis of that data is one of the most significant advantages of AI applications in the medical field. This is especially valuable in medical imaging, where AI algorithms can evaluate medical pictures like X-rays and MRI scans to discover diseases or anomalies. This can assist clinicians in making more precise diagnosis and improving patient outcomes. A neuron has a cell body, soma, a lot of fiber called dendritic, and a long fiber called axon, as shown in Figure 6.1. The human brain can be viewed as a multi-layered, nonlinear, parallel information processing system. The information processing and storage mechanisms in the brain are not fully independent; they may occur concurrently in the same neural network. In other words, these two procedures occur globally, not locally, in the neural network. The ability to learn is the most important feature of a biological neural network, and computers are used to replicate biological learning processes in order to fulfill the tasks that are required.

An artificial neural network is made up of a number of very simple processors known as neurons, which are analogous to biological neurons in the brain. There are weighted linkages between neurons that connect them as a full network. Signals are thus passed from one neuron to the next. The neurons' output signal will be split into multiple branches, each transmitting the same signal. The incoming connections of other neurons in the network end the outward branches. So far, neural networks have been widely used. In the field of computer science, there are a variety of algorithms and works based on neural networks that have been used to a variety of problems. Because of its self-learning and organizing abilities, neural networks are an excellent choice when several variables appear at random and the user needs to determine or clarify something based on them. In simple terms, a neural network (Moustafa & Slay, 2015) is a network that contains a number of neurons used to process information. A neural network consists of three main components, input layer, hidden layers, and output layer, as shown in Figure 6.2.

Another area in medicine where AI is being used is medication discovery and development. AI algorithms are able to analyze huge volumes of biological data, which can lead to the discovery of novel pharmaceuticals and a more effective

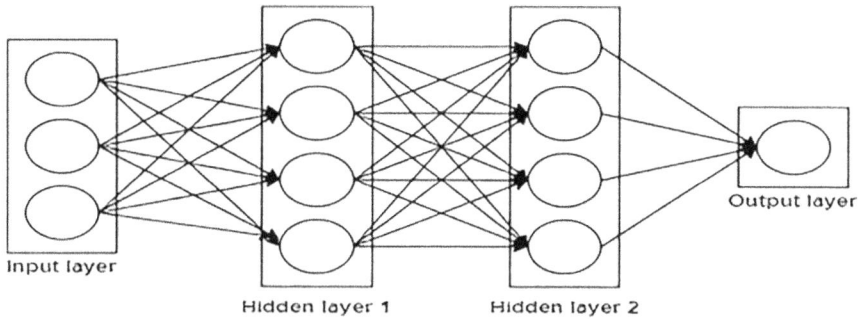

FIGURE 6.2 Neural network components.

treatment for diseases. AI may also help in the development of personalized medicine by identifying individualized treatment options based on a patient's genetic profile. This is one way AI can contribute to the field of personalized medicine.

AI can also be used to improve healthcare workflow by automating regular tasks and decreasing the time necessary for administrative chores. This could help healthcare providers concentrate more on providing care for patients and cut down on wait times. In addition, AI can be used to help clinical decision making by providing pertinent information and suggesting treatment alternatives based on a patient's medical history and present condition. This can be accomplished by providing information from AI systems that is relevant to the situation at hand.

The implementation of AI in medicine presents a number of formidable obstacles, one of the most significant of which is guaranteeing the safety and dependability of these systems. Since the quality of AI algorithms is directly proportional to the data they are trained on, it is of the utmost importance to guarantee that the data used for training is of a high standard, covers the relevant ground, and is devoid of any biases. In addition, before being implemented in therapeutic settings, AI algorithms must first be validated and subjected to extensive testing.

In conclusion, artificial intelligence has the potential to revolutionize the healthcare industry while also delivering major benefits to patients as well as to healthcare practitioners. However, it is essential to make certain that AI is applied in a responsible and ethical manner, and that its adoption is driven by a clear knowledge of the limitations of the technology as well as the possible threats it may pose. The medical field may pave the way for a brighter future for everyone if it accepts artificial intelligence (AI) and collaborates with others to speed up its research and development.

6.3 GROWTH FACTOR OF AI IN HEALTHCARE

The use of artificial intelligence (AI) has had a huge impact on the healthcare business as a result of the enormous improvements that it has made in a variety of fields, including healthcare. There are a number of variables that are driving the expansion of AI in the healthcare industry. These factors include the rising availability of data, advancements in computing power and algorithms, and the growing need for healthcare services that are more tailored and efficient.

The availability of more data is one of the key factors contributing to the expansion of AI in the healthcare industry. The proliferation of electronic health records, the expanding usage of wearable devices, and the expansion of telemedicine are all factors that have contributed to the expansion of the amount of data collected in the medical field. This data is being used to train artificial intelligence systems, which are then delivering diagnoses, treatment suggestions, and predictions about the progression of disease that are more accurate and dependable.

The development of more powerful computers and more efficient algorithms is also fueling the expansion of AI in the medical field. It is now possible, thanks to the rising availability of powerful computing systems, to process and analyze massive volumes of data in real time. This has resulted in the creation of new artificial intelligence algorithms that are capable of performing complicated tasks such as natural language processing, image recognition, and predictive analytics.

The rising need for healthcare services that are both more individualized and more efficient is another significant element that is driving the expansion of AI in the healthcare industry. Patients are increasingly interested in receiving care that is individualized to their specific requirements and preferences as a direct result of the growing influence of consumerism in the healthcare industry. AI is assisting in meeting this demand by offering patients individualized recommendations that take into account the patient's medical history, present state, and other relevant information.

In addition, the growing need for healthcare services that are also efficient financially is pushing the expansion of AI in the healthcare industry. Artificial intelligence has the potential to improve diagnostic and therapeutic precision, as well as streamline clinical workflow and cut the amount of time needed to complete administrative duties. Artificial intelligence has the potential to make medical treatment more accessible and inexpensive to patients by lowering the expenses associated with providing medical care.

Automating mundane processes and offering support for clinical decision making are two ways that artificial intelligence (AI) is assisting the healthcare industry in overcoming an issue that it is now facing: a shortage of competent healthcare personnel. This enables healthcare personnel to concentrate their efforts on more challenging and important activities, which has the potential to both increase the quality of care and lower the chance of making mistakes.

In spite of the numerous positive effects that artificial intelligence can have in the medical field, there are still many obstacles that need to be overcome before it can be used in a responsible and ethical manner. Before artificial intelligence systems can be utilized in clinical settings, they need to undergo exhaustive validation and testing beforehand. This is one of the most difficult tasks because it demands verifying the safety and reliability of AI systems. In addition, the ethical implications of utilizing AI in healthcare must be thoughtfully explored. This includes ensuring that algorithms are devoid of any biases and that patient privacy is safeguarded (Vyas & Gupta, 2022).

In conclusion, the growth factor of AI in healthcare is driven by a mix of reasons, some of which include the rising availability of data, advancements in computing power and algorithms, and the growing need for more personalized and efficient healthcare services. Even if there are problems that need to be solved, the potential benefits of AI in healthcare are tremendous, and its influence on the sector is anticipated to be significant in the years to come.

6.4 ETHICAL ISSUES IN AI-DRIVEN HEALTHCARE

6.4.1 Transparency in AI-Based Healthcare

Transparency in AI-based healthcare is critical for ensuring that AI systems are used in a responsible and ethical manner. Transparency refers to the degree to which AI systems are understandable, explainable, and accountable (Racine et al., 2019). In the context of healthcare, transparency is important for a number of reasons:

Patient safety: AI-based healthcare systems make decisions that can impact patient health and well-being. Transparency ensures that these decisions are based on accurate and reliable data and that patients are aware of the factors that influence these decisions.

Trust: Patients and healthcare providers must trust AI systems in order for them to be adopted and used effectively (Obermeyer, 2019). Transparency builds trust by allowing stakeholders to understand how AI systems make decisions and by providing a clear explanation of the factors that influence these decisions.

Compliance: Healthcare organizations must comply with a number of regulations, such as the General Data Protection Regulation (GDPR) and the Health Insurance Portability and Accountability Act (HIPAA), that require transparency and accountability in the use of patient data.

Ethical considerations: AI-based healthcare systems must operate in an ethical and responsible manner, taking into account the rights and well-being of patients. Transparency is an important tool for ensuring that AI systems are used in a manner that is consistent with ethical principles.

Transparency in AI-based healthcare can be achieved through a number of mechanisms, including (Vyas et al., 2023):

Explainable AI (XAI): XAI refers to AI systems that can provide clear and understandable explanations of their decision-making process.

Auditing: Regular auditing of AI systems can ensure that they are operating in a transparent and accountable manner.

Data transparency: Healthcare organizations must be transparent about how patient data is collected, processed, and used by AI systems (Lecher, 2018).

Regulation: Governments can play an important role in promoting transparency in AI-based healthcare by setting clear standards and guidelines for the use of AI in healthcare.

Overall, transparency is a critical component of AI-based healthcare and is necessary for ensuring that AI systems are used in a responsible and ethical manner.

6.4.2 JUSTICE AND FAIRNESS

Justice and fairness in AI-based healthcare are critical considerations for ensuring that AI systems are used in a responsible and ethical manner. These concepts refer to the degree to which AI systems treat patients equitably and without bias.

Justice in AI-based healthcare is concerned with ensuring that AI systems are designed and implemented in a manner that promotes equitable access to healthcare services and resources. This includes ensuring that AI systems do not discriminate based on factors such as race, gender, age, or socio-economic status (digital diagnostics).

Fairness in AI-based healthcare is concerned with ensuring that AI systems make decisions that are unbiased and equitable. This includes ensuring that AI systems are not influenced by factors such as personal preferences or prejudices and that they do not perpetuate existing inequalities or injustices.

There are a number of strategies for promoting justice and fairness in AI-based healthcare, including:

Data quality: Ensuring that the data used to train AI systems is accurate, representative, and free from bias is critical for promoting fairness in AI-based healthcare.

Model transparency: AI systems should be designed and implemented in a manner that is transparent and explainable, so that decisions made by the AI system can be understood and scrutinized.

Bias detection and correction: AI systems should be designed to detect and correct for biases in the data used to train the system and in the decisions made by the system.

Regular auditing: Regular auditing of AI systems can help to identify and correct biases and ensure that the AI system is operating in a fair and equitable manner.

Stakeholder engagement: Engaging with patients, healthcare providers, and other stakeholders can help to ensure that AI systems are designed and implemented in a manner that is consistent with the values and needs of these stakeholders.

Justice and fairness in AI-based healthcare are important for ensuring that AI systems are used in a manner that is responsible and ethical and that they do not perpetuate existing inequalities or injustices (Pasricha, 2022). By considering these concepts and implementing strategies to promote them, we can help to ensure that AI-based healthcare is accessible, equitable, and beneficial for all.

6.4.3 Privacy in AI-Based Healthcare

Privacy is a critical concern in AI-based healthcare. AI systems handle sensitive health information and personal data, and it is important to ensure that this data is protected and used in a manner that is consistent with privacy laws and regulations.

There are a number of factors that contribute to privacy concerns in AI-based healthcare, including:

Data collection: Healthcare organizations must be transparent about how they collect and store patient data, and they must ensure that this data is protected against unauthorized access, theft, or misuse.

Data sharing: AI-based healthcare systems often share patient data with other organizations, such as research institutions, healthcare providers, and data brokers. It is important to ensure that this data is shared in a manner that is consistent with privacy laws and regulations and that patients are informed about how their data is being used.

Data retention: AI-based healthcare systems must have clear policies for retaining and deleting patient data, and they must ensure that this data is deleted in a timely and secure manner.

Data security: AI-based healthcare systems must be designed and implemented in a manner that protects patient data from unauthorized access, theft, or misuse. This includes implementing appropriate security measures, such as encryption, firewalls, and access controls.

To address privacy concerns in AI-based healthcare, it is important to implement a privacy-by-design approach, which involves considering privacy at every stage of the development and implementation of AI systems. This includes:

Data protection: Implementing appropriate data protection measures, such as encryption, firewalls, and access controls, to protect patient data.

Data minimization: Minimizing the amount of personal data collected, stored, and shared by AI systems to reduce privacy risks.

Data governance: Implementing clear policies and procedures for how patient data is collected, stored, and shared to ensure that this data is used in a manner that is consistent with privacy laws and regulations.

Privacy impact assessments: Conducting regular privacy impact assessments to identify and mitigate privacy risks associated with AI systems.

Overall, privacy is a critical concern in AI-based healthcare, and it is important to implement appropriate measures to protect patient data and ensure that AI systems are used in a manner that is consistent with privacy laws and regulations.

6.4.4 AUTONOMY IN AI-BASED HEALTHCARE

Autonomy in AI-based healthcare refers to the degree to which AI systems can make decisions and act independently, without the need for human intervention. The use of AI in healthcare has the potential to greatly improve patient outcomes by enabling more accurate diagnoses, faster treatment decisions, and more efficient use of resources. However, there are also concerns about the ethical implications of granting AI systems greater autonomy in healthcare decision making.

There are several factors that contribute to the debate around autonomy in AI-based healthcare, including:

Accuracy of AI systems: While AI systems have the potential to be highly accurate, there is still the possibility of errors and misdiagnoses. It is important to ensure that AI systems are thoroughly tested and validated before they are used in healthcare decision making.

Bias in AI systems: There is a risk that AI systems may be biased based on the data used to train them, or based on the algorithms used to make decisions. It is important to address and correct biases in AI systems to ensure that they make fair and equitable decisions.

Ethical considerations: There are a number of ethical considerations that must be taken into account when granting AI systems greater autonomy in healthcare decision making. For example, there may be concerns about the potential for AI systems to perpetuate existing inequalities or to make decisions that are not in the best interests of patients.

Legal and regulatory considerations: There are also legal and regulatory considerations that must be taken into account when granting AI systems greater autonomy in healthcare decision making.

For example, there may be privacy concerns around the use of patient data in AI systems or concerns about the liability of healthcare providers in the event of a misdiagnosis or other error.

To address these concerns, it is important to approach the use of AI in healthcare in a responsible and ethical manner. This includes:

Transparency: Ensuring that AI systems are transparent and explainable, so that the decisions made by the AI system can be understood and scrutinized.

Human oversight: Ensuring that AI systems are subject to human oversight, so that any errors or biases can be detected and corrected.

Ethical review: Conducting regular ethical reviews of AI systems to ensure that they are used in a manner that is consistent with ethical principles.

Stakeholder engagement: Engaging with patients, healthcare providers, and other stakeholders to ensure that AI systems are designed and implemented in a manner that is consistent with the values and needs of these stakeholders.

In conclusion, autonomy in AI-based healthcare is a complex issue that requires careful consideration of both the benefits and the potential risks. By approaching the use of AI in healthcare in a responsible and ethical manner, we can ensure that AI systems are used in a way that benefits patients and promotes better health outcomes.

6.5 LEGAL ISSUES IN AI-DRIVEN HEALTHCARE

- Legal Liability in AI-Based Healthcare Can Be a Complex Issue, as It Involves Considerations of Both Medical Liability and Technology Liability.

From a medical liability perspective, healthcare providers and organizations may be held responsible for any harm or injury caused by AI-based diagnostic or treatment systems, just as they would be for harm caused by any other medical device or treatment. This means that they must ensure that the AI systems they use are safe and effective and that they provide adequate training and support for the users of these systems.

From a technology liability perspective, manufacturers of AI-based healthcare systems may be held responsible for any harm or injury caused by these systems. This could include liability for defects in the design or manufacturing of the system, or for failure to provide adequate warnings or instructions to users. Manufacturers may also be held liable for any harm caused by AI systems that are not properly maintained or updated, or for systems that are used in ways that they were not intended to be used.

To minimize legal liability in AI-based healthcare, it is important for healthcare providers and organizations to implement robust quality control and risk management processes, and to work closely with AI system manufacturers to ensure that the systems they use are safe, effective, and well supported. Additionally, it is important for manufacturers to be transparent about the limitations and potential risks of their systems, and to provide clear and comprehensive warnings and instructions to users.

Overall, it is important for all parties involved in AI-based healthcare to be proactive in addressing legal liability and to take steps to ensure that these systems are used in a responsible and ethical manner that prioritizes the well-being and safety of patients.

- Data Protection and Privacy Are Critical Considerations in AI-Based Healthcare, as AI Systems Often Process and Store Large Amounts of Sensitive Medical Information. Ensuring the Protection of This Information Is Essential for Maintaining Patient Trust and Confidence in These Systems, as Well as for Complying with Legal Requirements.

Here are some key steps that can be taken to ensure data protection and privacy in AI-based healthcare:

Implement strong data security measures: This may include encryption of data in transit and at rest, secure authentication and access controls, and regular backups and disaster recovery plans.

Adhere to privacy laws and regulations: Healthcare organizations must comply with relevant privacy laws and regulations, such as the General Data Protection Regulation (GDPR) in Europe and the Health Insurance Portability and Accountability Act (HIPAA) in the United States (Daniel et al., 2019).

Obtain patient consent: Before using AI systems to process or store patient data, it is important to obtain the patient's informed consent and provide clear information about how their data will be used, processed, and protected.

Ensure data minimization: AI systems should only collect and process the minimum amount of data necessary to perform their intended function. This reduces the risk of unauthorized access to or misuse of sensitive information.

Regularly monitor and audit the system: Regular monitoring and auditing of AI systems can help identify any security vulnerabilities or data breaches and allow for prompt remediation.

Implement data deletion policies: It is important to have clear policies in place for the deletion of patient data once it is no longer needed for the intended purpose. This helps prevent the accumulation of unnecessary or sensitive information.

Provide transparency: Patients have a right to understand how their data is being used, processed, and protected by AI systems (Racine et al., 2019). Healthcare organizations should provide clear and transparent information about their data protection and privacy practices.

By implementing these and other data protection and privacy measures, AI-based healthcare organizations can help ensure that sensitive medical information is properly protected and that patients can trust that their information is being used in a responsible and ethical manner.

- Cyber Security Is a Critical Concern in AI-Based Healthcare, as AI Systems Process and Store Large Amounts of Sensitive Medical Information and Are Vulnerable to Cyber Attacks. Ensuring the Security of These Systems Is Essential for Protecting Patients' Personal and Medical Information, as Well as Maintaining the Integrity and Reliability of AI-Based Healthcare Services.

Here are some key steps that can be taken to improve cyber security in AI-based healthcare:

Conduct risk assessments: Regularly conduct risk assessments to identify potential security vulnerabilities and prioritize the implementation of security measures that address the highest risk areas.

Implement strong access controls: Ensure that access to AI systems and sensitive medical information is controlled and restricted to authorized users only (Obermeyer, 2019). This may involve the use of strong authentication and authorization mechanisms, such as multi-factor authentication and role-based access controls.

Encrypt sensitive data: Encrypt sensitive medical information to protect it from unauthorized access or theft. This includes both data in transit and at rest.

Regularly update software and systems: Regularly update software and systems to address known vulnerabilities and prevent attackers from exploiting these weaknesses.

Implement backup and disaster recovery plans: Develop and implement backup and disaster recovery plans to ensure that critical data and systems can be restored in the event of a security breach or other disruption.

Monitor network activity: Regularly monitor network activity to detect and respond to potential security threats in a timely manner.

Provide security awareness training: Provide security awareness training to employees to help them understand the importance of cybersecurity and how they can help protect sensitive medical information (Garattini et al., 2019).

By implementing these and other cybersecurity measures, AI-based healthcare organizations can help reduce the risk of cyber-attacks and ensure the security and privacy of sensitive medical information.

6.6 CONCLUSION

In this chapter, various legal and ethical challenges that are faced by the application of AI in the health sector are reviewed and discussed. As in the future the use of AI in healthcare will increase, therefore it needs to be morally and ethically responsible for all the pros and cons. There has to be work done on improving the algorithms to work on unbiased data to give more accurate results. People who are building the algorithms need to be accountable for the results generated by it as the health sector is the most crucial sector for any economy to progress and develop the trust of people in new technologies like AI.

REFERENCES

Daniel, G. et al. (2019). *Current state and near-term priorities for AI-enabled diagnostic support software in health care*. White Paper, Duke-Margolis Center for Health Policy.

Dudley, J.T., Listgarten, J., Stegle, O., Brenner, S.E., & Parts, L. (2015). Personalized medicine: From genotypes, molecular phenotypes and the quantified self, towards improved medicine. *Pacific Symposium on Biocomputing*, 342–346. doi: 10.1142/9789814644730_0033

Garattini, C. et al. (2019). Big data analytics, infectious diseases and associated ethical impacts. *Philosophy & Technology*, 32(1), 69–85.

Ghebreyesus, T. (2018). *Artifcial intelligence for good global summit*. World Health Organization. http://www.who.int/dg/speeches/2018/artificial-intelligence-summit/en/.

Lecher, C. (2018). *What happens when an algorithm cuts your health care*. Verge.

Mehta, N., & Devarakonda, M.V. (2018). Machine learning, natural language programming, and electronic health records: The next step in the artificial intelligence journey? *The Journal of Allergy and Clinical Immunology, 141*, 201921. doi: 10.1016/j.jaci. 2018.02.025e1.

Moustafa, N., & Slay, J. (2015). UNSW-NB15: A comprehensive data set for network intrusion detection systems (UNSW-NB15 network data set). In *2015 Military Communications and Information Systems Conference (MilCIS)* (pp. 1–6). https:// www.researchgate.net/publication/287330529_UNSW-NB15_a_comprehensive_data_ set_for_network_intrusion_detection_systems_UNSW-NB15_network_data_set

Obermeyer, Z. (2019). Dissecting racial bias in an algorithm used to manage the health of populations. *Science, 366*, 447–453.

Pasricha, S. (2022). Ethics for digital medicine: A path for ethical emerging medical IoT design. *IEEE Computer. 56*, 32–40.

Racine, E. et al. (2019a). Healthcare uses of artificial intelligence: Challenges and opportunities for growth. *Healthcare Management Forum*.

Vyas, S., & Gupta, S. (2022). Case study on state-of-the-art wellness and health tracker devices. In *Handbook of research on lifestyle sustainability and management solutions using AI, big data analytics, and visualization* (pp. 325–337). IGI Global.

Vyas, S., Gupta, S., & Shukla, V.K. (2023, March). Towards edge AI and varied approaches of digital wellness in healthcare administration: A study. In *2023 International Conference on Computational Intelligence and Knowledge Economy (ICCIKE)* (pp. 186–190). IEEE.

Wang, T., Shen, F., Deng, H., Cai, F., & Chen, S. (2022). Smartphone imaging spectrometer for egg/meat freshness monitoring. *Analytical Methods, 14*, 508–517.

World Health Organization. (2018). A healthier humanity: The WHO investment case for 2019–2023. No. WHO/DGO/CRM/18.2.

Yu, K.H., Beam, A.L., & Kohane, I.S. (2018). Artificial intelligence in healthcare. *Nature Biomedical Engineering, 2*, 71931. doi: 10.1038/s41551-018-0305-z

7 The Future of the Healthcare System
A Meta-Analysis of Remote Patient Monitoring

Junaid Tantray, Sourabh Kosey, Jigna B. Prajapati, and Bhupendra Prajapati

7.1 INTRODUCTION

Remote patient monitoring includes various ways to take care of patients. It strives to enhance patient care through the digital transmission. The digital transformation in concern of health-related data which are collected on historical records. The accurate healthcare data enhances the patient-physician connection. The transparency in data will help a healthcare associate make an early diagnosis of illness. RPM (remote patient monitoring) is a subcategory of homecare tele-health. Telehealth enables patients to use various digital gazettes like mobile and other IoT-based medical devices to connect end-to-end note communication with healthcare associates. The latest technology is keeping track of patient health data (PGHD) by an active network and sending it to healthcare professionals on specified time intervals. Telemonitoring constantly keeps track of sudden changes in evaluating the variables discussed in clinical studies. Such changes also notify the respective people. RPM programs may be used to gather general information as routine changes in weight, other allergic affects, blood pressure, heart rate, and others. A customized telehealth computer system or software program that may be downloaded and installed on a computer, smartphone, or tablet is used to transmit patient data to a doctor's office once it has been gathered (Field & Grigsby, 2002).

Patients that require chronic, post-discharge, or geriatric care are typically helped by RPM. It can alert healthcare companies about possible health risks or maintain track of patient information between visits by connecting high-risk patients with remote monitoring. Additionally, organizations who wish to track workers' compensation claims and ensure that workers are on the appropriate route to returning to work can do so by using RPM.

DOI: 10.1201/9781003346289-7

7.1.1 EXAMPLES OF REMOTE PATIENT MONITORING TECHNOLOGY

RPM technology can include everything from mobile medical equipment to websites that let users enter their own data. Several instances include:

- Diabetes sufferers' glucose meters.
- Blood pressure or heart rate monitoring.
- IoT-based medical devices that can identify people with illnesses by monitoring from distance over the safe connected like dementia and notify medical personnel of a situation like a fall.
- Remote infertility treatment and monitoring.
- Drug misuse sufferers can be held responsible and kept on track with their objectives with the help of at-home testing.
- Caloric intake or diet logging programs.

In the delivery of routine medical treatment, online learning, and healthcare management, telehealth and telemedicine are currently becoming more pervasive. Many patients in underserved regions are obtaining services that they otherwise may not have been able to get without traveling long distances or getting through other transportation obstacles. Primary medical care to highly specialized treatment found in a top academic medical center are all services offered through telemedicine. Services are offered to patients of all ages, including the old and frail. Telehealth systems may be used in a variety of settings, including hospitals, clinics, nursing homes, rehab centers, assisted living residences, schools, jails, and health departments. The use of information technology to deliver healthcare services remotely is known as telehealthcare. It encompasses everything, such as inpatient or outpatient medical treatments. The physicians prefer continual assessment of the vital signs so they are constantly aware of the patient's history and how much has changed from yesterday to today. Once these results and data are accessible, a patient can receive treatment much sooner. Telemedicine has a lot to offer those who live in remote places and small towns. It is being utilized in almost all medical specialties (Nakamura et al., 2014).

Residents in these places have access to doctors or specialists who can do an accurate and thorough examination without the patient having to travel the usual distances to traditional hospitals. The capacity to see, debate, and evaluate patient concerns is now possible for healthcare workers thanks to new advancements in mobile collaboration technologies. With the help of telemedicine, skilled medical professionals may watch, demonstrate, and advise other medical professionals on how to do examinations more quickly and creatively. Patients in far-off places now have better access to healthcare. Through improved disease monitoring, pooled health professional personnel, and shorter travel distances, telemedicine lowers healthcare costs and increases efficiency. At both the patient and consultant doctor ends, the telemedicine system is made up of hardware and software, with certain diagnostic tools and a pathology microscope or camera available at the patient end (Farias et al., 2020).

The medical pictures and other info of the patients may be communicated to the consultant doctors over the satellite link using a telemedicine system made up of a basic computer and the required communication equipment. Since the day it became commercially available in every household, the use of handheld devices like the mobile phone for the purpose of receiving these data packets has been acknowledged. These packets are delivered to the doctor's office, where the data is rebuilt so the consultant doctor may examine the data, make the necessary diagnoses, discuss with the patient, and recommend the best course of action. The use of telemedicine and other communication technologies enables specialist doctors and patients who are thousands of kilometers apart to communicate and interact. Vital signs may now be sent from the hospital to a computer via a bio-signal collection device, according to current research (Blanchet, 2008).

In addition, more advanced cellular access technologies, including third generation (3G) and others, provide a consistent and high-speed Internet connection via smartphones and mobile modems than second-generation technology (2G). For our suggested technique, this relatively recent wireless communication technology is used. One of the fields where telemedicine is expanding significantly is cardiology (Mecklai et al., 2021). The leading cause of mortality worldwide is life-threatening cardiovascular illness, including heart attacks, arrhythmias, and strokes. Early identification and diagnosis are necessary for cardiovascular diseases (CVDs) prevention. A generic word for irregularities in the heart's rhythm is arrhythmia. By methodically analyzing the ECG, arrhythmias can be identified. One of the most widely utilized diagnostic methods for cardiac disorders nowadays is the electrocardiogram (ECG). In order to assess the significant data included in the ECG signal, a precise interpretation of the ECG (the electrical activity of the heart electrocardiogram) is generally necessary. By attaching electrodes to the body's surface, the ECG signal is obtained. It has been used for over a century as a diagnostic tool and is quite effective in detecting heart problems (Oleshchuk & Fensli, 2011; Ong et al., 2016; Sebastian et al., 2012). The traditional visual inspection of the ECG data is time-consuming and utterly ineffective. Instead, using changes in heart rate as a sign of a cardiac arrhythmia can be a quicker and more accurate way to identify heart disorders (Suh et al., 2011).

A person's heart rate in normal sinus rhythm ranges from 60 to 100 beats per minute. This rate could differ from person to person. Children often have a quicker heartbeat than older individuals, but the cardiac rhythm will always be regular (Ferrua et al., 2020). Bradycardia, for example, causes a heart rate that is less than 60 beats per minute, whereas tachycardia causes a heart rate that is higher than 100 beats per minute.

As a result, different cardiac arrhythmias can result from differences in heart rate. This work intends to provide a system for updating the patient's vital signs and generating an alarm signal based on differences in the patient's heart rate for the consulting doctor (Mohammed et al., 2014; Sharma et al., 2021). This warning is then sent to the consulting doctor's mobile phone, together with the patient's current vitals; enabling rapid decisions and prompt intervention on the part of the doctor should the patient's condition deteriorate. The historical backdrop of this particular undertaking is depicted in the following area. The data with respect to the arranged venture is remembered for the third part. The techniques used for the end result are portrayed in the fourth segment. The outcomes are made sense of in the fifth and sixth segments, separately.

7.2   ANDROID APPLICATION

Google and the Open Handset Alliance both provide the open-source TAndroid software stack for mobile devices. Every Android app is created using the Java programming language (Ondiege et al., 2017). A Software Creation Kit (SDK) includes the required tools and APIs are made available for the creation of applications.

An activity will transmit data to the server or request certain data from the server to show them to the user. The components can utilize the network adapter to do this. It enables both the doctor to view the image on the mobile phone and the user to create an account on the online application from the mobile device.

The system must have Android Market installed, and the user must log in to Android Market using their Google account. To log on to the C2DM server, you must use your account ID and password. Each gadget receives a unique ID when it has been registered. For communication between the server and C2DM server, the server provides this.

7.3   HOW REMOTE PATIENT MONITORING WORKS

The majority of technology uses comparable parts, even though RPM procedures might vary based on the gadget being utilized or the state being monitored. The first is a sensor that can record data and analyze specific physiological parameters wirelessly. In support of the connection, the storage carried a connection point for other sensors, databases of healthcare providers, and associated software (Lee et al., 1997). Applications usually provide consumers access to a user interface for tracking or analyzing data and displaying treatment suggestions.

The appropriate site receives and stores the data gathered by RPM devices in a various database. Healthcare facility analysis employs wireless telecommunications data in a single occurrence or as part of a comprehensive health history. As seen in Figure 7.1, the device frequently can alert patients when a medical practitioner has analyzed the data or finds a concern that requires the patient's visit.

7.4   BENEFITS OF REMOTE PATIENT MONITORING

Improved patient contribution: RPM devices let individuals take a functioning part in controlling and grasping their own clinical issues (Sharma et al., 2021). Better in general quality and worth based treatment on account of RPM's admittance to additional relevant patient information for patients and medical care experts. Further developed admittance to medical care – since RPM empowers people to do basic well-being tests all alone, more patients can get therapy from clinical

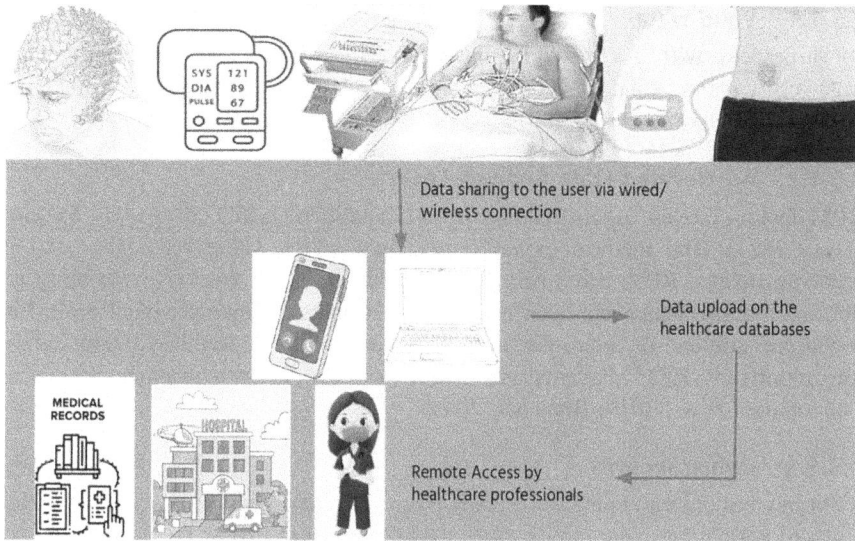

FIGURE 7.1 Diagrammatic sequence of how patient monitoring works.

specialists (El-Rashidy et al., 2021). More significant levels of schooling and backing RPM give patients data and criticism about their own circumstances every day, instructing them and offering help. Patient confirmation of steady checking can provide patients with a true serenity that any potential issues will be distinguished promptly (Khan et al., 2008).

7.5 RPM (REMOTE PATIENT MONITORING)

The advantages of patient checking to form good ways to incorporate the capacity to screen patients persistently, the capacity to distinguish sicknesses early and progressively, the counteraction of disease movement and troublesome demise, cost reserve funds from hospitalizations, a reduction in the quantity of hospitalizations, and the capacity to get more precise readings while as yet considering (Niyato et al., 2009; Taylor et al., 2021).

Work on the patient experience

Work on the soundness of populaces

Diminish the per capita cost of medical services

The advantages of patient remote checking include early and constant sickness location, capacity to screen patients persistently, avoidance of ailment deteriorating and sudden passing, cost reserve funds from hospitalizations, decrease in the quantity of hospitalizations, and capacity to get more exact appraisals as permitting.

7.5.1 IMPORTANT ASPECTS OF RPM

The second generation of remote monitoring technology is known as RPM. The main components of RPM are as follows.

7.5.1.1 Valid Data

RPM devices with AI (artificial intelligence) capabilities gather accurate and trustworthy patient data. RPM devices analyse, evaluate, and quantify reports using cutting-edge algorithms.

7.5.1.2 An Elevated Data Analysis Level

RPM devices provide automated responses if a patient's health changes (El-Rashidy et al., 2021). The medical expert reacts right away. These remedies assist in reducing delays. RPM can easily do repeated tasks like weekly or monthly reporting. Smart watches and other wearable electronics can measure intelligence like blood pressure and oxygen saturation. This information is seamlessly transformed into reports by RPM. Patients and health professionals can examine these data (Abdolkhani et al., 2019; Blanchet, 2008).

7.5.1.3 Cyber Security

RPM devices are vulnerable to privacy risks since they record the location, which contains patients' private information (Abdolkhani et al., 2019; Giger et al., 2015). To protect all stakeholders from security risks, RPM has a stronger cyber security network that works effectively.

7.5.1.4 Accessibly

The RPM record is accessible by both patients and caregivers at any time and from any location. Making correct health decisions and improving clinical operations are made easier with immediate access.

7.5.1.5 Cost-Efficient

RPM aids in lowering labor and administrative expenses.

7.5.2 APPLICATIONS

Peripheral devices with sensors collect subjective patient data and physiological measures like blood pressure. Pulse oximeters, glucometers, and blood pressure cuffs are a few examples of peripheral equipment. Wireless telecommunications equipment is used to communicate the data to healthcare organizations or other parties (Giger et al., 2015; Labrique et al., 2013). The data is well examined by a professional person associated with healthcare. The same can be done with a clinical decision support algorithm for all predicted issues, if any are discovered. The patient, caregivers, and healthcare professionals are immediately informed.

Therefore, prompt intervention guarantees successful patient outcomes. The more advanced programs also offer a way for the patient and the doctor to communicate with one another as well as alerts that remind you to take a test or fill a prescription.

The next section outlines RPM's applications; however, RPM is useful in a wide range of medical situations (Aalam et al., 2021).

7.5.2.1 Cancer

RPM use among cancer patients has been demonstrated to enhance overall results, with studies demonstrating reductions in re-hospitalization rates and increased efficiency in the use of healthcare resources. These remote monitoring techniques lessen pain intensity while also reducing depressive symptoms.

The RPM has increased cancer patients' life expectancy by 20%. Devices for remote patient observation support early treatments, prescriptions, chemotherapy adjustments, etc. (Hande et al., 2006). RPM has contributed to fewer cancer-related ER visits and shorter chemotherapy sessions. According to estimates, individuals with RPM had a hospitalization rate of 2.8%, compared to 13% for those without RPM (Ong et al., 2016).

7.5.2.2 COVID-19

Patients can get continuity of treatment from RPM after being released from the hospital if they have long-lasting COVID symptoms, medium to high oxygen desaturation levels (Coffey et al., 2021) without the need for hospitalization, or symptomatic COVID-19 (Muller et al., 2022). RPM is a crucial kind of treatment for individuals who are in danger, such as the elderly or those who have weakened immune systems, because of the nature of the pandemic (Tabacof et al., 2021).

Immunodeficiency is categorized by a diminished or absent immune system's ability to fight cancer and infectious diseases. Most instances are acquired ("secondary") due to factors outside the patient's control that have an effect on their immune system (Seshadri et al., 2020).

According to the study while pandemic, usage of RPM has lowered the demand for acute-care services along with the changed frequency of hospital admissions. The FDA has authorized the utilization of RPM technology in crisis situations to stop the spread of COVID-19 and avoid taxing healthcare professionals and resources (Fan et al., 2020).

7.5.2.3 Lung Diseases (COPD)

Chronic obstructive pulmonary disease (COPD), a progressive lung condition that affects people, is characterized by persistent respiratory symptoms and constrained airflow. The main symptoms are difficulty breathing and a cough that might or might not produce mucus (Palaniswamy et al., 2013). As COPD worsens over time, it becomes more difficult to do everyday tasks like dressing or walking. Although COPD is incurable, it may be prevented and treated. RPM may promote quicker healing times, more effective patient education, and collaborative decision making. Although there are contradictory facts, worries regarding cost and patients taking on more responsibility have also been addressed (Uddin et al., 2019).

7.5.2.4 Dementia & Falls

For individuals with dementia who are in danger of falling, RPM technology uses continuous monitoring to enhance safety and stop harm. The user or their adaptive equipment, also including canes and walkers, may have RPM sensors attached.

The sensors measure a person's position, gait, positive linear correlation, and initial acceleration (Hayajneh et al., 2016). They also employ a mathematical algorithm to detect changes in mobility, calculate fall risk, and alert caregivers in the event that a fall has taken place. Additionally, Wi-Fi, GPS, or radio frequency tracking technologies enable caregivers to locate senior wanderers (Alanazi & Daim, 2021).

7.5.2.5   Diabetes Mellitus

Blood pressure, weight, and blood glucose levels must all be under control when controlling diabetes. A collection of metabolic illnesses collectively referred to as diabetes are characterized by chronically elevated blood sugar levels (hyperglycemia). Incontinence, increased thirst, and increased hunger are typical symptoms (Alanazi & Daim, 2021). Diabetes can lead to a number of health issues if left untreated. Hyperosmolar hyperglycemia, diabetic ketoacidosis, and sometimes even mortality are examples of acute effects. Cardiovascular disease, stroke, chronic renal disease, foot ulcers, eye damage, nerve damage, and cognitive decline are some of the worst long-term complications (Su et al., 2019).

Patients and healthcare professionals may get early alerts and take necessary action due to the real-time dissemination of blood glucose and blood pressure data. There is proof that RPM diabetes treatment delivered every day is just as effective as therapy delivered through regular clinic visits every three months (Salehi et al., 2020).

7.5.2.6   Congestive Heart Failure (CHF)

Data on home monitoring for patients with heart failure underwent a thorough review, and it was discovered that RPM enhances quality of life, enhances patient-provider communication, reduces hospital visits, lowers death rates, and lowers healthcare system expenditures (Klersy et al., 2011).

Heart failure (HF), which has extremely high readmission rates, places a major financial strain on our healthcare resources. The care and prognosis of patients with HF can be greatly improved by remote monitoring (Ong et al., 2016). A period of subclinical hemodynamic decompensating that frequently precedes readmission for decompensated HF is when therapeutic measures might stop subsequent clinical decompensating and hospitalization. Several methods for remote patient monitoring include technological support with telephone, latest cutting-edge technology with implantable IoT-based devices, defibrillators, hemodynamic monitor, etc. Before this innovative technology is widely adopted, a number of medical, legal, and economical challenges must be resolved (Palaniswamy et al., 2013).

7.5.2.7   Infertility

For adequately screened people who were considering in vitro fertilisation (IVF) therapy, a recent study of a remote patient monitoring solution for infertility found that a six-month remote surveillance system had the same pregnancy rate as a round of IVF (Parimanam et al., 2022). The remote patient monitoring system and service used had a cost-per-patient of $800 as opposed to the typical cost of an IVF cycle, which is $15,000, demonstrating a 95% decrease in the cost of therapy for a similar outcome (Nirmala & Thanuja, 2014).

7.5.2.8   Surgery

According to a 2021 study, standard of care with no devices discovered 30% more medication errors, 10–14% fewer patient pain, and 5% fewer hospital readmissions than post-surgical remote patient monitoring. The RPM group spoke via text, chat, or video calls while being monitored remotely using at-home, clinical-grade vital sign monitoring equipment that transmitted their vital signs to a clinic portal (Bayliss et al., 2003). The standard of care group in the randomized experiment, which was conducted by PJ Devereaux and Michael McGillion, received the same standard of care as the other 905 patients (Parimanam et al., 2022). The researchers (Devereaux and McGillion) are now performing two further tests to look at the PVC RAM1's secondary results (PVC-RAM 2 and PVC-RAM 3) (McGillion et al., 2016).

7.5.2.9   Telemedicine in Prison Systems

In its jails throughout the latter 1980s, Florida conducted the first "basic" telemedicine experiments, which served as a precursor to RPM (Brunicardi, 1998). Glenn G. Hammack oversaw the creation of a ground-breaking telemedicine program at the University of Texas Medical Branch in Texas state jail from the early 1990s until 2007 in collaboration with Doctors Michael J. Davis and Oscar W. Boultinghouses (McCue et al., 2000).

7.5.2.10   Determine a Health Decline with Accuracy

RPM focuses on the prediction of sickness parameters more accurately. This has been challenging for populations with erratic disease progression. The measurement (e.g., spirometry, oximetry, or a hybrid) that would indicate the start of an exacerbation in the COPD group has been the subject of several investigations, but no definitive answer has been reached. RPM can be used in various demographic groups to provide metrics that can be looked at as potential predictors for future treatments and to track the progression of illness over time.

The population of people with heart failure may not receive adequate warning of decompensating from physiological signs (Mateo et al., 2019). In this group, readmission is typically not limited to physiological considerations and can often be a complex interaction of several factors. If the deterioration occurs too rapidly, there is not much time to take action. In order to accurately forecast patient health deterioration and define the optimal RPM strategy for this patient population, more study is required. The constant monitoring provided by implanted devices (like pacemakers) enables the early identification of comorbid conditions like atrial fibrillation, enabling proactive treatment. It can also improve the efficiency of clinical treatment for outside patient using IoT-based devices or historical issues (Lodenstein et al., 2013).

7.5.2.11   Deliver Prompt Treatment through a Responsive System

Patients must use the system (for example, timely data submission) in order for it to be beneficial, and clinicians must act appropriately and promptly when readings are outside of the normal range. Because they can minimize mistakes and delays brought on by manual entry, RPM systems that use automated data entry whenever possible are preferred (The bigger picture of planetary health, 2020). Standard RPM equipment is likely to be replaced by smartphone-based programs as technology advances, which could lead to patients providing data that is more consistently accurate and timely. RPM solutions must be straightforward to use (e.g., allowing effective data input, be transportable) and advantageous to patients in order to encourage long-term usage and engagement for inventions that depend on human data entry. Regular monitoring is also necessary. As an illustration, Srivastava et al. frequently checked data for anomalies or a loss of data; if a customer did not provide data for three days, nursing personnel made a call (Dahlgren & Whitehead, 2007).

7.5.2.12   Personalized Treatment

The effectiveness of an RPM intervention in lowering the utilization of acute care was also significantly influenced by the provision of a patient-centric and individualized approach. First, to ensure that the RPM innovation meets their needs and maximizes adoption and uptake, it must be codesigned with patients and their families (Frank et al., 2000). It will probably be necessary to tailor and occasionally repeat the instruction given to patients on how to operate the gadget. As indicated in Table 7.1, RPM alerts can also be customized by using individual data to establish alert levels.

7.5.2.13   Improve the Patient Experience

Meaningful use standards promote greater patient involvement, and it seems that remote monitoring of patients services can aid in this. By providing individuals new duties and supervised medical direction, RPM, together with further telehealth technologies, enables patients to participate more actively in the control of the medical conditions (Frank et al., 2000). Patients gain from an ongoing relationship with their medical team in addition to being more informed of their status (patients may check their vitals through their provider's patient portal). In reality, patients under RPM are often expected to report on a regular basis and provide essential medical information (Jamison et al., 2005).

TABLE 7.1

RPM Intervention

Context $\longrightarrow$	Mechanism $\longrightarrow$	Outcomes
RPM design (simple to use, co-design, automated data collection)	Two way interaction and communication between RPM team and patient	Staff respond alert
+	+	
Engaged and willing patients selected	Regular patient use and adherence to data collection	Timeliness of care
+	+	
RPM built into the workflow	Dedicated professional monitoring RPM	Optimized patient centered care

7.5.2.14   Improve the Health of Populations

Patients who are elderly or live in remote locations sometimes do not get the treatment they require. This could be as a result of a lack of specialized medical treatment in the area, problems with transportation, or difficulty getting about. The general health of both groups might considerably improve with the aid of RPM services (Brown et al., 2012). Patients in remote locations might get the care they require without having to go far. Elderly patients might be monitored remotely in the meantime to reduce needless doctor visits. This is significant because people who lack convenient access to healthcare frequently decide not to receive the essential medical care (Bradford et al., 2016).

7.5.2.15   Lower the Cost of Healthcare per Person

According to a 2008 analysis by Better Health Care Together based on Dr. Robert Litan's research, remote patient monitoring might save healthcare spending by $197 billion over the course of 25 years. It's true that recent telehealth projects have resulted in fewer hospital admissions and reduced total spending, but it's hard to determine with certainty at this time (Hanlon et al., 2017). For instance, a remote monitoring program used by the Veterans Hospital system reduced hospital admissions by 19% (McGillion et al., 2016). This implies that hundreds of people are burdening their healthcare system needlessly. Hospitals are more encouraged than ever to employ RPM solutions since CMS is fining healthcare organizations for high readmission rates (Jeddi & Bohr, 2020).

7.6   FACTORS PROMPTING THE EFFECTIVENESS OF RPM INTERVENTIONS

Nearly 70% of annual deaths worldwide are caused by non-communicable illnesses, including diabetes, chronic obstructive pulmonary disease (COPD), and heart disease. Healthcare systems throughout the world are under tremendous pressure to serve an increasing number of patients who are chronically ill due to the extra difficulty of an aging population. Due to their considerable resource needs, acute hospital admissions are one of the major causes of healthcare expenses for individuals with chronic diseases. Numerous innovative care models are being researched and tested in an effort to bring care into the home and decrease the need for unneeded acute care. A telehealth invention called RPM presents a number of options to improve healthcare outcomes, accelerate the administration of medicine, and maybe even shorten hospital stays and their related costs. Technology is used to remotely monitor a patient's physiological (such as heart rate and blood pressure) and behavioral (such as medication adherence and physical activity) data. With assistance, many individuals might be able to successfully manage their chronic conditions in the community. Despite the potential benefits of RPM, studies on its clinical and financial viability have so far shown conflicting findings. For instance, several thorough studies, meta-analyses, and reviews of reviews have been carried out in regard to the effects of RPM on the population with heart failure. Despite the bulk of these reviews being favorable regarding the potential advantages of RPM services for patients and healthcare professionals, some of them highlight that RPM services have little to no effect on lowering morbidity and mortality (Gupta et al., 2021).

A 2018 Cochrane analysis of patients with heart failure who were monitored remotely revealed no differences in hospitalizations, which varied from a 64% decline to a 60% rise, and no differences in all-cause death. In our most recent analysis, we summarized the information currently available on the effects of RPM on acute care utilization, including hospital admission events, hospital length of stay, and emergency department presentations.

RPM was shown to be associated with a reduction in the requirement for acute care in about 45% of trials. The remaining research largely found little or no change, while some studies indicated a rise in the demand for acute care (Tung et al., 2013).

The studies that were considered used a range of technologies and care models, and they examined a diversity of chronic diseases, countries, and healthcare organizations. Even though RPM might lessen the need for acute care, some enablers remain necessary.

Clinicians, researchers, and policymakers need more guidance to develop and implement RPM-facilitated models of care that will have the greatest effect (Fazio et al., 2015).

The fundamental causes of the significant variations in acute care consumption different RPM approaches must be further investigated. We made an attempt to comprehend the reasons behind the variations in the outcomes of RPM treatment. Analyzing how, why, and for whom remedies are helpful or ineffective allows for realist assessment. Consequently, the method has been used in a number of health initiatives (e.g., medical education programs, school feeding programs). According to the central principle of realism philosophy, an intervention's success is decided by the context in which it is carried out, which may cause processes that have both intended and unforeseen effects (Lodenstein et al., 2013). Realist evaluations are particularly helpful for complex therapies like RPM, because the efficacy is impacted by multiple linked elements including the intervention design, users, interpersonal relationships, institutions, and places where the intervention is administered, as shown in Table 7.2.

TABLE 7.2

Factors Including the Intervention Design, Users, Interpersonal Interaction

Early and Accurate Decline Detection	Selected Risk Points
Personalized care	Staff confident to discharge
Multiple component care (e.g., education)	Device easy to use
Timely response	RPM integrated into workflows
Self-management support	Observe effect

7.7   mHEALTH DEVELOPMENTS ARE WELLBEING FRAMEWORK FORTIFYING INSTRUMENTS

RPM solutions must be straightforward to use (e.g., allowing effective data input, be transportable) and advantageous to patients in order to encourage long-term usage and engagement for inventions that depend on human data entry (Hassan et al., 2018).

But their broad adoption or incorporation into wellness programs has been hindered by the lack of observational evidence proving the worth of these mHealth developments in terms of cost, execution, and wellbeing effects. The difficulties and competing demands that low- and middle-income legislative bodies must deal with prevent them from implementing innovations. To examine mHealth efforts in conjunction with fundamental health treatments and get recommendations on which mHealth arrangements to investigate in order to meet more challenging wellness framework goals, they need robust, trustworthy evidence concerning these initiatives. Their capacity to bear framework precariousness or disappointment can be low, in any event, when the norm might be similarly, or more, temperamental.

Greater scope viability and execution research activities are attempting to close the gaps in the evidence and demonstrate how the objectives for the healthcare system are impacted by mHealth hypotheses. There are several current projects to organize such discoveries.

7.7.1   mHEALTH AS A WELLBEING FRAMEWORK REINFORCING INSTRUMENT

Recent mHealth polls imply that rather than focusing on clear-cut mHealth solutions, innovators should focus on the overall wellness criteria that fundamental mHealth drives. International organizations and research associations have also attempted to situate mHealth interventions within the broader context of wellness framework aims or wellness outcomes. The phrase "wellbeing framework" recalls actions of all kinds whose primary goal is to improve, reestablish, or maintain health (Alanazi & Daim, 2021). We accept that a few previously proposed components of a structure for evaluating the effectiveness of health frameworks by connecting the aims of the framework to its essential capabilities can serve as a model for outlining and justifying mHealth drives and endeavors.

7.7.2   A COMMON SYSTEM TO MAKE SENSE OF mHEALTH DEVELOPMENTS

Attempts to identify, catalog, and organize evidence across this confusing scenario will continue to be hampered by the lack of a consistent vocabulary and method to portray mHealth mediations. Furthermore, it is challenging to explain mHealth innovations to partners in the traditional healthcare sector due to the lack of a uniform system.

The "mHealth and ICT System" (Figure 7.2) was created in collaboration with the World Health Organisation, the Johns Hopkins University Worldwide mHealth Drive, UNICEF, and frog Configuration. This aims to demonstrate how mHealth has advanced in the area of regenerative, maternal, infant, and child wellbeing

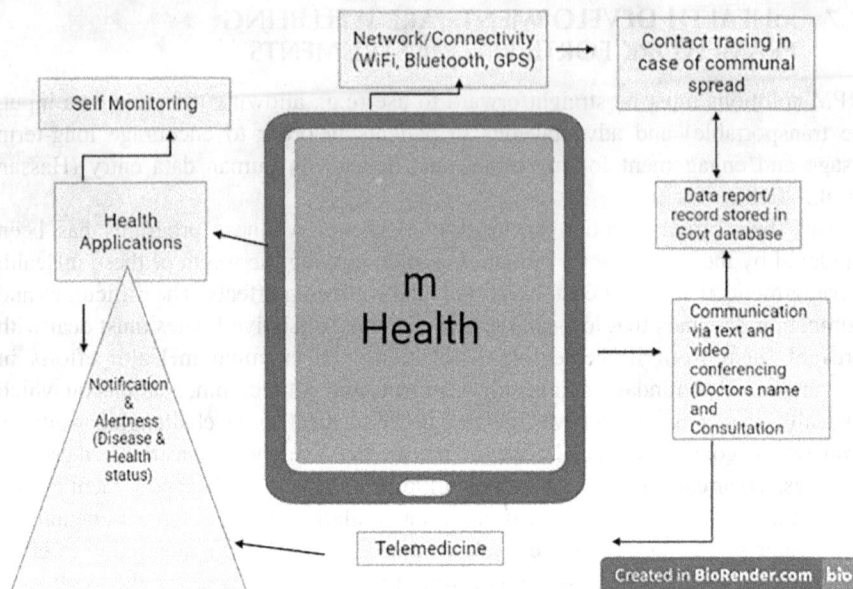

FIGURE 7.2   The mHealth and ICT structure.

(RMNCH), where multi-functional breakthroughs in wellbeing have been widely used over the past ten years around the globe. Unlike the proposed mHealth administration package, which is focused on the aims of the common health framework, our organizational design puts the intended recipients and consumers at the center of attention. The framework shows when, for whom, how is being lightened which imperatives, and the how of the approach by illustrating a specific mHealth system or strategy. Two crucial parts of the structure are as follows: A location to explain the specifics of the mHealth mediation, which is defined as one or more common mHealth or ICT apps used to address certain health framework requirements or challenges within a distinct section of the RMNCH continuum of care (Lee et al., 2013).

A visual portrayal of mHealth execution through the idea of "touch focuses," or resources, which depict the particular mHealth communications across wellbeing framework entertainers (for instance, clients, suppliers), areas (like centers or emergency clinics), and timings of connections and information trade.

7.7.3   NORMAL MHEALTH AND ICT APPLICATIONS

The initial part of the structure aims to address a previously acknowledged problem in mHealth: methodically illustrating the components of a mHealth methodology or platform. n order to achieve this, we establish connections between common ICT and mHealth usage and the wellness framework imperatives. The list of 12 typical mHealth applications has been assessed by a vast variety of mHealth partners and thought leaders, spanning from academic specialists using various emphases

TABLE 7.3

Client Education and Behavior Change and Communication between Them and Providers

Client Education and Behavior Change Communication (BCC)	Provider to Provider Communication User Groups, Consultation
Sensors and point of care diagnostics	Provider: work planning and scheduling
Data collection and reporting	Provider: training and education
Electronic health records	Supply chain, management

(Labrique et al., 2013). Even while a few mobile health programs only distribute one application, the majority always send at least two (Figure 7). Also, mHealth projects utilize at least one cell phone capability, for example, short message administration (SMS) and intuitive voice reaction (IVR), to achieve the normal applications (Table 7.3).

7.8 LIMITATIONS

RPM is significantly impacted by someone's desire to keep their health fit. RPM implementation is probably doomed to failure even in the absence of patient motivation to actively engage in their care. The lack of defined reimbursement guidelines for RPM services may make it challenging to deploy them in clinical settings. Questions concerning liability are raised by the RPM responsibility transfer. There are no clear guidelines on when doctors must act after receiving an alarm, irrespective of the urgency. The continual input of patient data must be managed by a trained team of healthcare experts, which may potentially require more work. Both RPM and health informatics technologies struggle with the same issues. RPM employs a variety of methods, regardless of the comorbidities being evaluated. In remote places, it might not be possible or practical to implement RPM since it requires a sizable wireless telecommunications infrastructure. Since RPM requires communicating sensitive patient data across telecommunications networks, information security is a concern. The potential for RPM's cyber security issues, particularly the potential for cyber assaults that might steal private medical data, is under debate. The vast majority of remote monitoring tools are also only suitable for single-user applications. Future growth might combine multi-user technology more effectively (Perl et al., 2013).

7.9 CONTROVERSY

The *New England Journal of Medicine* published the results of a randomized controlled trial including patients with congestive heart failure and found no evidence that telemonitoring was more effective than standard therapy. The telemonitoring patient group was given instructions to dial a specific number each day and enter their answers to a series of keypad questions regarding their symptoms.

The technique described differs noticeably from the RPM approach illustrated in the overview, which entails the actual collection and exchange of physiological data utilizing point-of-care equipment. It might be challenging to dispel the myth that telemonitoring and remote patient monitoring are comparable, particularly in light of *Forbes* headlines linking RPM to the bad findings. Despite the fact that remote patient monitoring is a more practical method of treating elderly patients at home, particularly in an ongoing situation like the current COVID-19 pandemic, researchers at the Semnan University of Medical Science have discovered that it is difficult for doctors to maintain control over their care while not directly responsible for it.

It is difficult to discern between the various technologies aided patient monitoring techniques since RPM nomenclature and descriptions are inconsistent. Researchers assert that different RPM procedures have varying degrees of effectiveness and that more investment should go toward developing technologies that do away with the shortcomings of existing methods (Hassan et al., 2018).

7.10 SOME ORGANIZATIONS THAT ARE SURPRISING TELEMEDICINE

India achieved a significant milestone in its digital health project in Walk 2022 by finishing 170,000 teleconsultations in a single day, utilizing its national telemedicine service, "eSanjeevani." The epidemic has significantly impacted India's medical facilities, despite the fact that the majority of Indians have received the coronavirus vaccination. The number of persons admitted to clinical offices in cities and towns declined as resources were redirected to coronavirus patients. Getting face-to-face therapy was hazardous and challenging for persons with chronic diseases or other moderate conditions. Telemedicine has existed in some form since the 1950s. The majority of patients have always favored in-person encounters with their primary care providers, despite the computer technology of many businesses. Despite the fact that by 2020 telemedicine use was on the rise, the RAND Organization ("research and development") data estimated that just 4% of Americans were using it.

After the coronavirus hit, telemedicine usage shot through the roof. For instance, UCLA Wellbeing in Los Angeles reported an increase in the daily average of telehealth visits from 100 to between 3,000 and 4,000 between spring 2020 and May 2020. Consumers' recently discovered preference for telehealth will persist, just like other aspects of our "new normal." It is anticipated that interest in telemedicine will rise by 38% over the next five years as more and more individuals are eager to benefit from the accessibility of contemporary medical treatments.

To address this tendency, innovative telemedicine firms are pushing forward. Here is a look at some of the new and established organizations that are broadening the scope of telemedicine and improving what it can achieve.

7.10.1 TELADOC

Teladoc might be the most notable telemedicine supplier. It offers every minute of every day admittance to specialists by means of telephone or video call, so clients can get clinical consideration any place they are and at whatever point they need it. Teladoc just stands apart for doing it especially well, with reliably high evaluations in application stores and an expansive client base (Palaniswamy et al., 2013).

7.10.2 Livongo

Livongo addresses the side of telehealth that is centered around creating customized wellbeing improvement plans. As opposed to tending to intense medical problems, the organization gives clients equipment that assists them with following their own wellbeing information, and afterward utilizes that information to propose ways they can make in the way of life enhancements (Uscher-Pines & Mehrotra, 2014).

7.10.3 Exploring Malignant Growth

Not all ailments can be treated online, and not all circumstances can be handled by patients alone. Innovative businesses are creating solutions for these more difficult circumstances as well. One example is Exploring Malignant Growth, a business that develops electronic phases especially made to assist sufferers through the cancer therapy with venturing. Patients benefit from better individualized therapy as a result of being able to share their side effects and unexpected outcomes. In the meantime, suppliers may utilize the platform to track patient progress, concentrate on understanding requirements, and tempt patients with useful information. Due to the stage's affiliation and relationship, a lot of businesses have committed to it, and 86% of them have endorsed it (Tung et al., 2013).

7.10.4 Amazon

A considerable lot of the more prominent organizations in telehealth are youthful new businesses, yet that doesn't mean laid-out organizations don't need a slice of the developing pie. Numerous tech goliaths are starting to put down wagers on telemedicine, including Amazon.

Telehealth doesn't seem like a field for an online business stage to play in; however, Amazon's development objectives aren't precisely conventional. It as of late constructed a telehealth administration called Amazon Care that has previously opened up to Amazon representatives. The organization desires to start growing Amazon Care to organizations around the nation over this mid-year.

Despite how it shows, doubtlessly telemedicine is the eventual fate of medical services. Computerized specialist visits and high-level clinical programming make medical care more proficient and more compelling. Throughout the following couple of years, this area will probably keep filling in new and fascinating ways (Bollyky et al., 2018).

7.10.5 Netmeds

Netmeds is an authorized e-drug store situated in Chennai. Dependence Retail procured a 60% stake in Net meds' parent Vitalic for roughly ₹620 crore.

7.11 CONCLUSION

While telemedicine cannot be the answer to every problem, it is crucial in treating a wide range of problems. Services like tele-wellness, tele-training, and tele-home medical care are becoming more and more popular in the medical profession. When all earthbound communication channels are disrupted in a disaster situation, the importance of satellite correspondences is highlighted. Global telemedicine initiatives are bringing people closer together, and at this point, geography is not a barrier to providing quality medical treatment. Telemedicine has not yet made the "blast" that it was expected to make, despite having so much promise. Lack of awareness and acceptance of new innovation, both from the general public and professionals, is what's holding it back. States are now starting to show a distinct interest in developing telemedicine initiatives to bring about a gradual but steady rise in its application in general wellbeing. Ideally, telemedicine methods will reach their full potential in a few years.

REFERENCES

Aalam, A.A., Hood, C., Donelan, C., Rutenberg, A., Kane, E.M., & Sikka, N. (2021, March 1). Remote patient monitoring for ED discharges in the COVID-19 pandemic. *Emergency Medicine Journal, 38*(3), 229–231.

Abdolkhani, R., Gray, K., Borda, A., & DeSouza, R. (2019, December). Patient-generated health data management and quality challenges in remote patient monitoring. *JAMIA Open, 2*(4), 471–478.

Alanazi, H., & Daim, T. (2021, August 1). Health technology diffusion: Case of remote patient monitoring (RPM) for the care of senior population. *Technology in Society, 66*, 101662.

Bayliss, E., Steiner, J.F., Fernald, D.H., Crane, L.A., & Main, D.S. (2003). Descriptions of barriers to self-care by persons with comboid chronic diseases. *Annals of Family Medicine, 1*(1), 15–21. 10.1370/afm.

Blanchet, K.D. (2008, March 1). Remote patient monitoring. *Telemedicine and e-Health, 14*(2), 127–130.

Bollyky, J., Lu, W.E., Schneider, J., & Whaley, C. (2018, July 1). Cost savings associated with usage and blood glucose control for members of the Livongo for diabetes program. *Diabetes, 67*(Supplement_1).

Bradford, N.K., Caffery, L.J., & Smith, A.C. (2016). Telehealth services in rural and remote Australia: A systematic review of models of care and factors influencing success and sustainability. *Rural Remote Health, 16*, 245.

Brown, R.S., Peikes, D., Peterson, G., Schore, J., & Razafindrakoto, C.M. (2012, June). Six features of medicare coordinated care demonstration programs that cut hospital admissions of high-risk patients. *Health Affairs (Millwood), 31*(6), 1156–1166.

Brunicardi, B.O. (1998). Financial analysis of savings from telemedicine in Ohio's prison system. *Journal of Telemedicine and Telecare, 4*, 49–54.

Coffey, J.D., Christopherson, L.A., Glasgow, A.E., Pearson, K.K., Brown, J.K., Gathje, S.R., Sangaralingham, L.R., Carmona Porquera, E.M., Virk, A., Orenstein, R., & Speicher, L.L. (2021, August 13). Implementation of a multisite, interdisciplinary remote patient monitoring program for ambulatory management of patients with COVID-19. *NPJ Digital Medicine, 4*(1), 1-1.

Dahlgren, G., & Whitehead, M. (2007). Tackling inequalities in health: What can we learn from what has been tried? Working paper prepared for the king's fund international seminar on tackling inequalities in health. In *European strategies for tackling social inequities in health* (Vol. 2). World Health Organization.

El-Rashidy, N., El-Sappagh, S., Islam, S.M.R., El-Bakry, H., & Abdelrazek, S. (2021, March 29). Mobile health in remote patient monitoring for chronic diseases: Principles, trends, and challenges. *Diagnostics*, *11*(4), 607.

Fan, K.G., Mandel, J., Agnihotri, P., & Tai-Seale, M. (2020, May 21). Remote patient monitoring technologies for predicting chronic obstructive pulmonary disease exacerbations: Review and comparison. *JMIR mHealth and uHealth*, *8*(5), e16147.

Farias, F.A., Dagostini, C.M., Bicca, Y.D., Falavigna, V.F., & Falavigna, A. (2020, May 1). Remote patient monitoring: A systematic review. *Telemedicine and e-Health*, *26*(5), 576.

Fazio, M., Celesti, A., Márquez, F.G., Glikson, A., & Villari, M. (2015, July 6). Exploiting the FIWARE cloud platform to develop a remote patient monitoring system. In *2015 IEEE Symposium on Computers and Communication (ISCC)* (pp. 264–270). IEEE.

Ferrua, M., Minvielle, E., Fourcade, A., Lalloué, B., Sicotte, C., Di Palma, M., & Mir, O. (2020, December). How to design a remote patient monitoring system? A French case study. *BMC Health Services Research*, *20*(1), 1–6.

Field M.J., & Grigsby J. (2002, July 24). Telemedicine and remote patient monitoring. *JAMA, 288*(4), 423–425.

Frank, C., Mohamed, M.K., Strickland, G.T., Lavanchy, D., Arthur, R.R., Magder, L.S. et al. (2000). The role of parenteral antischistosomal therapy in the spread of Hepatitis C virus in Egypt. *Lancet*, *355*(9207), 887–891.

Giger, J.T., Pope, N.D., Vogt, H.B., Gutierrez, C., Newland, L.A., Lemke, J., & Lawler, M.J. (2015, March 1). Remote patient monitoring acceptance trends among older adults residing in a frontier state. *Computers in Human Behavior*, *44*, 174–182.

Gupta, D., Gupta, M., Bhatt, S., & Tosun, A.S. (2021, August 10). Detecting anomalous user behavior in remote patient monitoring. In *2021 IEEE 22nd International Conference on Information Reuse and Integration for Data Science (IRI)* (pp. 33–40). IEEE.

Hande, A., Polk, T., Walker, W., & Bhatia, D. (2006, September 22). Self-powered wireless sensor networks for remote patient monitoring in hospitals. *Sensors*, *6*(9), 1102–1117.

Hanlon, P., Daines, L., Campbell, C. et al. (2017). Telehealth interventions to support self-management of long-term conditions: A systematic metareview of diabetes, heart failure, asthma, chronic obstructive pulmonary disease, and cancer. *Journal of Medical Internet Research*, *19*, e172.

Hassan, M.K., El Desouky, A.I., Elghamrawy, S.M., & Sarhan, A.M. (2018, August 1). Intelligent hybrid remote patient-monitoring model with cloud-based framework for knowledge discovery. *Computers & Electrical Engineering*, *70*, 1034–1048.

Hayajneh, T., Mohd, B.J., Imran, M., Almashaqbeh, G., & Vasilakos, A.V. (2016, March 24). Secure authentication for remote patient monitoring with wireless medical sensor networks. *Sensors*, *16*(4), 424.

Jamison, D.T., Lau, L.J., & Wang, J. (2005). Health's contribution to economic growth in an environment of partially endogenous technical progress. In G. Lopez-Casasnovas, B. Rivera, & L. Currais (Eds.), *Health and economic growth: Findings and policy implications* (pp. 67–91). Cambridge, MA: MIT Press.

Jeddi, Z., & Bohr, A. (2020, January 1). Remote patient monitoring using artificial intelligence. In *Artificial intelligence in healthcare* (pp. 203–234). Academic Press.

Khan, J.Y., Yuce, M.R., & Karami, F. (2008, August 20). Performance evaluation of a wireless body area sensor network for remote patient monitoring. In *2008 30th Annual International Conference of the IEEE Engineering in Medicine and Biology Society* (pp. 1266–1269). IEEE.

Klersy, C., De Silvestri, A., Gabutti, G., Raisaro, A., Curti, M., Regoli, F., & Auricchio, A. (2011, April). Economic impact of remote patient monitoring: An integrated economic model derived from a meta-analysis of randomized controlled trials in heart failure. *European Journal of Heart Failure, 13*(4), 450–459.

Labrique, A.B., Vasudevan, L., Kochi, E., Fabricant, R., & Mehl, G. (2013). mHealth innovations as health system strengthening tools: 12 common applications and a visual framework. *Global Health: Science and Practice, 1*(2), 160–171. 10.9745/GHSP-D-13-00031.

Lee, H.S., Park, S.H., & Woo, E.J. (1997, November 2). Remote patient monitoring service through World-Wide Web. In *Proceedings of the 19th Annual International Conference of the IEEE Engineering in Medicine and Biology Society. Magnificent Milestones and Emerging Opportunities in Medical Engineering (Cat. No. 97CH36136)* (Vol. 2, pp. 928–931). IEEE.

Lee, S.I., Ghasemzadeh, H., Mortazavi, B., Lan, M., Alshurafa, N., Ong, M., & Sarrafzadeh, M. (2013, November 1). Remote patient monitoring: What impact can data analytics have on cost?. In *Proceedings of the 4th Conference on Wireless Health* (pp. 1–8).

Lodenstein, E., Dieleman, M., Gerretsen, B. et al. (2013). A realist synthesis of the effect of social accountability interventions on health service providers' and policymakers' responsiveness. *Systematic Reviews, 2*, 98.

Mateo, M., Álvarez, R., Cobo, C., Pallas, J.R., López, A.M., & Gaite, L. (2019, May). Telemedicine: Contributions, difficulties and key factors for implementation in the prison setting. *Revista Espanola de Sanidad Penitenciaria, 21*(2), 95.

McCue, M.J., Hampton, C.L., Malloy, W., Fisk, K.J., Dixon, L., & Neece, A. (2000). Financial analysis of telecardiology used in a correctional setting. *Telemedicine Journal and e-Health, 6*, 385–391.

McGillion, M., Yost, J., Turner, A., Bender, D., Scott, T., Carroll, S., Ritvo, P., Peter, E., Lamy, A., Furze, G. et al. (2016, July). Technology-enabled remote monitoring and selfmanagement—vision for patient empowerment following cardiac and vascular surgery: User testing and randomized controlled trial protocol. *JMIR Research Protocols, 5*(3), e5763.

Mecklai, K., Smith, N., Stern, A.D., & Kramer, D.B. (2021, April 15). Remote patient monitoring: Overdue or overused? *The New England Journal of Medicine, 384*(15), 1384–1386.

Mohammed, J., Lung, C.H., Ocneanu, A., Thakral, A., Jones, C., & Adler A. (2014 September 1). Internet of Things: Remote patient monitoring using web services and cloud computing. In *2014 IEEE International Conference on Internet of Things (IThings), and IEEE Green Computing and Communications (GreenCom) and IEEE Cyber, Physical and Social Computing (CPSCom)* (pp. 256–263). IEEE.

Muller, A.E., Berg, R.C., Jardim, P.S., Johansen, T.B., & Ormstad, S.S. (2022, July 1). Can remote patient monitoring be the new standard in primary care of chronic diseases, post-Covid-19?. *Telemedicine and e-Health, 28*(7), 942–969.

Nakamura, N., Koga, T., & Iseki, H. (2014, January). A meta-analysis of remote patient monitoring for chronic heart failure patients. *Journal of Telemedicine and Telecare, 20*(1), 11.

Nirmala, S.G., & Thanuja, K. (2014, April). Wireless technology to monitor remote patients: A survey. *International Journal of Computer Networking, Wireless and Mobile Communications (IJCNWMC), 4*, 65–76.

Niyato, D., Hossain, E., & Camorlinga, S. (2009, May 5). Remote patient monitoring service using heterogeneous wireless access networks: Architecture and optimization. *IEEE Journal on Selected Areas in Communications, 27*(4), 412–423.

Oleshchuk, V., & Fensli, R. (2011, April). Remote patient monitoring within a future 5G infrastructure. *Wireless Personal Communications, 57*(3), 431–439.

The bigger picture of planetary health. (2020). Lancet Planet Health; 3. https://www. thelancet.com/journals/lanplh/article/PIIS2542-5196(19)30001-4/fulltext.

Ondiege, B., Clarke, M., & Mapp, G. (2017, February 24). Exploring a new security framework for remote patient monitoring devices. *Computers*, 6(1), 11.

Ong, M.K., Romano, P.S., Edgington, S., Aronow, H.U., Auerbach, A.D., Black, J.T., De Marco, T., Escarce, J.J., Evangelista, L.S., Hanna, B., & Ganiats, T.G. (2016, March 1). Effectiveness of remote patient monitoring after discharge of hospitalized patients with heart failure: The better effectiveness after transition–heart failure (BEAT-HF) randomized clinical trial. *JAMA Internal Medicine*, 176(3), 310–318.

Palaniswamy, C., Mishkin, A., Aronow, W.S., Kalra, A., & Frishman, W.H. (2013, May 1). Remote patient monitoring in chronic heart failure. *Cardiology in Review*, 21(3), 141–150.

Parimanam, K., Lakshmanan, L., & Palaniswamy, T. (2022, March 30). Hybrid optimization based learning technique for multi-disease analytics from healthcare big data using optimal pre-processing, clustering and classifier. *Concurrency and Computation: Practice and Experience*, 34, e6986.

Perl, S., Stiegler, P., Rotman, B., Prenner, G., Lercher, P., Anelli-Monti, M., Sereinigg, M., Riegelnik, V., Kvas, E., Kos, C., & Heinzel, F.R. (2013, November 30). Socioeconomic effects and cost saving potential of remote patient monitoring (SAVE-HM trial). *International Journal of Cardiology*, 169(6), 402–407.

Salehi, S., Olyaeemanesh, A., Mobinizadeh, M., Nasli-Esfahani, E., & Riazi, H. (2020, June). Assessment of remote patient monitoring (RPM) systems for patients with type 2 diabetes: A systematic review and meta-analysis. *Journal of Diabetes & Metabolic Disorders*, 19(1), 115–127.

Sebastian, S., Jacob, N.R., Manmadhan, Y., Anand, V.R., & Jayashree, M.J. (2012, September 1). Remote patient monitoring system. *International Journal of Distributed and Parallel Systems*, 3(5), 99.

Seshadri, D.R., Davies, E.V., Harlow, E.R., Hsu, J.J., Knighton, S.C., Walker, T.A., Voos, J.E., & Drummond, C.K. (2020). Wearable sensors for COVID-19: A call to action to harness our digital infrastructure for remote patient monitoring and virtual assessments. *Frontiers in Digital Health*, 8.

Sharma, N., Mangla, M., Mohanty, S.N., Gupta, D., Tiwari, P., Shorfuzzaman, M., & Rawashdeh, M. (2021, July 1). A smart ontology-based IoT framework for remote patient monitoring. *Biomedical Signal Processing and Control*, 68, 102717.

Su, D., Michaud, T.L., Estabrooks, P., Schwab, R.J., Eiland, L.A., Hansen, G., DeVany, M., Zhang, D., Li, Y., Pagán, J.A., & Siahpush, M. (2019, October 1). Diabetes management through remote patient monitoring: The importance of patient activation and engagement with the technology. *Telemedicine and e-Health*, 25(10), 952–959.

Suh, M.K., Chen, C.A., Woodbridge, J., Tu, M.K., Kim, J.I., Nahapetian, A., Evangelista, L.S., & Sarrafzadeh, M. (2011, October). A remote patient monitoring system for congestive heart failure. *Journal of Medical Systems*, 35(5), 1165–1179.

Tabacof, L., Kellner, C., Breyman, E., Dewil, S., Braren, S., Nasr, L., Tosto, J., Cortes, M., & Putrino, D. (2021, June 1). Remote patient monitoring for home management of coronavirus disease 2019 in New York: A cross-sectional observational study. *Telemedicine and e-Health*, 27(6), 641–648.

Taylor, M.L., Thomas, E.E., Snoswell, C.L., Smith, A.C., & Caffery, L.J. (2021, March 1). Does remote patient monitoring reduce acute care use? A systematic review. *BMJ Open*, 11(3), e040232.

Tung, H.Y., Tsang, K.F., Tung, H.C., Chui, K.T., & Chi, H.R. (2013, December 23). The design of dual radio ZigBee homecare gateway for remote patient monitoring. *IEEE Transactions on Consumer Electronics*, *59*(4), 756–764.

Uddin, M.A., Stranieri, A., Gondal, I., & Balasubramanian, V. (2019, October 21). A decentralized patient agent controlled blockchain for remote patient monitoring. In *2019 International Conference on Wireless and Mobile Computing, Networking and Communications (WiMob)* (pp. 1–8). IEEE.

Uscher-Pines, L., & Mehrotra, A. (2014, February 1). Analysis of Teladoc use seems to indicate expanded access to care for patients without prior connection to a provider. *Health Affairs*, *33*(2), 258–264.

8 Artificial Intelligence for Healthcare Delivery System
Future Prospective

Manish Soni and Sunita Rao

8.1 INTRODUCTION

Artificial intelligence (AI) initiatives are becoming popular as a means of achieving improvements in the health sector, particularly around the global coronavirus (COVID-19) pandemic, where new information, remote monitoring, and models are being developed at a rapid pace. AI research has led to the development of methods that can actually perceive radiographs, detect abnormal heart rhythms using only a wearable device, recognize media reports about contagious diseases automatically, identify heart-related risk factors from retinal images, and discover new drug and treatment targets. These models are successful as they have been targeted at regulated, tagged, and organized data points (Feldman et al., 2019; Liu et al., 2019; Majkowska et al., 2020; Perez et al., 2019; Poplin et al., 2018; Topol, 2019; Vamathevan et al., 2019). AI's ability to provide continuous, unwavering, and quick data analysis has the potential to revolutionize the current societal aspect of health promotion, prevention, and management, as shown in Figure 8.1. During the pandemic witnessed by people all over the world, people had developed a habit of writing and asking health-related queries on various platforms. Trending words like "read" were mentioned by almost 1.3 million research articles indexed by PubMed each year; "examine" posts from 1.5 crore facebook users; or "monitor" by 500 million tweets of people regularly talking about psychological poor health, foodborne diseases, or the common cold by the means social platform is obviously not a good way to deal with the health problems. But it somehow helps the community seeking answers related to their health, concerns, and troubles, and confronts all of them at the same time (Friedman, 2014; Medline, 2019; Twitter usage statistics, 2020).

Speech-based tools are frequently used in software agents and virtual assistants (VAs), including chatbots and "voice assistants" (VA). Chatbots and VAs are related in both their definition and use. By utilizing rule-based discourse or machine learning, both enable end users to occasionally speak with healthcare experts in their native language. Users allow multimedia communication (e.g., screen, text,

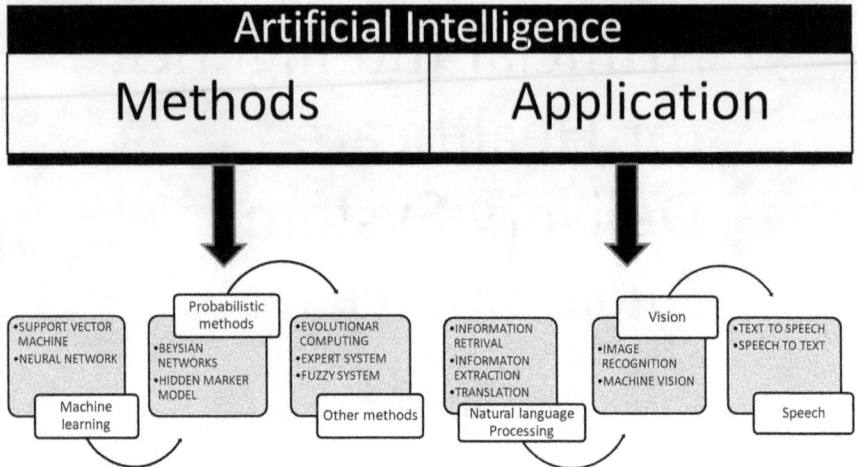

FIGURE 8.1   Artificial intelligence as methods and applications for a better approach.

speech, and sound) with VAs by engaging them primarily via voice interface and chatbots using text. They are generally thought to be system conversational agents that participate in multi-turn dialogues like Woebot, a treatment chatbot, whereas VAs are found to be an advanced technology that communicates and executes tasks using voice commands like Amazon's Alexa. Therefore, in hospitals, AI can automate and optimize tasks. Diagnostic tools powered by AI could be simpler, cheaper, and much more precise than the present methods. Machine learning algorithms could help with triage and scheduling. AI-enabled solutions may be able to increase efficiency by 30–50% in the nursing field alone. In developed economies, framework improvements, such as automating the repetitive processes, might reduce health spending as a percentage of gross domestic product (GDP).

The present chapter gives an overview of the functionalist AI more efficiently worldwide so that the developed and developing nations could work together to help each other by timely and effective data sharing so that the losses could be managed easily. With the advent of science, there is no doubt evidence of improving human lives. Therefore, AI and the latest medical care delivery system can change the world by facing the challenges in each and every situation of the world.

8.2   ROLE OF SENSORS IN THE HEALTHCARE SECTOR

As a result, in the change of technology over a period of time for wireless functionalities in sensors, two methods to sensor system development have emerged: (1) incorporation of Wi-Fi service or power transferring circuits in traditional sensors, and (2) production of motion detectors body signals by detecting radio frequency (Vyas & Gupta, 2022). The first approach uses a Bluetooth module to transmit signals to a smartphone while measuring the drain current of transistor-type pressure sensors. This approach can also be utilized with other kinds of sensors (Cheong et al., 2019; Shin et al., 2016). The passive sensing strategy is the second method. For instance, the development of split-ring resonator-type pressure sensors alters in response to pressure and can be applied to a variety of sensors and formats (Nie et al., 2019; Tseng et al., 2018). As an outcome, numerous studies of wearable sensing systems have demonstrated that they can closely touch the surface of body parts or be inserted into the body with little to no physiological tissue damage. Examples include stretchable electronic skin, soft neural interfaces with tissue-level stiffness, and smart corrective lenses made of soft contact lens materials that cause minimal eye strain (Chen & Pei, 2017; Yetisen et al., 2018). In recent times, non-visible or transparent wearable sensors reduce the user's pain in addition to wearable sensing devices that continuously collect biological signals while being attached to or fixed into the human body (An et al., 2016; Choi et al., 2017; Kim et al., 2019; Park et al., 2019).

8.3 ROLE OF SOFTWARE-BASED MOBILE DEVICES IN THE HEALTHCARE SECTOR

Smartphones with sensor technology in operating systems make a user more interested in online surfing, and software program execution according to personal interest. These are termed "smart" because these can deliver information at the push of a button when needed. Nowadays, smartphones are equipped with nearly all features, including cameras, video recorders, GPS navigation, games, email sending and receiving, and web search tools for a range of purposes. Modern-day utilities like start-ups, health industry, social life, education, banking, and other spheres are smartphones-user friendly by the development of application (Vyas et al., 2023). There are many smartphone applications for solving health-related issues and those have been proven to be useful in many clinical situations. The scientific application of the program varies in degrees of success, and this is due to their ease of learning and user efficiency (Pandey et al., 2013). An early study conducted ten years ago found that the usage of smartphones provided practitioners with immediate access to medical and health information, which had a positive effect on the healthcare system. This technology has enhanced telemedicine communication among hospital workers by reducing medical errors and improving decision making (Payne et al., 2012). The percentage of health professionals that use smartphones increased from 66% to 90% (Dubey et al., 2014). More than 85% of respondents to a digital survey conducted by the California-based Accreditation Council for Graduate Medical Education (ACGME) on the use of cell phones and software reported using a smartphone, with the iPhone being among the most popular (56%). Over half of the respondents gave an insight of different apps in their professional practice, with pharmacological guides (79%), medical calculators (18%), and coding and billing apps (4%). Textbooks/reference materials (average response: 55%), classification/treatment algorithms (46%), and general medical knowledge (43%) were the most frequently selected program categories (Franko & Tirrell, 2012). The findings also showed that all healthcare professional groups have an overwhelmingly favorable opinion regarding using mobile devices to coordinate patient care, which is corroborated by previous research that looked at coordination of care and communication among healthcare professionals (Krebs & Duncan, 2015; Ventola, 2014). Personal usage of such gadgets, on the other hand, may cause distraction, which can harm team cohesion and the quality of patient experience and care. This could explain less positive perceptions concerning influence on patient safety and ED collaboration and cohesion were reported in this study, particularly among attending physicians. It could also be due to a generation effect, in which attending physicians, who are older than the other professional groups, are less receptive to technology (Franko & Tirrell, 2012; Lane & Manner, 2011). But the situation has now changed due to global lockdowns faced by each corner of the world from March 2020; it was a necessity of people to be connected for their regular health check-up and mobile phones made it possible.

8.4   NATURAL LANGUAGE PROCESSING (NLP)

Machine learning (ML), a branch of artificial intelligence (AI) that aims to replicate human intellect, is the process of teaching computer systems to make predictions based on prior performance. The goal of NLP, a subfield of AI, is to develop algorithms and models that can use language in a manner similar to that of a human. It is commonly found in virtual assistants such as "Siri" and "Alexa," as well as Google searches and translations. It can be broadly divided into supervised and unsupervised learning. Supervised learning involves giving the computer access to research results, whereas unsupervised learning involves not providing any actual result data. Both techniques search for data patterns to predict outcomes, such as the presence or absence of cancer, survival rates, or risk groups. A method that is commonly used in the study of cancer and in the analysis of unstructured clinical data is natural language processing (NLP) (Yim et al., 2016). It translates free-text that isn't structured into a format that computers can understand, allowing labor-intensive procedures to be automated.

In order to achieve human-like language processing for a variety of activities or applications, NLP is defined as a series of potentially supercomputing approaches for evaluating and expressing naturally occurring texts (in oral or written human language) at one or more levels of language analysis (Liddy, 2003). NLP allows humans to converse with computers in natural languages. Multidisciplinary best describes the nature of NLP. Specifically in the area of human–computer interaction, it is a branch of computer science with research connections to the field of artificial intelligence (HCI). Languages, behavioral biology, psychiatry, ontology, and logical reasoning are all relevant to NLP (Copestake, 2003). NLP enables the analysis and extraction of information from unorganized inputs, as well as automated information retrieval, sentiment analysis, and text summarization (Topol, 2019). Since natural language (communication) is the only method for collecting knowledge and exchanging information in public health and medicine, NLP holds the key to releasing AI's full potential in the biomedical sciences. It is evident that the popularity of consumer voice assistants began with the release of "i-Phone Siri" in 2011. The technology behind speech recognition, "text-to-speech," "speech-to-text," NLP techniques, and conversational AI has been steadily advancing since Amazon debuted Alexa Voice Assistant and Echo devices in 2014 (Microsoft Voice Report 2019; Olson et al., 2019). Healthcare institutions have adopted VAs to interact with users seeking care services and knowledge and provide medical tips and guidelines, wellness headlines, updates on operational processes, correspondence, connectivity, first aid guidance, and more.

8.5 MEDICAL IMAGING TECHNOLOGY UTILIZING AI

The most popular catchphrase in recent years has been AI in medical imaging. Articles on the benefits of AI in medical imaging have emerged in the medical literature. AI in diagnostics can analyze imaging data to digital algorithms with the primary objective of detecting anomalies and quantifying disease processes. Computer-aided detection (CAD) has so far demonstrated exceptional accuracy and precision in several disease processes, including the diagnosis of breast carcinoma (microcalcification), colon cancers (polyps), and pneumonia. They have not ever, unfortunately, performed much better with the proper intervention of the experts. The primary flaw in CAD is that it just aims to detect lesions; it offers no information on how biologically active the lesions are, but the problem is AI algorithms cannot replace the inherent knowledge of diagnostic imaging specialists unless they are educated to differentiate between benign anomalies and clinically significant lesions. Is AI therefore useless for medical imaging? In actuality, it has great value. Support comes in the form of aided imaging in massive quantities.

8.6 ROLE OF AI IN CANCER MANAGEMENT

Cancer is a fatal condition that is frequently driven by several germline abnormalities and structural abnormalities. Cancer in its early-stage detection can lower the mortality rate because it allows more effective therapy before the tumor to be spread to other parts of the body. Early cancer detection, though, is not without its challenges. AI developments have opened up new possibilities for the effective and early identification of this lethal disease, as shown in Figure 8.2. Numerous studies have developed cancer prediction models with machine learning and deep learning assistance to identify cancer from previously available information with greater accuracy, sensitivity, and specificity. Numerous organizations, including the World Health Organization (WHO) and the International Alliance for Cancer Early Detection, recognize timely detection as the highest concern on a global scale (Vyas et al., 2023). Even in disease groups where screening programs are well established, such as breast cancer, there are continuous discussions regarding patient selection and risk-benefit trade-offs. Studies have shown that scanning can raise cancer in its early stage and can prevent death. Additionally, there are concerns raised about a perceived "one size fits all" strategy that is inconsistent with the goals of targeted therapies (Esserman, 2017; Maroni et al., 2021; Sasieni, 2003).

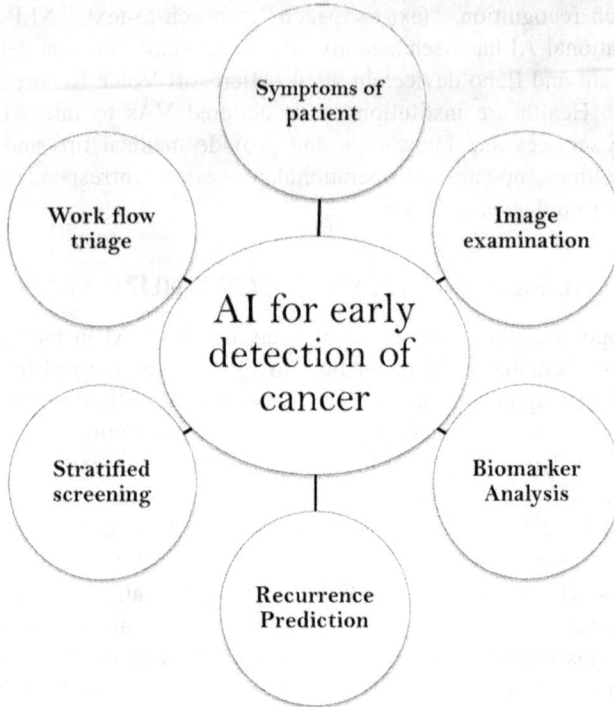

FIGURE 8.2 Role of artificial intelligence in cancer management for early detection of cancer.

8.7 REMOTE-CONTROLLED ROBOTIC SURGERY

The two primary AI technologies that currently support surgical practice are augmented reality and anatomical profiling. Machine learning and computer vision (CV) enable computers and systems to extract useful information from digital images, videos, and other visual inputs and to take actions or make suggestions based on that data (Liu et al., 2022; Mascagni, 2021). The performance of surgeons can be improved by applying CV software, which focuses on the sub-components of the human visual system including object detection, identification, motion extraction, or spatial understanding. These software have three main applications: the understanding of the surgical process, computer-aided detection, and computer-assisted navigation. For instance, tele-manipulative surgery can be performed from a remote booth using surgical robots like the da Vinci Surgical System (Chadebecq et al., 2020).

8.8 PRECISION MEDICINE

Akin to other fields, precision medicine is expanding quickly. The National Research Council previously referred to the development of "a new classification of human pathogens based upon molecular biology" as precision medicine.

Another way to put it is a revolution in healthcare brought on by the knowledge gained from sequencing the human genome (National Research Council, 2011). Since then, the field has made progress in comprehending how the fusion of data with medical history, social/behavioral characteristics, and environmental knowledge effectively characterizes health states, disease states, and therapy alternatives for persons who are afflicted (Ziegelstein, 2017). We will refer to the healthcare way-of-life and research agenda mentioned previously as precision medicine for the duration of this work and customized care to reflect the importance of the traditional way of individual care method.

8.9 EARLY SEPSIS DETECTION USING A DEEP NEURAL NETWORK

When a human being's body reacts incorrectly to an infection, it results in the deadly illness known as sepsis. For persons diagnosed with infections, sepsis is a rather frequent cause of death. The mortality rate is still too high and must be reduced. Proper assessment methods for sepsis prognostic evaluation could improve clinical judgment and reduce mortality (Wernly et al., 2021). A deep neural network (DNN) can assess a patient's clinical condition following ICU treatment and indicate the mortality rate within 96 hours of admission. The model may help physicians recognize patients with poor prognosis, enabling for "re-triage" and modification of treatment strategies. The creation of decision support systems that were dependent on advances in machine learning, as seen in Figure 8.3, is one area where healthcare techniques have advanced. One of the active topics is predicting the onset of septic shock (Yee et al., 2019). To enhance clinical outcomes and advance real-time resource optimization in

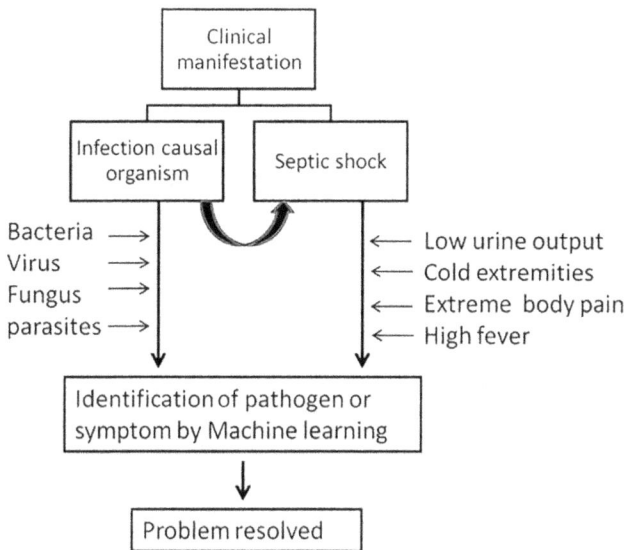

FIGURE 8.3 Role of machine learning in early diagnosis of sepsis.

healthcare, numerous studies have created intelligent decision support technologies for septic shock. In order to create a prediction model of septic shock, one study examined eight different machine learning algorithms (Misra et al., 2021).

8.10   IMPACT OF AI ON EMPLOYMENT IN DEVELOPED AND DEVELOPING NATIONS

During the United States (US)-European Commission (EC) Trade and Technology Council in late September 2021, the United States-EC both expressed a strong interest in working together on a joint study to assess the potential impact of AI on workforces. The Pittsburgh Declaration pledged a collaborative "economic research analyzing the influence of AI on the future of our workforces, with regard to results in employment, pay, and the distribution of labor market opportunities" (The White House, 2021). Definitely, it is clearly visible that a skilled workforce with a specialization in AI is required to deal with the situations. This can lead to problems and risks for patients that definitely require a completely different perspective of treatment that requires the human brain and promptness of the line of treatment.

8.11   DEPENDENCY OF DOCTORS OVER ARTIFICIAL INTELLIGENCE IN CLINICAL TERMS

Nowadays, a relatively limited use of clinical AI may be a reflection of clinicians need to change, as well as potential misunderstandings and adverse views (He et al., 2019; Scott et al., 2021). Naturally, clinicians will probably be the "earliest" believers of AI and will unavoidably operate it directly. Their perspectives need to be explored and comprehended because of their critical role in the acceptance and implementation of clinical AI. Future generations of doctors, who will be affected by AI-driven innovations, are medical students. Therefore, in order to create efficient education and health programs, research should be designed to comprehend their views. The body of research regarding how doctors and medical students feel about AI is expanding. However, there are variations across the world, but the majority of research and development was done in developed countries (Park et al., 2021; Sarwar et al., 2019). Even if there have been a few comprehensive evaluations on this subject, it is estimated that it only offers a limited insight (Santomartino et al., 2022; Scott et al., 2021). The opinions of doctors and medical students around the world must be understood while developing the tools by the leading healthcare industry partners that take a patient as a subject of monetary benefit.

8.12   CONCLUSION

The chapter can be summarized by stating that there is still a substantial gap between the creation of AI algorithms and their use in quantifiable settings. The clinical management function of infection specialists cannot be replaced by AI. AI-based algorithms should only be employed as development tools until they can include behaviors that are consistent with understood physiology and show that the results may be modified prospectively in a range of scenarios.

8.13 FUTURE PERSPECTIVES

In the healthcare sector, including surgeries, the possibility of developing AI-based healthcare apps may be on par with or better than that of clinicians. However, the sector of healthcare is intricate and constantly evolving, with strict requirements for security. Any technical malfunction could endanger patients. Doctors and safety engineers essentially have no control over the decision-making process when an AI system makes one, and it is challenging to completely comprehend how the AI system makes an accurate choice. AI-based tools lack ethical restraints and safety standards in comparison to conventional healthcare treatment. Nevertheless, all clinical settings specifications as displayed in the computer are very tough to achieve in the technical stage of tool designing. Therefore, it is required to make a road map of the healthcare delivery system for ethical conduct of practices with the changing needs of a growing population.

REFERENCES

An, B.W., Gwak, E., Kim, K., Kim, Y., Jang, J., Kim, J., & Park, J. (2016). Stretchable, transparent electrodes as wearable heaters using nanotrough networks of metallic glasses with superior mechanical properties and thermal stability. *Nano Letters*, *16*(1), 471–478. 10.1021/acs.nanolett.5b04134

Chadebecq, F., Vasconcelos, F., Mazomenos, E., & Stoyanov, D. (2020). Computer vision in the surgical operating room. *Visceral Medicine*, *36*(6), 456–462. 10.1159/000511934

Chen, D., Pei, Q. (2017). Electronic muscles and skins: A review of soft sensors and actuators. *Chemical Reviews*, *117*(17), 11239–11268. 10.1021/acs.chemrev.7b00019

Cheong, W.H., Oh, B., Kim, S., Jang, J., Ji, S., Lee, S., Cheon, J., Yoo, S., Lee, S., & Park, J. (2019). Platform for wireless pressure sensing with built-in battery and instant visualization. *Nano Energy*, *62*, 230–238. 10.1016/j.nanoen.2019.05.047

Choi, T.Y., Hwang, B., Kim, B., Trung, T.Q., Nam, Y.H., Kim, D., Eom, K., & Lee, N. (2017). Stretchable, transparent, and stretch-unresponsive capacitive touch sensor array with selectively patterned silver nanowires/reduced graphene oxide electrodes. *ACS Applied Materials & Interfaces*, *9*(21), 18022–18030. 10.1021/acsami.6b16716

Copestake, A.. (2003). *Natural language processing: Part 1 of lecture notes*. Cambridge: Ann Copestake Lecture Note Series.

Dubey, D., Amritphale, A., Sawhney, A., Amritphale, N., Dubey, P., & Pandey, A. (2014). Smart phone applications as a source of information on stroke. *Journal of Stroke*, *16*(2), 86–90. 10.5853/jos.2014.16.2.86

Esserman, L.J., WISDOM Study and Athena Investigators (2017). The wisdom study: Breaking the deadlock in the breast cancer screening debate. *NPJ Breast Cancer*, *3*, 34. 10.1038/s41523-017-0035-5

Feldman, J., Thomas-Bachli, A., Forsyth, J., Patel, Z.H., & Khan, K. (2019). Development of a global infectious disease activity database using natural language processing, machine learning, and human expertise. *Journal of the American Medical Informatics Association*, *26*(11), 1355–1359. 10.1093/jamia/ocz112

Franko, O.I., & Tirrell, T.F. (2012). Smartphone app use among medical providers in ACGME training programs. *Journal of Medical Systems*, *36*(5), 3135–3139. 10.1007/s10916-011-9798-7

Friedman, C., & Elhadad, N. (2014). Natural language processing in health care and biomedicine. In: E. Shortliffe, & J. Cimino (Eds.), *Biomed informatics*. London: Springer.

He, J., Baxter, S.L., Xu, J., Xu, J., Zhou, X., & Zhang, K. (2019). The practical implementation of artificial intelligence technologies in medicine. *Nature Medicine*, *25*(1), 30–36. 10.1038/s41591-018-0307-0

Kim, K., Park, Y., Hyun, B.G., Choi, M., & Park, J. (2019). Recent advances in transparent electronics with stretchable forms. *Advanced Materials*, *31*(20), e1804690. 10.1002/adma.201804690

Krebs, P., & Duncan, D.T. (2015). Health app use among US mobile phone owners: A national survey. *JMIR mHealth & uHealth*, *3*(4), e101. 10.2196/mhealth.4924

Lane, W., & Manner, C. (2011). The impact of personality traits on smartphone ownership and use. *International Journal of Business & Social Science*, *2*, 22–28.

Liddy, E.D. (2003). *Natural language processing. New York: Encyclopedia of library and information science*. Marcel Dekker Inc.

Liu, R., Greenstein, J.L., Sarma, S.V., & Winslow, R.L. (2019). Natural Language Processing of Clinical Notes for Improved Early Prediction of Septic Shock in the ICU. *Annual International Conference of the IEEE Engineering in Medicine & Biology Society. IEEE Engineering in Medicine & Biology Society. Annual International Conference* 2019 (pp. 6103–6108). 10.1109/EMBC.2019.8857819.

Liu, R., An, J., Wang, Z., Guan, J., Liu, J., Jiang, J., Chen, Z., Li, H., Peng, B., & Wang, X. (2022). Artifcial intelligence in laparoscopic cholecystectomy: Does computer vision outperform human vision? *Artificial Intelligence Surgery*, *2*(2), 80–92. 10.20517/ais.2022.04

Majkowska, A., Mittal, S., Steiner, D.F., et al. (2020). Chest radiograph interpretation with deep learning models: Assessment with radiologist-adjudicated reference standards and population-adjusted evaluation. *Radiology*, *294*(2), 421–431. 10.1148/radiol.2019191293

Maroni, R., Massat, N.J., Parmar, D., et al. (2021). A case-control study to evaluate the impact of the breast screening programme on mortality in England. *British Journal of Cancer*, *124*(4), 736–743. 10.1038/s41416-020-01163-2.

Mascagni, P., Alapatt, D., Urade, T., et al. (2021). A computer vision platform to automatically locate critical events in surgical videos: Documenting safety in laparoscopic cholecystectomy. *Annals of Surgery*, *274*(1), e93–e95. 10.1097/SLA.0000000000004736

MEDLINE. updated (2019)/11/19. PubMed Production Statistics; Accessed 2021/12/27. https://www.nlm.nih.gov/bsd/medline_pubmed_production_stats.html. Bethesda, (MD): United States National Library of Medicine.

Microsoft Voice Report 2019: From Answers to Action: Customer Adoption of Voice Technology and Digital Assistants. (2019). https://voicebot.ai/2017/07/14/timeline-voiceassistants-short-history-voice-revolution/.

Misra, D., Avula, V., Wolk, D.M., et al. (2021). Early detection of septic shock onset using interpretable machine learners. *Journal of Clinical Medicine*, *10*(2), 301. 10.3390/jcm10020301

National Research Council (2011). *Toward precision medicine: Building a knowledge network for biomedical research and a new taxonomy of disease*. Washington: The National Academies Press.

Nie, B., Huang, R., Yao, T., Zhang, Y., Miao, Y., Liu, C., Liu, J., & Chen, X. (2019). Textile-based wireless pressure sensor array for human-interactive sensing. *Advanced Functional Materials*, *29*(22), 1808786. 10.1002/adfm.201808786

Olson, C., & Kemery, K. (2019) https:. Microsoft Voice Report 2019: From Answers to Action: Customer Adoption of Voice Technology and Digital Assistants, Accessed June 1 2020. http://about.ads.microsoft.com/en-us/insights/2019-voice-report.

Pandey, A., Hasan, S., Dubey, D., & Sarangi, S. (2013). Smartphone apps as a source of cancer information: Changing trends in health information-seeking behavior. *Journal of Cancer Education*, *28*(1), 138–142. 10.1007/s13187-012-0446-9

Park, C.J., Yi, P.H., & Siegel, E.L. (2021). Medical student perspectives on the impact of artificial intelligence on the practice of medicine. *Current Problems in Diagnostic Radiology*, *50*(5), 614–619. 10.1067/j.cpradiol.2020.06.011

Park, Y., An, H.S., Kim, J., & Park, J. (2019). High-resolution, reconfigurable printing of liquid metals with three-dimensional structures. *Science Advances*, *5*(6), eaaw2844. 10.1126/sciadv.aaw2844.

Payne, K.F.B., Wharrad, H., & Watts, K. (2012). Smartphone and medical related app use among medical students and junior doctors in the United Kingdom (UK): A regional survey. *BMC Medical Informatics & Decision Making*, *12*(1), 121. 10.1186/1472-6947-12-121

Perez, M.V., Mahaffey, K.W., Hedlin, H., et al. (2019). Large-scale assessment of a smartwatch to identify atrial fibrillation. *New England Journal of Medicine*, *381*(20), 1909–1917. 10.1056/NEJMoa1901183

Poplin, R., Varadarajan, A.V., Blumer, K., et al. (2018). Prediction of cardiovascular risk factors from retinal fundus photographs via deep learning. *Nature Biomedical Engineering*, *2*(3), 158–164. 10.1038/s41551-018-0195-0.

Santomartino, S.M., & Yi, P.H. (2022). Systematic review of radiologist and medical student attitudes on the role and impact of AI in radiology. *Academic Radiology*, *29*. https://doi.org/S1076-6332(21)00624-3. 10.1016/j.acra.2021.12.032.

Sarwar, S., Dent, A., Faust, K., et al. (2019). Physician perspectives on integration of artificial intelligence into diagnostic pathology. *npj Digital Medicine*, *2*, 28. 10.1038/s41746-019-0106-0

Sasieni, P. (2003). Evaluation of the UK breast screening programmes. *Annals of Oncology*, *14*(8), 1206–1208. 10.1093/annonc/mdg325.

Scott, I.A., Carter, S.M., & Coiera, E. (2021). Exploring stakeholder attitudes towards AI in clinical practice. *BMJ Health & Care Informatics*, *28*(1), e100450. 10.1136/bmjhci-2021-100450

Shin, K., Lee, J.S., & Jang, J. (2016). Highly sensitive, wearable and wireless pressure sensor using free-standing ZnO nanoneedle/PVDF hybrid thin film for heart rate monitoring. *Nano Energy*, *22*, 95–104. 10.1016/j.nanoen.2016.02.012

Topol, E.J. (2019 Jan). High-performance medicine: The convergence of human and artificial intelligence. *Nature Medicine*, *25*(1), 44–56. 10.1038/s41591-018-0300-7, PubMed: 30617339

Tseng, P., Napier, B., Garbarini, L., Kaplan, D.L., & Omenetto, F.G. (2018). Functional, RF-trilayer sensors for tooth-mounted, wireless monitoring of the oral cavity and food consumption. *Advanced Materials*, *30*(18), e1703257. 10.1002/adma.201703257

Vamathevan, J., Clark, D., Czodrowski, P., et al. (2019). Applications of machine learning in drug discovery and development. *Nature Reviews Drug Discovery*, *18*(6), 463–477. 10.1038/s41573-019-0024-5.

Ventola, C.L. (2014). Mobile devices and apps for health care professionals: Uses and benefits. *Pharmacy and Therapeutics*, *39*, 356.

Vyas, S., & Gupta, S. (2022). Case Study on state-of-the-art wellness and health tracker devices. In *Handbook of research on lifestyle sustainability and management solutions using AI, big data analytics, and visualization* (pp. 325–337). IGI Global.

Vyas, S., Gupta, S., & Shukla, V.K. (2023, March). Towards Edge AI and Varied Approaches of Digital Wellness in Healthcare Administration: A Study. In *2023 International Conference on Computational Intelligence and Knowledge Economy (ICCIKE)* (pp. 186–190). IEEE.

Wernly, B., Mamandipoor, B., Baldia, P., et al. (2021). Machine learning predicts mortality in septic patients using only routinely available ABG variables: A multi-centre evaluation. *International Journal of Medical Informatics*, *145*, 104312. 10.1016/j.ijmedinf.2020.104312

The White House. (2021). U.S. – EU Trade and Technology Council Inaugural Joint Statement. Briefing Room, September 29, 2021.

Yee, C.R., Narain, N.R., Akmaev, V.R., & Vemulapalli, V. (2019). A data-driven approach to predicting septic shock in the intensive care unit. *Biomedical Informatics Insights*, *11*, 1178222619885147. 10.1177/1178222619885147

Yetisen, A.K., Martinez-Hurtado, J.L., Ünal, B., Khademhosseini, A., & Butt, H. (2018). Wearables in medicine. *Advanced Materials*, *30*(33), e1706910. 10.1002/adma.201706910

Yim, W., Yetisgen, M., Harris, W.P., & Kwan, S.W. (2016). Natural language processing in oncology: A review. *JAMA Oncology*, *2*(6), 797–804. 10.1001/jamaoncol.2016.0213

Ziegelstein, R.C. (2017). Personomics and precision medicine. *Transactions of the American Clinical & Climatological Association*, *128*, 160–168.

9 Contemporary Practice of Automated Machine Learning for Clinical Repository in the Medicinal Field

Sonali Vyas, Shaurya Gupta, and Vinod Kumar Shukla

9.1 INTRODUCTION

Healthiness besides happiness stands as the central apprehension of everyone nowadays, which is being verified by the constant development of universal healthcare businesses, and it is predicted to touch a mark of $20 trillion by 2024 (Azghadi et al., 2020). The hopeful technical expertise to attain a competitive edge in this ever-growing business involves artificial intelligence (Rong et al., 2020) and its implementation with machine learning. Current expansions in machine learning technology provide a wonderful prospect for overall improvement in terms of healthcare consequences of patients as it has been used broadly in terms of healthcare besides medicinal uses for varied illnesses documentation (Beam & Kohane, 2018; Li et al., 2020), oral illness analysis besides extrapolation (Leite et al., 2020), and determining cancer lumps from radiology imageries (Feurer et al., 2015). In recent times, auto machine teaching (Hutter et al., 2019; Warring et al., 2020; Yao et al., 2018) is projected in terms of expansion of application dominion of various machine learning algorithms besides simplifying their disposition to several examples of healthcare (Alsalem et al., 2022; Ooms & Spruit, 2020). Though it is still an emergent technology, it is already being applied in the medicinal imaging field (Tsamardinos et al., 2022), diabetes analysis (Karaglani et al., 2020), and Alzheimer's judgment, besides electronic (Gehrmann et al., 2017) health record investigation. Specifically, earlier developments (Nigam, 2016; Venkataraman et al., 2020) in terms of clinical notes use standard machine learning procedures for processing in terms of disease detection (Huang et al., 2019; Pineda et al., 2019). The above readings have been formulated with the deployment of machine learning algorithms in amalgamation with other procedures like natural language processing (Zheng et al., 2012) and perception abstraction resolutions (Liu et al., 2013; Mullenbach et al., 2018). Section 9.1 discusses the background, besides the motivation of the manuscript. Section 9.2

DOI: 10.1201/9781003346289-9

discourses regarding essential perception of automated machine learning besides accessible tools and procedures. Section 9.3 reviews usage of automated machine learning expertise in healthcare business. Section 9.4 delivers specifics on dissimilar machine learning phases responsible for excerpting inferences from clinical notes involving pre-processing, feature extraction, and selection besides the assessment phases of an automated machine learning platform.

9.2 AUTOMATED MACHINE LEARNING

Though machine learning algorithms don't necessitate anthropological interfering while learning, arranging information, which is going to be used by some algorithms, narrowing down to accurate process, besides alteration for the attainment of desired results. Data scientists' custom dissimilar rehearses for the purpose of pre-processing varied machine learning algorithms. Such procedures are human-reliant, besides necessitating distinctive assistance in varied verticals of computer science. The concept of automated machine learning is solely responsible for minimizing the overall human interference to a bare minimum level (Bisong, 2019); apart from this, it systematizes the foremost procedures of machine learning. The following steps are involved in the complete process:

- **Data Preparation:** It comprises data incorporation, conversion, cleaning, and finally data lessening. It is extended development that involves the maximum time of data scientists (Alaa & Schaar, 2018).
- **Feature Extraction and Selection:** It involves the selection of a sub-section of the data set that focuses on information protection of a certain data set, although refining the education or learning simplification (Khalid et al., 2014).
- **Algorithm Selection:** It inculcates the period wherein a process is involved in choosing the paramount procedure that will provide the most precise consequences. Regulating the procedures and its associated circumstances will enhance the further outcomes, termed *hyperparameter optimization* (Yao et al., 2020).

Automated machine learning methods use dissimilar approaches besides optimization procedures in attaining anticipated correctness besides enactment. Bayesian optimization is a process that focuses on the optimization of hyperparameters for machine learning procedures centered on probability theory, called Bayes' theorem (Gupta & Katarya, 2021; Zhang et al., 2020), whereas other more naive methods like grid search and random search are also used. Meta-learning is an additional process for hyperparameter optimization, wherein an automated machine learning system absorbs from past understandings that in turn is entitled "learning to learn" (Thornton, 2014).

Out of the numerous platforms accessible for automated machine learning, some of them are open source, whereas others are saleable or commercial in other words. The contrasts among some of the prevalent automated machine-learning platforms are as follows:

- Coding Necessities.
- Processing Site.
- Acknowledged input information, besides budget.

The platforms that are quite acceptable for open-source automated machine learning libraries are Auto-Sklearn and Sklearn library (Hutter et al., 2019), which are responsible for automating procedure selection besides hyperparameter optimization with the help of Bayesian optimization methods plus meta-learning. An additional open-source environment is Auto-WEKA (Ahmad et al., 2018; Kotthoff et al., 2019), i.e., a computerized Waikato setting for knowledge investigation that is very much like Auto Sklearn. These automated machine learning environments use statistical algorithms that can practice regulated data like stock market prices, etc. RapidMiner is another commercial automated machine learning platform (Yu et al., 2022), which is in fact a platform wherein entire machine learning procedure is computerized from the point of information preparation to procedure, besides choosing the hyperparameter and is centered on the channel that is responsible for guiding procedures besides resolving procedures in addition to procedures that should be used, apart from considering the factors like data scrutiny, imagining, and plus transcript excavating. Google itself has its cloud-centered automated machine learning platform (Perotte et al., 2014), which is distributed into dissimilar prototypes centered on sorts of input data sets. Automated machine learning language is widely used for the following:

- Structured information.
- Text cataloging plus objects documentation.
- Automated machine learning translation is being used for translations.
- Automated machine learning vision is widely used for image cataloging besides object recognition.
- Automated machine learning video intelligence is used in audiovisual cataloging besides object trailing.

Varied structures mechanize machine learning procedures though every system works inversely, aiming at diverse data sets, stages, and algorithms; besides, it has exceptional compensations plus shortcomings (Luo et al., 2017). Auto-WEKA is facilitated by graphical user interface that is also used as Java command, but its use is quite restricted to arithmetical procedures. RapidMiner has information scrutiny competence, though it necessitates human supervision at times. Google automated machine learning supports varied data sets, though it is cloud centered in addition to its custodies or charges for data processing (Starmans et al., 2021).

9.3 AUTOMATED MACHINE LEARNING IN THE HEALTHCARE INDUSTRY

Though automated machine learning is being used in a multiplicity of solicitations like fraud discovery (Varoquaux & Cheplygina, 2022), besides illness analyses (Rajkomar et al., 2019), apart from many applications that are still using customary machine learning procedures moderately to automated machine learning. It is because of varied applications types that necessitate procedures like data scrubbing besides feature selection, which are not at all supported by varied automated machine learning platforms. Considering healthcare commerce, there are many studies that have concentrated critically on the implementation of automated machine learning systems specific to the well-being domain. Bearing in mind resource constraints for the purpose of scientific coding (Erickson et al., 2017) besides extraordinary salaries being demanded by data scientists (Abbas et al., 2022), it became quite important to discover a cost-saving process that permits healthiness establishments to extract assistance from machine learning proficiencies minus enormous expenditures. Additionally, such kinds of approaches may advance patient consequences, which finally aims at dominant prominence in the healthcare tool development. In developing technical expertise, automated machine learning helps in the attainment of such kinds of goalmouths for fitness administrations, particularly for mining analyses from clinical notes that in turn protect the well-being of employees' time, apart from advancement in patient conclusions with the means of fast-tracking patient cure forecasting, besides refining the diagnose accuracy (Quazi, 2022). The all-purpose methods that are being used in studying automated machine learning in terms of the healthcare industry are as follows:

- Building novel automated machine learning gears for facilitating medical data sets.
- The second method involves the usage of prevailing automated machine learning libraries, besides platforms in performing extrapolative demonstrating or cataloging in the case of a quantifiable clinical data set.

Underneath, these methods are being deliberated in added detail.

9.3.1 Constructing Automated Machine Learning for Medical Data Sets

Here, an automated machine learning tool is quite focused in terms of examining medical or clinical information, taking into consideration the careful supervision strategy; besides forecasting patient's medical budgets or categorizing medicinal archives, besides all the medicinal imageries (Jardim et al., 2022) and the data sets may be moreover structured or unstructured. Taking into consideration a regulated data set, information is signified in the form of rows and columns, which is effortlessly administered by workstations apart from the data sets that exist in catalogs, spreadsheets, and varied platforms that support data in the tabular arrangement. An instance of structured data sets may entail patient info like name, age, and gender, besides test outcomes that may contain hemoglobin or cholesterol intensities (Javeed et al., 2022). Unstructured data sets generally form or compromise unformatted data that entails any format of data that is not at all in tabular format. Many machine learning procedures like linear regression or support vector machines require evolution or any sort of transition procedure that is converting formless data to an arranged layout (Hill et al., 2019). With time and recent developments, an automated machine learning system that is quite focused in terms of bioinformatics applications besides translational medicine has been developed and is named JADBIO, which supports an in-built prognostic plus diagnostic experimental prototype. Another platform is AHPS, which practices parameters like:

- Data set scope.
- Feature dimensionality.
- Besieged value type that is used in recognized among pertinent procedures.
- Identifying practices for feature selection besides data pre-processing.

AHPS's production is served as input in engendering a list of pipelines with obtainable hyperparameters and thereby serving as a configuration generator (Wang et al., 2019). Formerly, conformation assessment procedures k-fold cross authentication in regulating finest information pre-processing approaches, which features:

- Engineering algorithms.
- Hyperparameters.
- Assessing the model's enactment.

CEP assortment afterwards is functional to unique data sets besides building prognostic. JADBIO automated machine teaching model functions together (Ngiam & Khor, 2019) and includes the following entities:

- Linear ridge regression.
- Support vector machine and decision trees.
- Random forests.
- Gaussian kernel support vector machines.

While comparing it to the Auto-Sklearn arrangement, which uses around 748 data sets, it failed completely in processing of nearby 39.44% data sets because of timeout and inner inaccuracies.

Jamieson and Goldfarb (2019) projected approaches that are an entity of progressive development of an automated machine learning structure helping healthcare experts and in turn accomplishing extrapolations, besides groupings on vast experimental information minus any contribution from data scientists. They recognized three challenges that are overall affecting the automated machine learning procedure that are as follows:

- The first problem is features, besides procedure assortment and hyperparameters optimization. It centrally leads to the functioning of all conceivable amalgamation or combinations that are feasible and available.

- The second issue involves consortium besides amassing of information, required by data scientists. It becomes a fragment of data training, besides feature abstraction procedures.
- Lastly, the challenge involves simplifying machine learning models. Individually, every medical data set possesses individual physiognomies besides requiring a diverse model in attaining efficient accuracy.

The authors (Ehrmann et al., 2022) deliberated regarding collective algorithm assortment besides hyperparameter optimization plus consecutive model-centered algorithmic outline, which is not at all effectual for big data. Consequently, a novel technique using Bayesian optimization was taken into consideration. An auto prognosis model that was being discussed was responsible for automating the medical extrapolative model, besides the FLASH model (Vandemeulebroucke et al., 2022), which was using double coatings of Bayesian optimization in successfully selecting algorithms besides optimizing hyperparameters. Alanazi (2022) used an automated machine learning method and neural architecture search that uses optimization of neural network models for high-resolution 3D medicinal imageries, requiring additional storage spaces.

9.3.2 Consuming Prevailing Automated Machine Learning Tools for Clinical Data Sets

Prevailing automated machine learning tools may have their indulgence in a diversity of applications, comprising medicinal information investigation. The authors (Suneetha et al., 2020) have associated two automated machine knowledge platforms:

a. Google's automated machine learning
b. Apple creates machine learning.

Experimentation was conducted on lungs, besides colon cancer imageries data sets wherein the size ranges between 250 to 750 images. Though estimated consequences were near, Apple's machine language proved to deliver the best optimal results in terms of data sets. Google's automated machine learning tool sustains a certain price for medicinal operators to process data sets besides categorizing therapeutic imageries (Dash et al., 2019). Additionally, Google's model supplies medicinal imageries plus data that is available on a cloud platform, but useless for subtle patient information. Apple's model is charge free, besides stocks information locally. Other automated machine learning archives were used in the classification of data sets for illnesses like breast cancer, diabetes, etc. There are numerous preceding tasks that used dissimilar automated machine learning platforms for varied therapeutic applications and these are classified into two groups centered on data set type, which may be:

i. Structured
ii. Unstructured.

Varied kinds of unstructured information were administered by saleable automated machine learning gears, though the structured information may be frequently handled with the help of open-source gears besides therapeutic automated machine learning interfaces. The key cause for this is that dealing with controlled information is quite easier (Supriya & Deepa, 2020). The viable corporations that attain monetary welfare by extending or providing automated machine learning platforms with time have developed more progressive tools that are quite capable of processing imageries besides script (Albahri et al., 2021). Consequently, therapeutic specialists may use any of the tools in identifying dissimilar illnesses that are centered on obtainable data sets from laboratory fallouts and the medicinal history of patients.

9.4 CHALLENGES AND BENEFITS OF WORKING WITH CLINICAL NOTES

The experimental script comprises countless therapeutic terms besides non-medical verses that later confuses the computer in terms of their classification. Taking an instance, the word "bad" has a generic meaning, whereas in medicinal terms it means "bipolar affective disorder" (Haq et al., 2020). Additionally, it is a challenge to decide what NLP practices must be used for the information training phase of automated machine learning, which requires or necessitates the need of a data scientist. As the records are generated, numerous features from the medical script are produced. The larger the medicinal data set, the more it affects the performance of the model considerably. The medicinal acronym stands as one other obstacle, which is usually applied in clinical notes; however, these are sifted out by stop-word elimination procedures, as medicinal acronyms might be quite like stop words. For instance, stop words like "AND" signify the allowance in terms of normal demise; "IS" signifies inducement spirometry. A typographical error in medicinal reports stands as an additional problem because predicting a word in the wrong manner may designate alternative illnesses or analyses. Nowadays, quite prominently, noteworthy enhancement in terms of patient results is attained with the help of quicker supplementary precise detection of illness. Though individual healthcare associations have dissimilar configurations besides medical notes size, a possible automated machine wisdom contrivance will help in finding the preeminent algorithm besides delivering great accuracy minus any kind of human intervention; therefore, it will result in endowment saving with the help of plummeting medical coders in any hospital or infirmaries (Umer et al., 2022). Furthermore, such sorts of tools would be of great assistance to medicinal doctors in better patient administration, further adding to better-quality medicinal source administration besides decreasing overall encumbrance nationwide besides global medicinal organizations. Additionally, hospitals will benefit from an increase in the number of clinical coders, which will help in the reduction of coding mistakes and progress coding quality, thereby supporting medical coders considerably apart from helping subordinate medical coders in recognizing diseases and infections with their possible remedies (Gupta et al., 2023).

9.5 CONCLUSION AND FUTURE SCOPE

To augment in terms of patient consequences and progress in terms of healthcare business, novel procedures besides technical expertise are under progressive development. Machine learning generally necessitates social acquaintance in terms of training procedures, besides relying on the proficiency of human inventors in order to achieve utmost performance. The necessity of human intervention may result in somewhat lesser acceptance in the healthcare business, even though machine learning is an inordinate possibility in terms of enhancement in patient health's consequence, besides saving while plummeting overall load on the therapeutic structure. This automatic learning mechanism is called automated machine learning technology, and today has a wider acceptance in the medicinal field, with time assures a promising future in healthcare purview with improved analyses of disease, condensed budgets, and reduced cure time of patients.

REFERENCES

Abbas, A., O'Byrne, C., Fu, D.J., Moraes, G., Balaskas, K., Struyven, R., & Keane, P.A. (2022). Evaluating an automated machine learning model that predicts visual acuity outcomes in patients with neovascular age-related macular degeneration. *Graefe's Archive for Clinical and Experimental Ophthalmology*, 1–13.

Ahmad, M.A., Eckert, C., & Teredesai, A. (2018, August). Interpretable machine learning in healthcare. In *Proceedings of the 2018 ACM international conference on bioinformatics, computational biology, and health informatics* (pp. 559–560).

Alaa, A., & Schaar, M. (2018, July). Auto prognosis: Automated clinical prognostic modeling via Bayesian optimization with structured kernel learning. In *International conference on machine learning* (pp. 139–148). PMLR.

Alanazi, A. (2022). Using machine learning for healthcare challenges and opportunities. *Informatics in Medicine Unlocked*, 100924.

Albahri, A.S., Zaidan, A.A., Albahri, O.S., Zaidan, B.B., Alamoodi, A.H., Shareef, A.H., ... & Mohammed, K.I. (2021). Development of IoT-based mhealth framework for various cases of heart disease patients. *Health and Technology*, *11*(5), 1013–1033.

Alsalem, M.A., Mohammed, R., Albahri, O.S., Zaidan, A.A., Alamoodi, A.H., Dawood, K., ... & Jumaah, F. (2022). Rise of multiattribute decision-making in combating COVID-19: A systematic review of the state-of-the-art literature. *International Journal of Intelligent Systems*, *37*(6), 3514–3624.

Azghadi, M.R., Lammie, C., Eshraghian, J.K., Payvand, M., Donati, E., Linares-Barranco, B., & Indiveri, G. (2020). Hardware implementation of deep network accelerators towards healthcare and biomedical applications. *IEEE Transactions on Biomedical Circuits and Systems*, *14*(6), 1138–1159.

Beam, A.L., & Kohane, I.S. (2018). Big data and machine learning in health care. *JAMA*, *319*(13), 1317–1318.

Bisong, E. (2019). *Building machine learning and deep learning models on Google cloud platform: A comprehensive guide for beginners*. Apress.

Dash, S., Shakyawar, S.K., Sharma, M., & Kaushik, S. (2019). Big data in healthcare: Management, analysis and future prospects. *Journal of Big Data*, *6*(1), 1–25.

Ehrmann, D.E., Gallant, S.N., Nagaraj, S., Goodfellow, S.D., Eytan, D., Goldenberg, A., & Mazwi, M.L. (2022). Evaluating and reducing cognitive load should be a priority for machine learning in healthcare. *Nature Medicine*, 1–2.

Erickson, B.J., Korfiatis, P., Akkus, Z., & Kline, T.L. (2017). Machine learning for medical imaging. *Radiographics, 37*(2), 505.

Feurer, M., Klein, A., Eggensperger, K., Springenberg, J., Blum, M., & Hutter, F. (2015). Efficient and robust automated machine learning. *Advances in Neural Information Processing Systems, 28.*

Gehrmann, S., Dernoncourt, F., Li, Y., Carlson, E.T., Wu, J.T., Welt, J., & Celi, L.A. (2017). Comparing rule-based and deep learning models for patient phenotyping. ArXiv preprint arXiv: 1703.08705.

Gupta, G., & Katarya, R. (2021). EnPSO: An Automated machine learning technique for generating ensemble recommender system. *Arabian Journal for Science and Engineering, 46*(9), 8677–8695.

Gupta, S., Sharma, H.K., Kapoor, M. (2023). Digital medical records (DMR) security and privacy challenges in smart healthcare system. In: *Blockchain for secure healthcare using internet of medical things (IoMT)*. Cham: Springer. 10.1007/978-3-031-18896-1_6

Haq, A.U., Li, J.P., Khan, J., Memon, M.H., Nazir, S., Ahmad, S., & Ali, A. (2020). Intelligent machine learning approach for effective recognition of diabetes in E-healthcare using clinical data. *Sensors, 20*(9), 2649.

Hill, B.L., Brown, R., Gabel, E., Rakocz, N., Lee, C., Cannesson, M., & Halperin, E. (2019). An automated machine learning-based model predicts postoperative mortality using readily-extractable preoperative electronic health record data. *British Journal of Anesthesia, 123*(6), 877–886.

Huang, J., Osorio, C., & Sy, L.W. (2019). An empirical evaluation of deep learning for ICD-9 code assignment using MIMIC-III clinical notes. *Computer Methods and Programs in Biomedicine, 177*, 141–153.

Hutter, F., Kotthoff, L., & Vanschoren, J. (2019). *Automated machine learning: methods, systems, challenges* (p. 219). Springer Nature.

Jamieson, T., & Goldfarb, A. (2019). Clinical considerations when applying machine learning to decision-support tasks versus automation. *BMJ Quality & Safety, 28*(10), 778–781.

Jardim, P.S.J., Rose, C.J., Ames, H.M., Echavez, J.F.M., Van de Velde, S., & Muller, A.E. (2022). Automating risk of bias assessment in systematic reviews: A real-time mixed methods comparison of human researchers to a machine learning system. *BMC Medical Research Methodology, 22*(1), 1–12.

Javeed, A., Khan, S.U., Ali, L., Ali, S., Imrana, Y., & Rahman, A. (2022). Machine learning-based automated diagnostic systems developed for heart failure prediction using different types of data modalities: A systematic review and future directions. *Computational and Mathematical Methods in Medicine, 2022.*

Karaglani, M., Gourlia, K., Tsamardinos, I., & Chatzaki, E. (2020). Accurate blood-based diagnostic bio signatures for Alzheimer's disease via automated machine learning. *Journal of Clinical Medicine, 9*(9), 3016.

Khalid, S., Khalil, T., & Nasreen, S. (2014, August). A survey of feature selection and feature extraction techniques in machine learning. In *2014 science and information conference* (pp. 372–378). IEEE.

Kotthoff, L., Thornton, C., Hoos, H.H., Hutter, F., & Leyton-Brown, K. (2019). Auto-WEKA: Automatic model selection and hyper parameter optimization in WEKA. In *Automated machine learning* (pp. 81–95). Cham: Springer.

Leite, A.F., Vasconcelos, K.D.F., Willems, H., & Jacobs, R. (2020). Radiomics and machine learning in oral healthcare. *Proteomics–Clinical Applications, 14*(3), 1900040.

Li, J.P., Haq, A.U., Din, S.U., Khan, J., Khan, A., & Saboor, A. (2020). Heart disease identification method using machine learning classification in e-healthcare. *IEEE Access, 8*, 107562–107582.

Liu, H., Wagholikar, K.B., Jonnalagadda, S., & Sohn, S. (2013, September). Integrated cTAKES for concept mention detection and normalization. In CLEF (Working Notes).

Luo, G., Stone, B.L., Johnson, M.D., Tarczy-Hornoch, P., Wilcox, A.B., Mooney, S.D., ... & Nkoy, F.L. (2017). Automating construction of machine learning models with clinical big data: proposal rationale and methods. *JMIR Research Protocols*, 6(8), e7757.

Mullenbach, J., Wiegreffe, S., Duke, J., Sun, J., & Eisenstein, J. (2018). Explainable prediction of medical codes from clinical text. ArXiv preprint arXiv: 1802.05695.

Ngiam, K.Y., & Khor, W. (2019). Big data and machine learning algorithms for health-care delivery. *The Lancet Oncology*, 20(5), e262–e273.

Nigam, P. (2016). Applying deep learning to ICD-9 multi-label classification from medical records. Technical report, Stanford University.

Ooms, R., & Spruit, M. (2020). Self-service data science in healthcare with automated machine learning. *Applied Sciences*, 10(9), 2992.

Perotte, A., Pivovarov, R., Natarajan, K., Weiskopf, N., Wood, F., & Elhadad, N. (2014). Diagnosis code assignment: models and evaluation metrics. *Journal of the American Medical Informatics Association*, 21(2), 231–237.

Pineda, A.L., Bear, O.J., Venkataraman, G.R., Zehnder, A.M., Ayyar, S., Page, R.L., ... & Rivas, M.A. (2019). FasTag: automatic text classification of unstructured medical narratives. bioRxiv, 429720.

Quazi, S. (2022). Artificial intelligence and machine learning in precision and genomic medicine. *Medical Oncology*, 39(8), 1–18.

Rajkomar, A., Dean, J., & Kohane, I. (2019). Machine learning in medicine. *New England Journal of Medicine*, 380(14), 1347–1358.

Rong, G., Mendez, A., Assi, E.B., Zhao, B., & Sawan, M. (2020). Artificial intelligence in healthcare: Review and prediction case studies. *Engineering*, 6(3), 291–301.

Starmans, M., van der Voort, S.R., Phil, T., Timbergen, M.J., Vos, M., Padmos, G.A., & Klein, S. (2021). Reproducible radiomics through automated machine learning validated on twelve clinical applications. ArXiv preprint arXiv: 2108.08618.

Suneetha, K.C., Shalini, R.S., Vadladi, V.K., & Mounica, M. (2020). Disease prediction and diagnosis system in cloud based IoT: A review on deep learning techniques. *Materials Today: Proceedings*.

Supriya, M., & Deepa, A.J. (2020). Machine learning approach on healthcare big data: A review. *Big Data and Information Analytics*, 5(1), 58–75.

Thornton, C. (2014). Auto-WEKA: combined selection and hyper parameter optimization of supervised machine learning algorithms (Doctoral dissertation, University of British Columbia).

Tsamardinos, I., Charonyktakis, P., Papoutsoglou, G., Borboudakis, G., Lakiotaki, K., Zenklusen, J.C., & Lagani, V. (2022). Just Add Data: Automated predictive modeling for knowledge discovery and feature selection. *NPJ Precision Oncology*, 6(1), 1–17.

Umer, M., Sadiq, S., Karamti, H., Karamti, W., Majeed, R., & Nappi, M. (2022). IoT based smart monitoring of patients' with acute heart failure. *Sensors*, 22(7), 2431.

Vandemeulebroucke, T., Denier, Y., & Gastmans, C. (2022). The need for a global approach to the ethical evaluation of healthcare machine learning. *The American Journal of Bioethics*, 22(5), 33–35.

Varoquaux, G., & Cheplygina, V. (2022). Machine learning for medical imaging: methodological failures and recommendations for the future. *NPJ Digital Medicine*, 5(1), 1–8.

Venkataraman, G.R., Pineda, A.L., Bear Don't Walk IV, O.J., Zehnder, A.M., Ayyar, S., Page, R.L., & Rivas, M.A. (2020). FasTag: Automatic text classification of unstructured medical narratives. *PLoS One*, 15(6), e0234647.

Wang, H.L., Hsu, W.Y., Lee, M.H., Weng, H.H., Chang, S.W., Yang, J.T., & Tsai, Y.H. (2019). Automatic machine-learning-based outcome prediction in patients with primary intracerebral hemorrhage. *Frontiers in Neurology*, 10, 910.

Waring, J., Lindvall, C., & Umeton, R. (2020). Automated machine learning: Review of the state-of-the-art and opportunities for healthcare. *Artificial Intelligence in Medicine, 104*, 101822.

Yang, C., Akimoto, Y., Kim, D.W., & Udell, M. (2019, July). OBOE: Collaborative filtering for Automated machine learning model selection. In *Proceedings of the 25th ACM SIGKDD international conference on knowledge discovery & data mining* (pp. 1173–1183).

Yao, Q., Wang, M., Chen, Y., Dai, W., Li, Y.F., Tu, W.W., & Yu, Y. (2018). Taking human out of learning applications: A survey on automated machine learning. ArXiv preprint arXiv: 1810.13306.

Yao, Q., Chen, X., Kwok, J.T., Li, Y., & Hsieh, C.J. (2020, April). Efficient neural interaction function search for collaborative filtering. In *Proceedings of The web conference 2020* (pp. 1660–1670).

Yu, T.H., Su, B.H., Battalora, L.C., Liu, S., & Tseng, Y.J. (2022). Ensemble modeling with machine learning and deep learning to provide interpretable generalized rules for classifying CNS drugs with high prediction power. *Briefings in Bioinformatics, 23*(1), bbab377.

Zhang, J., Sun, Z., & Qi, Y. (2020, August). AutoIDL: Automated Imbalanced Data Learning via Collaborative Filtering. In *International Conference on Knowledge Science, Engineering and Management* (pp. 96–104). Springer, Cham.

Zheng, J., Chapman, W.W., Miller, T.A., Lin, C., Crowley, R.S., & Savova, G.K. (2012). A system for reference resolution for the clinical narrative. *Journal of the American Medical Informatics Association, 19*(4), 660–667.

10 Smart Innovative Medical Devices Based on Artificial Intelligence

*Nayankumar C. Ratnakar,
Beenkumar R. Prajapati, Bhupendra G. Prajapati,
and Jigna B. Prajapati*

10.1 INTRODUCTION OF AI-ENABLED MEDICAL DEVICES

Researchers and the biomedical industry have been interested in artificial intelligence (AI), owing to its power to practice huge amounts of data, provide exact outcomes, and then switch procedures to achieve optimal effect. AI is not new since computers remain in existence and are used to decide and predict the long-term effect of illnesses. In today's world, machines and algorithms assist with most of regular tasks (Cfr - Code of Federal Regulations Title 21). In AI, the Internet of Medical Things (IoMT) remains a division of the Internet of Things (IoT) technologies and consists of health devices connected and composed with the drive of monitoring enduring care. IoMT devices integrate automated, internal devices, and artificial intelligence based on machine learning to enable healthcare monitoring without human involvement (Fda Ai-Machine Learning Strategy Remains Work in Progress). IoMT tools enable distant access to acquire, progression, and convey health data through a safe system via integrating patients and clinicians through medical devices. By providing remote monitoring of health parameters, IoMT systems aid in reducing unwanted clinic halts with health expenses. The IoMT health technology section comprises wearables devices with real-time health monitoring and point-of-care systems based in hospitals or clinical settings (POC) (Smart Medical Device Market Is Expected to Witness Massive Growth with a Significant Cagr of 7. 22%, Delveinsight | Medgadget). The global market for smart medical devices is anticipated to rise as chronic diseases like diabetes and asthma are diagnosed more frequently. The World Health Organization (WHO) projects that by 2045, diabetes will affect nearly 10% of people worldwide between the ages of 20 and 79 (What Is the Future of the Smart Medical Devices Market?). In 2020, the market for smart medical devices was estimated to be worth US$48.6 billion. From 2021 to 2028, foresee toward growth CAGR of 7.5%. Through the completion of 2028, arcade aimed at smart medical devices expected a value of $70.1 billion (Alemzadeh et al., 2013). In recent years, smartphones and other smart devices have firmly established

DOI: 10.1201/9781003346289-10

themselves in everyone's lives. The accessibility and availability of medical facilities and services have improved thanks to recent breakthroughs in smartphone technology (Allen et al., 2021). Mobile devices that offer medical services make it possible for patients, cargivers, and healthcare practitioners to communicate efficiently and effectively. With usage of these movable health apps, healthcare physicians may monitor their patients more quickly than ever before, identify illnesses, and get test results (Allen et al., 2021).

According to the most recent figures from GSMA Intelligence, there are 5.34 billion individual mobile phone users worldwide. The software for mobile applications is designed to do specific tasks inside a mobile phone, such as a smartphone. These IoT-based gadgets have the potential to be very productive in an extensive diversity of industries, with telemedicine, healthcare, manufacturing, and agricultural production. Mobile apps with intelligent automation are widely employed during the COVID-19 epidemic (Athilingam et al., 2016). During the COVID pandemic, smart medical devices with IoT applications were set up to track social distancing (Benjamens et al., 2020) and isolate patients, monitor the virus, sanitize the containment area, and support an e-learning facility (Bohr & Memarzadeh, 2020a).

The Internet of Things (IoT) enables doctors to reach patients outside of the usual clinical location. To avoid needless and expensive excursions in a consulting room, patients plus doctors can check a person's health at home with the use of smart medical devices. The Internet of Things is progressively rising to influence healthcare, aiding patients and clinicians. Patients now monitor their fitness through the use of linked USG, temperature monitoring, glucose monitoring, ECG, and other super smart medical devices (Bohr & Memarzadeh, 2020b). IoT deployment in the medical sector is centered on a variety of sensors, medical equipment, artificial intelligence, testing, and advanced imaging technologies. The productivity and standard of living are increased by these technologies in both traditional and modern economies and cultures (Dente et al., 2017). Smart beds are now being used in many hospitals. These beds may detect the existence of patients and then automatically acclimate the right pressure and viewpoint to deliver the right sustenance of a healthcare professional (Ekins et al., 2019). IoT strength possibly recovers at domiciliary care. Miserably, certain patients nosedive toward taking their medicine on correct intervals or in the exact quantities. Intelligent medicine dispensers in the home convey data to the server mechanically and inform doctors once patients don't take their medicine. Additionally this thoughtful machinery focused health specialists to some possibly dangerous enduring behavior (Fernando et al., 2020). Some of top trending FDA-approved mobile apps supporting smart medical devices are as follows: Mobile MIM, HealthTap, WebMD, Pingmd, Epocrates, Talkspace, UpToDate, ResolutionMD, etc. The relation of smart medical devices, IoMT, and cloud computing is shown in Figure 10.1.

10.2   DEVELOPMENT STAGES OF AI MEDICAL DEVICES

The effectiveness and safety of AI technologies in healthcare will be dependent on good AI development procedures and practices. These methods must take into account factors such as the resilience of the product, algorithms train, testing and

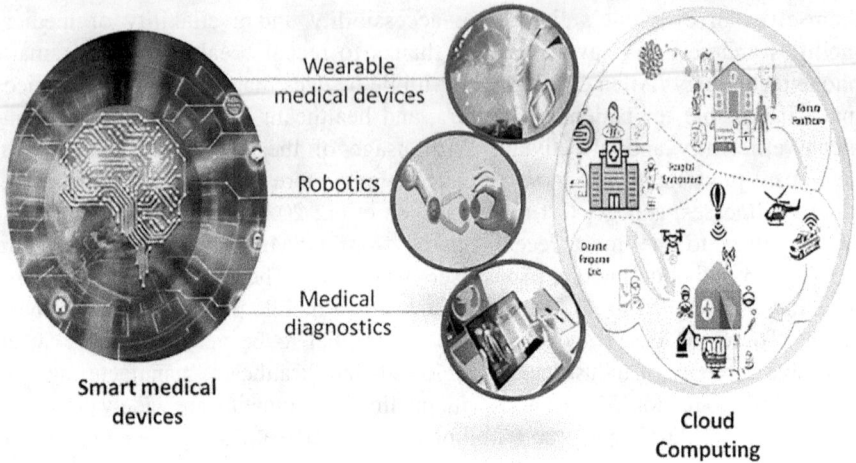

FIGURE 10.1   Smart medical devices with IoMT and cloud computing.

validation modification processes, and identification and documenting of various AI solution iterations (Ghersi et al., 2018). The type, nature, and hazards associated with the algorithm will determine how regulatory methods for AI-based medical device software are modified. Consideration of these various strategies ought to start with existing smart medical devices classification criteria. The classification of international medical device regulators forum software depends on the severity of the healthcare issue (critical, serious, or not serious) and the importance of the data it provides to guide clinical treatment. The ten guiding principles highlight areas in which the IMDRF, global standards bodies, and other cooperative groups should cooperate to develop GMLP. IEC 82304–1 outlines the general system standards for the security and safety of health software. These specifications, which cover the whole life cycle of the product, including design, production, validation, installation, maintenance, and disposal of products, are intended for software products used on common computing platforms without specialized hardware (Haleem et al., 2021). The IEC 62304 standard also addresses the software life span for medical device applications. It is applicable to software that is governed by medical device rules and can be applied to both independent software and software that is integrated into a physical device. IEC 62304 specifies criteria for "software of unknown provenance" (SOUP), which refers to available commercially software that has not been specifically created aimed at usage in a health device acceptable documents of the development cycle are not readily available. The extra controls for SOUP within IEC 62304 might serve as a good place to start when addressing the AI's "black box" characteristics. It is worth giving thought to whether they are appropriate or if more instruction is needed (Health, Center for Devices, and Radiological, 2022a). There might be some helpful AI development techniques out there. For instance, ISO/IEC JTC1/SC42, a global standardization body for general AI, is presently creating best practices for risk management, bias, reliability, and governance implications. ITU/WHO and IEEE are developing more standards, best practices, and guidelines (Health, Center for Devices, and Radiological, 2022b). To create a U.S. National Evaluation System for health technology, the FDA is currently working with stakeholders (NEST). This aims to more effectively produce better evidence in support of medical technologies. It will make use of empirical data acquired from many sources as well as sophisticated analytics (Health, Center for Devices, and Radiological, 2022c). Similar to this, new evidence criteria have been created in the United Kingdom to guarantee that digital health solutions are both clinically and financially beneficial. This helps innovators and commissioners better grasp what high evidence standards should entail (Health, Center for Devices, and Radiological, 2021a).

In a white paper published in 2019, the FDA suggested a potential strategy for overcoming this adaptive learning issue. It is built on the following four guiding ideas:

a. *Unblemished anticipations on eminence arrangements and decent machine learning performs.*

The FDA anticipates that software as a medical device developers will put a system in place to ensure that their product conforms with laws and relevant quality standards, just like any other device maker would. In addition, an algorithm developer would have to follow recognized best practices, or GMPs, when creating the algorithm (GMLP). The present Good Manufacturing Practice regulations for devices may eventually need to be amended to accommodate this set of standards, as they are still under development (Health, Center for Devices, and Radiological, 2019). In order to report what unanswered queries about best performances aimed at algorithm strategy, training, analysis look like, the FDA recently declared that it seeks input from the industry and stakeholders (Health, Center for Devices, and Radiological, 2022d).

b. *Premarket valuation of software as a medical devices.*

In the preliminary pre-market evaluation of software as a medical device that uses machine learning, programmers have a choice to submit a predetermined modification control plan, which is a strategy for future adjustments. This strategy would include the kinds of possible revisions as well as the method the developer would employ to put those adjustments into place and lower the risks involved (Health, Center for Devices, and Radiological, 2021b).

c. *Repetitive checking of software as a medical devices through producers to govern once an algorithm modification needs FDA evaluation.*

Many modifications to a software as a medical devices product would probably necessitate a new premarket application underneath the current regulatory system, according to the developer. Developers would only need to document changes performed under the suggested method if they fall within the parameters of the established change control plan. For example, if the development company tends to make software as a medical devices consistent through other data sources or integrates different kind of facts, changes go beyond purview to start changing the governing program and nevertheless result in a new intentional practice for the device. The FDA may review or start changing the control plan alone but approve a new edition (Hetrick).

   d. *Transparency and real-world performance monitoring*.

The FDA would anticipate that developers will agree to follow specific transparency guidelines and participate in continuing performance evaluation as part of this plan. As a result, creators are required to deliver regular reporting to organizations regarding proceeding upgrades that have been deployed and performance data, among other obligations (Hirai et al., 2009).

10.3 REGULATORY ASPECTS AND GUIDELINE

The FDA takes severe regulatory standards aimed at medical device certification owing to high-risk characteristics of medical devices and unknown consequences of applying AI/ML for medical management and statistics. The creation of AI/ML-centered health devices plus algorithms requires creators to undergo laborious, interval, and reserve exhaustive methods. It could remain a crucial barrier for the use of AI/ML in medication. The parent company must acquiesce health devices or else software to the FDA evaluation prior to making it legally available in the U.S. market. Regulatory organizations take three degrees of authorization for AI/ML-based medical algorithms, together with 510(k) (Hitti et al., 2021) premarket endorsement with a de novo pathway (Jain et al., 2021), as shown in Figure 10.2.

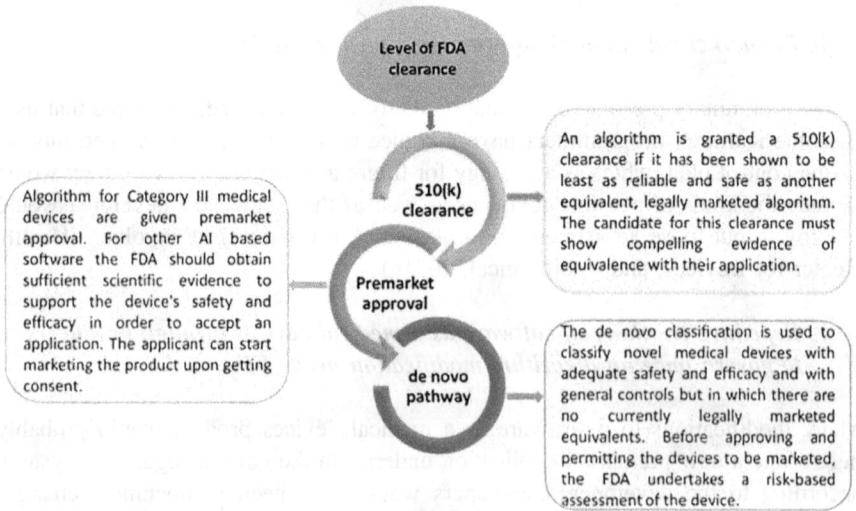

Level of FDA clearance

510(k) clearance

Premarket approval

de novo pathway

Algorithm for Category III medical devices are given premarket approval. For other AI based software the FDA should obtain sufficient scientific evidence to support the device's safety and efficacy in order to accept an application. The applicant can start marketing the product upon getting consent.

An algorithm is granted a 510(k) clearance if it has been shown to be least as reliable and safe as another equivalent, legally marketed algorithm. The candidate for this clearance must show compelling evidence of equivalence with their application.

The de novo classification is used to classify novel medical devices with adequate safety and efficacy and with general controls but in which there are no currently legally marketed equivalents. Before approving and permitting the devices to be marketed, the FDA undertakes a risk-based assessment of the device.

FIGURE 10.2 FDA approval process for AI-based smart medical devices.

The FDA has currently certified or approved a multitude of medical devices that make use of "locked" algorithms. A "locked" algorithm is one that consistently produces similar output when identical contribution is realistic, according toward their definition. The FDA refers to these as adaptive algorithms, which are used by many modern medical devices, notably those that are AI/ML based, although regulatory framework were not created for them (Javaid & Khan, 2021).

10.4   MERITS AND DEMERITS OF MEDICAL DEVICES

Meanwhile, tools' potentials to increase efficiency of healthcare facilities then diminish the weight of healthcare workers; it is difficult to exaggerate the general meaning of healthcare software reactions. It is vitally proved that persons remain alive lengthier and that enduring illnesses continue flattering additional predominant. Key benefits of intelligent, cutting-edge medical devices include the following.

a. **Remote surveillance**

In the occasion of a therapeutic emergency, actual distant checking by linked smart devices might sense diseases, treat diseases, and protect lives. In addition to opening up options for on-demand medical facilities utilizing fitness following applications and search platforms, advancements in wireless communication and cell phones have also made remote interactions, which are accessible anywhere and anytime, a new method of delivering healthcare (Kaur et al., 2022). Those with chronic diseases and the elderly may benefit greatly from remote monitoring and spotting early disease indications. The patient's determination remains to interrelate with their healthcare providers; this situation is carried out by a smart device manually entering statistics aimed at an extended age of time without having to interfere with their regular life. This is an excellent illustration of how algorithms might work with medical experts to create results that are good for people (Kaushik et al., 2021). Assistive technologies encourage user involvement in communication and information technology devices deliver remote care facilities that help deliver information to health practitioners, increasing the self-dependence of patients. The use of assistive devices is expanding quickly, particularly among those 65 to 74 years old (Kaushik et al., 2021). Nowadays, many health-related requests use AI and syndicate it with capabilities for remote doctors to provide answers to some of the straightforward issues that might not require an in-person visit to the doctor (Kaur et al., 2022).

b. **Prevention**

Patients typically cope well with the basic feeling of feeling poorly as well as the many difficulties that come with moderate illnesses. For some illnesses, it remains crucial toward control of these indications in command to stop afterward attainment inferior and ultimately relieve more severe symptoms. Smart sensors evaluate these health issues and lifestyle decisions and suggest preventative steps that will lower the likelihood of developing illnesses and acute states (Konieczny & Roterman, 2019).

c. Reduction of healthcare costs

Smart, cutting-edge medical equipment lowers the cost of testing and cuts back on expensive doctor appointments and hospital stays. For illustration, the aptitude of AI to decrease prices for a therapeutic scheme is an important factor in the acceptance of AI requests. In 2026, it is foreseen that AI applications resolve U.S. healthcare expenditures by $150 billion yearly. The change since unique sensitive proactively healthcare prototypical that highlights patient care relatively than disease management is mainly answerable aimed at these price investments. This is anticipated to lead to a decrease in hospital stays, medical visits, and treatments (Miller & Brown, 2018).

d. Medical data accessibility

Since the advent of electronic medical records, and the substantial records of data scheduled for all patients that, when combined, can remain utilized toward spot healthcare drifts across various illness categories. The EMR databases include clinical narratives, lab test results, records of diagnosis and interventions, and histories of hospital contacts. Prognostic replicas can remain shaped together by these data set contributions of clinicians with diagnosis and numerous cure executives. As AI techniques advance, it will power feasibility toward a variety of data, including correlations between past and present medical events and linked illness effects (Nunan & Di Domenico, 2013). In the previous ages, there have remained an important upsurge in quantity of evidence available for assessing the action of pharmacological complexes and biomedical material. This is a consequence of increasing mechanization and growth of new investigational approaches like parallel synthesis and concealed Markov model-based text-to-speech fusion. The availability toward electronic medical records helps healthcare professionals make wise medical decisions and avoid difficulties while also enabling patients to obtain high-quality care (Okuzaki et al., 2014). Results extracted after a script narrative, predictive algorithms founded on test data, and scientific choice created on distinct therapeutic pasts are instances of existing submissions. Additionally, AI consumes a portion enabling the combination of electronic medical record data with dissimilar health applications (Pak & Park, 2012).

e. Improved healthcare and treatment management

Smart, cutting-edge medical equipment makes it easier to monitor medicine delivery, patient reaction, and medical error. Healthcare authorities can gather useful data about the efficiency of workers and equipment using smart, new medical gadgets, and use that knowledge to suggest innovations (Patel et al., 2007a).

f. Precision medication

Precision medication lets healthcare activities remain custom-made to persons or clusters of patients liable on their illness outline, predictive or investigative evidence, or treatment retort. The modified treatment choice will take into explanation of genomic changes as well as medical management underwriting issues such as age, sex, topography, race, family history, immunological outline, metabolic contour, micro-biome, and ecological vulnerability. Precision medication pursues to service individual ecology in its place of populace biology through all phases of a patient's therapeutic journey. This involves getting facts from patients, such as genetic data, physiological monitoring information, EMR data, and personalizing their treatment founded on progressive models. Precision medication welfare includes inferior healthcare expenses, less adverse drug reactions, and better drug action efficacy (Patel et al., 2007b).

Although smart revolutionary medical gadgets have the potential to improve healthcare, there are significant difficulties to overcome before full-scale application. The subsequent risks and disadvantages of using these technologies in healthcare follow.

a. Security plus privacy

Users are deterred from utilizing smart devices for health reasons owing to security plus privacy concerns, as healthcare watching devices take probable to remain cooperated or hewed. The revelation of delicate information concerning a patient's well-being, as well as interfering by sensor records, can have grave consequences. Electronic health records typically contain privacy-sensitive information, which might cause privacy issues when joined or dispersed to a third party in the cloud. The majority of known privacy-preserving approaches are intended for small-scale databases and are not adequate for handling large amounts of data (Pham et al., 2017). Security issues, such as hacking and other unwanted access, are examples of privacy challenges. Several databases that were not before viewed as having privacy consequences are mingled or integrated, potentially jeopardizing privacy. Due to the existing limitations of analysis, merging data sets without even any analysis may result in future privacy breaches. When data is collected autonomously and without human intervention, ethical considerations arise (Quezada et al., 2021).

b. Risk of failure

Electronic and computer-based gadgets are increasingly being used in clinical and personalized settings, thanks to reduced hardware and enhanced portability and interconnectivity. However, with ease of deployment comes an increase in device complexity and substantial hurdles in reliability, improved safety, and security. Medical equipment are frequently prone to non-negligible failures with potentially disastrous consequences for patients. Failure or defects in hardware can all influence the functionality of devices and linked gear, leaving healthcare services in danger. Also, delaying software upgrades might be much dodgier than hopping a doctor's appointment (Rahman et al., 2020).

c. **Amalgamation**

Incorporating gadgets with electronic medical records necessitates work at multiple levels, from software development to the establishment of a network of data conversation gateways. The concluding creation may employ numerous procedures to interface with hospital plans and middleware to connect well with electronic medical records. There is also no contract on smart medical device procedures and values; thus, devices from numerous constructers might not work efficiently (Research, 2022).

10.5 APPLICATIONS

Medical device businesses are creating artificial intelligence (AI) medical devices to support purposes such as chronic disease management, medical imaging, and Internet of Things as technology develops. Medical gadgets utilizing artificial intelligence could track patients with chronic conditions and administer care or medication as necessary. For example, diabetic patients might wear sensors that monitor their levels of blood sugar and give insulin to manage them. To undergo medical imaging with greater image quality and clarity, companies are creating medical devices using artificial intelligence. Additionally, these tools would lessen a patient's radiation exposure. Medical practitioners can handle data, inform patients, save money, monitor the patient, and work better effectively and efficiently by using a network of wireless, interconnected, and networked digital devices. To improve patient outcomes, businesses are combining the Internet of Things with intelligent medical devices (Tsiukhai, 2021). An overall product entire life sequence controlling outline for artificial intelligence skills is being considered by the FDA's center for devices and radiological health. This framework would certify for revisions to be made from real-world erudition and variation while confirming software's security and efficiency as a medical device are preserved. The FDA has examined and approved an increasing number of devices that have been lawfully marketed (through 510(k) approval, approved De Novo request, as well as approved PMA) utilizing machine learning over the previous ten years in a variety of medical specialties. The initial list of medical devices with AI/ML capabilities that are marketed in the United States is provided by the FDA as information for the general public on these devices. Various AI software/devices with the names of the inventors and application area are mentioned in Table 10.1.

10.6 FUTURE OF AI-DRIVEN MEDICAL DEVICES AND CONCLUSION

Smart medical equipment is in demand for various reasons and its commercial scope is expected to rise for 2022–2027 at a CAGR of 7.22%. It can be USD $42,558.17 million by the near future, probably the year 2027. Medical devices will be in demand for many reasons but especially rapid increasing because of bad living styles such as improper diet, imbalance routine, exercise, etc. This type of bad habit creates the causes for chronic illnesses. Generally, chronic illnesses

TABLE 10.1

AI Software/Devices, Their Inventors, and Applications (Udell et al., 2019)

Sr No.	Date of US FDA Approval	510(k) Clearance Number	AI Device/Software	Inventors	Category	Applications/Indications
1	01-Feb-20	K183089	Air Next	NuvoAir AB	Anesthesiology	Air Next is designed to measure fundamental lung function and spirometry. In a forced expiratory manoeuvre, it assesses characteristics like the forced expiratory in one second (FEV1) and the forceful vital capacity (FVC). These parameters can be used to detect, assess, and monitor lung function problems such as bronchial asthma, COPD, and cystic fibrosis.
2	05-May-17	K163665	MATRx Plus	Zephyr Sleep Technologies	Anesthesiology	The various Health Care Professional (HCP) are using the such devices for enhancing the diagnosis and assessment. The use of device working for sleep disordered breathing specially for adult patients. It works with checking the nostril airflow, pulses, respiratory system, snoring habits and posture of body while sleeping.
3	06-Nov-21	K210484	LINQ_II_Insertable_ Cardiac_Monitor Zelda AI_ECG_ Classification_ System	Medtronic, Inc.	Cardiovascular	The device-ICM-LINQ_II_Insertable_Cardiac_Monitors is a programmable for monitoring and recording ECG. It also keep track of physiological parameters by arrhythmias & patient-initiated triggering/markings.
4	04-Jan-21	DEN200038	Gili-Pro-Biosensor/ Gili-Biosensor-System	Continuse-Biometrics- Ltd.	Cardiovascular	The Gili Pro BioSensor (Gili BioSensor System) features an optical module designed to gather motion-vibration signals from a lighted surface in order to calculate heart rate and respiratory rate.

(Continued)

TABLE 10.1 (Continued)
AI Software/Devices, Their Inventors, and Applications (Udell et al., 2019)

Sr No.	Date of US FDA Approval	510(k) Clearance Number	AI Device/Software	Inventors	Category	Applications/Indications
5	10-Sep-20	K193631	Stethee-Pro-1/Stethee-Pro-Software-System	M3DICINE-Pty-Ltd.	Cardiovascular	The StetheeTM Pro 1 electronic stethoscope detects and amplifies sound from heart, lungs, arteries, veins, and other internal organs regardless of the use of selected frequencies.
6	26-Mar-20	K192732	Bodyguardian Remote Monitoring System	Preventice Technologies, Inc.	Cardiovascular	The Preventice BodyGuardian Remote Monitoring System is designed for use with paediatric patients in clinical and non-clinical situations to collect and send health parameters for monitoring and evaluation to healthcare professionals.
7	20-Mar-20	K200036	AI ECG Tracker	Shenzhen-Carewell-Electronic-Co-Ltd.	Cardiovascular	The AI-ECG Tracker is intended for use in hospitals and healthcare institutions by trained healthcare professionals for the detection of arrhythmias utilizing ECG data obtained from individuals (age 22 and above) without pacemakers. The product also allows you to download and analyze data captured from electrode in compatible formats from whatever device used to diagnose arrhythmias.
8	15-Aug-19	K190925	HeartfFow FFRct Analysis	HeartFlow, Inc.	Cardiovascular	This coronary physiological simulation software is analyzing Heart Flow. It allows quantitative and qualitative analysis for clinic purpose. It works on the basis of historical data. The dataset is related to coronary artery disease clinically stable symptomatic patients.

9	03-Nov-19	K181823	KardiaAI	AliveCor, Inc.	Cardiovascular	KardiaAI is a software analytic package designed to evaluate adult patients' ambulatory electrocardiogram (ECG) rhythms. The gadget may analyse data from every ambulatory ECG device, such as event recorders or other comparable devices, that is recorded in a suitable format.
10	17-Jan-19	K182456	StudyWatch	VerilyLifeSciencesLlc	Cardiovascular	The purpose of the StudyWatch is to record, save, transfer, and show single-channel electrocardiogram (ECG) patterns. Medical practitioners, adult patients (aged 22 and up) with suspected or known heart issues, and health-conscious individuals use it.
11	14-Dec-18	K182790	Cardio-TriTest v6.5	Cardio-Phoenix Inc.	Cardiovascular	The device's intended application is to acquire and record the three various types of heart biosignals (ECG, PCG, and MCG) and integrate the results into a contiguous representation as an assistance to diagnostic assessment by a Physician in a clinical context.
12	01-Apr-18	K171056	WAVE Clinical Platform	Excel Medical Electronics, LLC	Cardiovascular	A WAVE Clinical System is software developed to route, save, and display data, alerts, findings, and clinical findings from medical devices, EMRs, and Clinical Information Management. The WAVE Clinical System is a telemonitoring platform that shows physiological data, waveforms, alarms, outcomes, and diagnostic data.
13	22-Dec-17	K172507	SOZO	ImpediMed Limited	Cardiovascular	The SOZO Fluid Status Monitoring is designed for adult heart failure patients. Under the supervision of a physician, this device is designed for noninvasive monitoring of individuals with fluid management difficulties due to heart failure.

(Continued)

TABLE 10.1 (Continued)
AI Software/Devices, Their Inventors, and Applications (Udell et al., 2019)

Sr No.	Date of US FDA Approval	510(k) Clearance Number	AI Device/Software	Inventors	Category	Applications/Indications
14	17-Aug-17	K163339	Spectral_MD_DeepView_Wound_Imaging_System 2.0	SpectralMDInc.	Cardiovascular	The SpectralMDTM DeepViewTM Wound Imaging Technology 2.0 is an optics imaging system used to study blood flow in the microcirculation. It has clinical uses in plastic surgery, diabetes, cosmetics, vascular surgery, wound repair, neurology, physiology, neurosurgery, and anaesthesia. It can be used to assess the perfusion of both healthy and wounded skin, such as burn injuries, epidermis flaps, chronic wounds, pressure sores, and diabetic ulcers.
15	19-Oct-16	K160401	Tyto Stethoscope	Tyto Care Ltd.	Cardiovascular	The Tyto Stethoscope is an electronic stethoscope which allows for the transmission of auscultation sound data, allowing a clinician at one location on an IP network to hear to the auscultatory audio of a patient on premises or at a different location on the IP network via an IP connection between the two locations.
16	15-Jul-16	K160016	Steth IO	Stratoscientific, Inc.	Cardiovascular	The StratoScientific Steth IO Stethoscope as well as Phonocardiogram Model 1.0 detects and amplifies sounds from the heart and lungs by employing selective frequency ranges.
17	01-Nov-18	K171122	CEREC Ortho Software	Dentsply Sirona	Dental	In order to produce 3D virtual models for data collecting and modelling evaluation for orthodontic patients and circumstances, CEREC Ortho Software is designed to be used with image data gathered from handheld intraoral 3D camera and desktop laboratory scanners.

18	04-Sep-21	DEN200055	GI Genius	Cosmo Artificial Intelligence - AI, Ltd.	Gastroenterology-Urology	The tool is a piece of software that uses AI-ML algorithms for assisting in the real-time detection of colonic mucosal lesions during endoscopy.
19	25-May-18	K180163	TransEnterix Senhance Surgical System	TransEnterix, Inc	General And Plastic Surgery	This System is created to assist the high precision of laparoscopic instruments. The Senhance Surgical System focus on visualization and endoscopic tissue manipulation. It also include gripping, slicing, rough and keen dissection, approximations and retraction.
20	28-Oct-20	K202089	Lenshooke_X1_Pro_Semen_Quality_Analyzer	Bonraybio Co., Ltd.	Hematology	The Lens_Hooke_X1_Semen_Quality Analyzer delivers direct and computed quantitative readings for Sperm counts (106-per-ml), Total-Motility (PR+NP,%), Sperm_Morphology (normal_forms,%), and pH value. It is an optical instrument for human semen examination.
21	08-Jan-17	K171315	Advanced RBC Application	Cella Vision AB	Hematology	The Advanced RBC Program on the CellaVision DM1200 automatically detects and displays pictures of blood cells on smears of peripheral blood. The operator recognises and confirms the suggested type-based classification of each cell.
22	27-Nov-19	K192945	7D_Surgical_System: Cranial_Biopsy and Ventricular_Catheter_Placement_Application	7DSurgicalInc.	Neurology	A stereotaxic image guidance system called the 7D Surgical System is used by surgeons to spatially position and orient neurosurgery equipment. During image-guided surgery, the device is also meant to serve as the main surgical lighting.
23	09-Nov-19	K190815	Brainscope TBI	BrainScope Company, Inc.	Neurology	An adjunct to routine clinical practice, BrainScope TBI is a multi-modal, multiple parameter evaluation that can help with the assessment of patients who have suffered a closed head injury.

(Continued)

TABLE 10.1 (Continued)
AI Software/Devices, Their Inventors, and Applications (Udell et al., 2019)

Sr No.	Date of US FDA Approval	510(k) Clearance Number	AI Device/Software	Inventors	Category	Applications/Indications
24	28-Dec-18	DEN170091	EyeBOX	Oculogica, Inc.	Neurology	In addition to a standard neurological evaluation of concussion, the EyeBOX is designed to measure and analyse eye movements to assist in the detection of traumatic event within one week following a brain injury in patients aged 5 to 67.
25	19-Jul-19	K182798	KIDScore D3	Vitrolife A/S	Obstetrics And Gynecology	The KIDScore D3 tool scores embryos in accordance with their statistical viability to provide decision support for the chance of embryos progressing to the blastocyst stage.
26	06-Oct-21	K203629	IDx-DR	Digital Diagnostics Inc.	Ophthalmic	The IDx-DR AI diagnostic tool analyses patient data to autonomously identify whether a patient has diabetic retinopathy (including macular edema)
27	19-Jun-18	K180038	NuVasive Pulse System	NuVasive, Incorporated	Orthopedic	The Pulse NVM5 is designed for intraoperative neurophysiologic monitoring during thoracic surgery, upper and lower extremity surgery, neck dissection, and spinal surgery.
28	17-Jun-21	K203514	Precise Position	Philips Healthcare (Suzhou) Co., Ltd.	Radiology	With the use of Artificial Intelligence (AI)*, Philips Precise Position helps swiftly and accurately situate patients for successful CT scans.
29	16-Jun-21	K202718	QmentaCarePlatform-Family	MintLabs, Inc., D/B/A.QMENTA	Radiology	The imaging platform from QMENTA is an all-inclusive, fully integrated platform for end-to-end solutions. Every component necessary for an imaging study is included, including secure data gathering and storage, quantified image processing, centralized reviewing procedures, and study closeout.

#	Date	K-number	Product	Company	Specialty	Description
30	19-May-21	K210237	CINA CHEST	Avicenna.AI	Radiology	It detect and triage of pulmonary embolism and aortic dissection with AI tool as CINE-PE and CINE-AD
31	30-Jun-21	K210001	Hyper_AiR	Shanghai_United_Imaging_Healthcare_Co._Ltd.	Radiology	Fluorodeoxyglucose (FDG) PET scans' noise and contrast can be improved with the help of the image processing tool HYPER AiR, which is designed to be employed by radiologist and nuclear medicine specialists.
32	23-Apr-21	K203314	Cartesion_Prime_V10.8	Canon_Medical_Systems_Corporation	Radiology	AiCE-i for PET uses deep learning artificial neural network techniques that can analyse the statistical features of the signal and noise of PET data in order to enhance image quality and lower image noise for FDG whole body data. The AiCE algorithm can be used to denoize and enhance PET image quality.
33	23-Apr-21	K203502	MEDO-Thyroid	MEDO DX Pte. Ltd.	Radiology	Medo then analyses these to present the most relevant image of the lobes and nodules. Furthermore, it segments the lobes and the nodules and calculates the TI-RADS scoring.
34	20-Apr-21	K203610	Automatic Anatomy Recognition (AAR)	Quantitative Radiology Solutions, LLC	Radiology	Designed to work independently of a specific treatment planning system, AAR is deployed on a cloud-based platform.
35	19-Apr-21	K203469	AI Segmentation	Varian Medical Systems	Radiology	AI segmentation is designed to produce high-speed, correct, and perceptive contouring to duplicability of important form description. It reduces 80% contouring time than manual contouring
36	16-Apr-21	K203517	Saige-Q	DeepHealth, Inc.	Radiology	The Saige-Q code represents the software's suspicion of any potential findings in mammography. Saige-Q additionally creates a preview image of the exam's most suspect image for tests that are flagged as suspicious. Radiologists can sort and rank their cases appropriately by viewing the Saige-Q codes and preview images in a worklist on their workstation.

(Continued)

TABLE 10.1 (Continued)
AI Software/Devices, Their Inventors, and Applications (Udell et al., 2019)

Sr No.	Date of US FDA Approval	510(k) Clearance Number	AI Device/Software	Inventors	Category	Applications/Indications
37	14-Apr-21	K202992	BriefCase, RibFx	Aidoc_Medical_Ltd.	Radiology	The rib fractures solution from Aidoc is a treatment and notification tool designed for use with CT image processing. It flags and notifies users of potential positive rib fracture findings.
38	04-Feb-21	K202441	Eclipse-II-Smart_Noise_Cancellation	Carestream Health_Inc.	Radiology	Cares Stream is Smart Noise Cancellation uses deep, convolutional neural network technology to give exceptional CNN-based noise reduction, improving image quality, preserving fine detail, improving contrast-to-noise ratio, and making radiographs simpler to read.
39	31-Mar-21	K203258	Syngo_CT Lung_CAD-VD20	Siemens_Healthcare_GmbH	Radiology	In order to help radiologists identify solid pulmonary nodules while reviewing chest CT scans, Syngo CT Lung CAD was created.
40	31-Mar-21	K203443	MAGNETOM Vida_Sola_Lumina_Altea:syngo_MR_XA31A	Siemens_Medical_Sol._Inc.	Radiology	With the 1.5 T Open Bore MAGNETOM Altea system, you can perform MRI services with complete assurance in terms of output, reproducibility, and patient happiness.
41	31-Mar-21	K210071	SIS_System: 5.1.0	Surgical Information Sciences, Inc.	Radiology	SIS System is designed for viewing, presenting, and documenting medical imaging. It has a variety of features for image analysis, image compression, and intraoperative functional planning. The 3D outputs can be used with orthogonal image guided surgery or other devices for further processing and visualization.

42	24-Mar-21	K203225	Aquilion ONE (TSX-306 A/3) V10.4 with Spectral Imaging System	Canon Medical Systems Corporation	Radiology	Utilize the strength of one-beat cardiac and AI increased image sharpness in combination with PIQE to overcome the difficulties of coronary artery visualization in stents and badly calcified vessels.
43	23-Mar-21	K210209	Viz ICH	Viz.Ai, Inc.	Radiology	Automated Detection and Triage of Suspected ICH
44	19-Mar-21	K203235	VBrain	Vysioneer Inc.	Radiology	Vbrain is the first AI device to receive FDA clearance for tumor auto contouring in radiation therapy.
45	03-Sep-21	K203256	Imbio RV/LV Software	Imbio, LLC	Radiology	Imbio RV/LV Analysis helps physicians quickly assess potential ventricular dilation by automatically processing CTPA scans to assess the most diameters of the right. It also scan the left ventricles along with diameters of heart, and reporting the resulting RV/LV ratio.
46	03-May-21	K202300	Optellum Virtual Nodule Clinic, Optellum Software	Optellum Ltd	Radiology	The world's first AI-based early lung cancer decision support software.
47	25-Feb-21	K202990	NinesMeasure	Nines, Inc.	Radiology	The novel lung nodule assessment tool Nines Measure, created with artificial intelligence (AI), can hasten the diagnosis of several respiratory illnesses.
48	01-Jul-21	K202414	BrainInsight	Hyperfine Research, Inc.	Radiology	Hyperfine BrainInsight offers automated artificial intelligence (AI) tools for brain imaging that give clinicians useful quantitative biomarker information. These tools may help to improve patient care and lessen the workload of neuroradiological specialists by eliminating the need for manual measurements.

(Continued)

TABLE 10.1 (Continued)
AI Software/Devices, Their Inventors, and Applications (Udell et al., 2019)

Sr No.	Date of US FDA Approval	510(k) Clearance Number	AI Device/Software	Inventors	Category	Applications/Indications
49	12-Jul-20	K201039	HepaFat-AI	Resonance_Health_ Analysis_ Services_ Pvt_Ltd.	Radiology	HepaFat-AI analyses MRI datasets automatically to determine the amount of liver fat in patients, giving medical professionals a comprehensive, multi-metric solution to utilize in the evaluation of patients with confirmed or suspected fatty liver disease. Doctors get access to several variables about a patient's liver function via a single short, non-invasive MRI scan.
50	12-Apr-20	K202487	HealthJOINT	Zebra Medical Vision Ltd.	Radiology	The Zebra Health Joint is intended to assist clinicians in the preoperative planning of the surgical procedure
51	20-Nov-20	K200873	HALO	NICo-Lab B.V.	Radiology	HALO is to process and analyze contrast-enhanced CT angiograms of the brain obtained in an acute situation in order to make it easier for patients suspected of having a stroke to have their brain vasculature evaluated
52	10-Nov-20	K202501	Quantib Prostate	Quantib BV	Radiology	The MRI prostate reporting procedure is improved by Quantib Prostate, which is readily accessible from the radiologist's reading station and makes use of deep learning algorithms.

| 53 | 16-Sep-21 | K201369 | Augmented Vascular Analysis | See-Mode Technologies Pte. Ltd. | Radiology | AVA (Augmented Vascular Analysis) analyses each image in a vascular ultrasound scan using deep learning with a single click, delivering highly accurate vascular reports in under a minute. No need for hand drawings. |
| 54 | 30-Jul-20 | K193283 | AI-Rad Companion Prostate MR | Siemens Medical Solutions USA Inc | Radiology | MRI scans of the prostate are automatically segmented using the AI-Rad Companion Prostate MR for Biopsy Support, which also allows radiologists to mark any areas they believe to be suspicious. |

such as asthma, diabetes, and high BP (blood pressure) may be based on historical medical states, cardiovascular disorders, etc. In addition, advanced Internet facility in various medical devices, some of which are also accessible from remote places, are used widely day by day as intelligent medical devices for heterogenous in the healthcare sub-sectors. The medical devices are in acceptance due to their function on large historical data, which assist physicians, doctors, lab technicians, and other healthcare service staff to monitor a patient's health on prescribed time intervals. It also makes sure about the routine care of patients. Looking to all facilities, there is a huge demand for smart medical devices. Furthermore, rising patient knowledge and need for real-time data will assure proper patient care. Furthermore, increased demand for residential therapeutic and diagnostic settings globally is expected to strengthen the market. There will be wide growth of such smart medical tools in the overall market in the near future, such as years 2022–2027 (Unsworth et al., 2021). With the introduction of disruptive new competitors, personalized technology, and constantly altering regulatory constraints, traditional R&D tactics used by medical device manufacturers are proving ineffective in maintaining a competitive advantage. As a result, many corporations are increasingly outsourcing their R&D operations to companies outside of the industry that specialize in value-driven product creation. Using outside assistance in R&D (and thereby having access to innovative technology) can reduce risk, shorten time to market, and ensure high-quality innovation. Outsourcing can be especially advantageous for medical device firms that lack the resources or skills to produce a product from concept to commercialization safely (Vyas, 2023). Big internet companies like Google, Amazon, and Apple, which have a long history of emphasizing innovation and speed, are now making inroads into the "digital health sector." However, such an approach frequently clashes with the stringent development process required in the med-tech business. Apart from Apple, most big tech companies have often struggled to translate product ideas into viable medical equipment. In regard to the ethical concerns raised by the 510(k) FDA approval procedure, producers of medical devices and their collaborators are increasingly adopting methods to guarantee worldwide compliance standards are met, notably in regions with less stringent rules than the EU (Vyas & Gupta, 2022; Wang & Alexander, 2020).

REFERENCES

Cfr - Code of Federal Regulations Title 21. Print.
Fda Ai-Machine Learning Strategy Remains Work in Progress. MedTech Dive. Print.
Smart Medical Device Market Is Expected to Witness Massive Growth with a Significant Cagr of 7. 22%, Delveinsight | Medgadget. Web.
What Is the Future of the Smart Medical Devices Market? SRG. Print.
Alemzadeh, H., et al. (2013). Analysis of safety-critical computer failures in medical devices. *IEEE Security & Privacy, 11*(4), 14–26. Print.
Allen, B., et al. (2021). Evaluation and real-world performance monitoring of artificial intelligence models in clinical practice: Try it, buy it, check it. *Journal of the American College of Radiology, 18*(11), 1489–1496. Print.

Athilingam, P., et al. (2016). Features and usability assessment of a patient-centered mobile application (heartmapp) for self-management of heart failure. *Applied Nursing Research*, *32*, 156–163. Print.

Benjamens, S., Dhunnoo, P., & Meskó, B. (2020). The state of artificial intelligence-based FDA-approved medical devices and algorithms: An online database. *NPJ Digital Medicine*, *3*(1), 118. Print.

Bohr, A., & Memarzadeh, K. (2020a). The rise of artificial intelligence in healthcare applications. *Artificial Intelligence in Healthcare*, 25–60. doi: 10.1016/B978-0-12-818438-7.00002-2. Epub 2020 Jun 26. Print.

Bohr, A., & Memarzadeh, K. (2020b). Chapter 2 – The rise of artificial intelligence in healthcare applications. In A. Bohr, & K. Memarzadeh (Eds.), *Artificial intelligence in healthcare* (pp. 25–60). Academic Press. Print.

Dente, C.J., et al. (2017). Towards precision medicine: Accurate predictive modeling of infectious complications in combat casualties. *Journal of Trauma and Acute Care Surgery*, *83*(4), 609–616. Print.

Ekins, S., et al. (2019). Exploiting machine learning for end-to-end drug discovery and development. *Nature Materials*, *18*(5), 435–441. Print.

Fernando, W.K.A.U.K., Samarakkody, R.M., & Halgamuge, M.N. (2020). Smart transportation tracking systems based on the internet of things vision. In Z. Mahmood (Ed.), *Connected vehicles in the internet of things: Concepts, technologies and frameworks for the Iov* (pp. 143–166). Cham: Springer International Publishing. Print.

Ghersi, I., Mariño, M., & Miralles, M.T. (2018). Smart medical beds in patient-care environments of the twenty-first century: A state-of-art survey. *BMC Medical Informatics and Decision Making*, *18*(1), 63. Print.

Haleem, A., et al. (2021). Telemedicine for healthcare: Capabilities, features, barriers, and applications. *Sensors International*, *2*, 100117. Print.

Health, Center for Devices, and Radiological. (2022a). Artificial Intelligence and Machine Learning (Ai/Ml)-Enabled Medical Devices. FDA. Print.

Health, Center for Devices, and Radiological. (2022b). Artificial Intelligence and Machine Learning in Software as a Medical Device. FDA. Print.

Health, Center for Devices, and Radiological. (2022c). De Novo Classification Request. FDA. Print.

Health, Center for Devices, and Radiological. (2021a). Good Machine Learning Practice for Medical Device Development: Guiding Principles. FDA. Print.

Health, Center for Devices, and Radiological. (2019). National Evaluation System for Health Technology(Nest). FDA. Print.

Health, Center for Devices, and Radiological. (2022d). Premarket Approval(Pma). FDA. Print.

Health, Center for Devices, and Radiological. (2021b). Premarket Notification 510(K). FDA. Print.

Hetrick, C. The Use of Artificial Intelligence in Medical Devices. Sterling Medical Devices. 2021/12/20/T20:21:09+00:00 2021. Web.

Hirai, T., et al. (2009). Electrically Active Artificial Pupil Showing Amoeba-Like Pseudopodial Deformation, *21*(28), 2886–2888. Print.

Hitti, E., et al. (2021). Mobile device use among emergency department healthcare professionals: Prevalence, utilization and attitudes. *Scientific Reports*, *11*(1), 1917. Print.

Jain, S., et al. (2021). Internet of medical things (IomT)-integrated biosensors for point-of-care testing of infectious diseases. *Biosensors and Bioelectronics*, *179*, 113074. Print.

Javaid, M., & Khan, I.H. (2021). Internet of things (IoT) enabled healthcare helps to take the challenges of Covid-19 pandemic. *Journal of Oral Biology and Craniofacial Research*, *11*(2), 209–214. Print.

Kaur, D., et al. (2022). Trustworthy artificial intelligence: A review. *ACM Computing Surveys*. Article 39. Print.

Kaushik, A., et al. (2021). From nanosystems to a biosensing prototype for an efficient diagnostic: A special issue in honor of professor Bansi D. *Malhotra, 11*(10), 359. Print.

Konieczny, L., & Roterman, I. (2019). Personalized precision medicine, *15*, 4, Print.

Miller, D.D., & Brown, E.W. (2018). Artificial intelligence in medical practice: The question to the answer? *The American Journal of Medicine, 131*(2), 129–133. Print.

Nunan, D., & Di Domenico, M.L. (2013). Market research and the ethics of big data. *International Journal of Market Research, 55*(4), 505–520. Print.

Okuzaki, H., et al. (2014). Ionic liquid/polyurethane/pedot: Pss composites for electro-active polymer actuators. *Sensors and Actuators B: Chemical, 194*, 59–63. Print.

Pak, J.G., & Park, K.H. (2012). Construction of a smart medication dispenser with high degree of scalability and remote manageability. *Journal of Biomedicine and Biotechnology, 2012*, 381493. Print.

Patel, V.M., et al. (2007a). Mucoadhesive bilayer tablets of propranolol hydrochloride, *8*(3), E203–E208. Print.

Patel, V.M., Prajapati, B.G., & Patel, M.M.. (2007b). Formulation, evaluation, and comparison of bilayered and multilayered mucoadhesive buccal devices of propranolol hydrochloride. *Aaps Pharmscitech, 8*(1), E147–E154. Print.

Pham, T., et al. (2017). Predicting healthcare trajectories from medical records: A deep learning approach. *Journal of Biomedical Informatics, 69*, 218–229. Print.

Quezada, R., et al. (2021). Technological aspects for pleasant learning: A review of the literature, *8*(2), 25. Print.

Rahman, Md S., et al. (2020). Defending against the Novel Coronavirus (Covid-19) Outbreak: How can the internet of things (Iot) help to save the world? *Health Policy and Technology, 9*(2), 136–138. Print.

Research, S. (2022). Diabetes care devices market size is projected to reach Usd 87.15 Billion by 2030, growing at a Cagr of 6%: Straits Research. *GlobeNewswire News Room*. Print.

Tsiukhai, T. (2021). Iot and Medical Device Integration: Putting Healthcare Data into Action. *The Healthcare Guys*. Print.

Udell, M., et al. (2019). Towards a smart automated society: Cognitive technologies, knowledge production, and economic growth. *Economics, Management, and Financial Markets, 14*(1), 44–50. Print.

Unsworth, H., et al. (2021). The nice evidence standards framework for digital health and care technologies - Developing and maintaining an innovative evidence framework with global impact. *Digit Health, 7*, 20552076211018617. Print.

Vyas, S. (2023). Extended reality and edge AI for healthcare 4.0: Systematic study. In *Extended reality for healthcare systems* (pp. 229–240). Academic Press.

Vyas, S., & Gupta, S. (2022). Case study on state-of-the-art wellness and health tracker devices. In *Handbook of research on lifestyle sustainability and management solutions using AI, big data analytics, and visualization* (pp. 325–337). IGI Global.

Wang, L., & Alexander, C.A. (2020). Big data analytics in medical engineering and healthcare: Methods, advances and challenges. *Journal of Medical Engineering & Technology, 44*(6), 267–283. Print.

11 Virtual Consultation

Scope and Application in Healthcare

Sheetal Yadav, Rakhi Ahuja, Puneeta Ajmera, and Sheetal Kalra

The accurate meaning of telemedicine is "healing from remote." Telemedicine has been defined by the World Health Organization (WHO) as "The delivery of healthcare services, where distance is a critical factor, by all healthcare professionals using information and communication technologies for the exchange of valid information for diagnosis, treatment and prevention of disease and injuries, research and evaluation, and for continuing education of healthcare providers, all in the interests of advancing the health of individuals and their communities."

11.1 TECHNOLOGY AND TELEMEDICINE

An essential component of telemedicine is telenursing and robotics. Telenursing is a crucial component of mobile health. Telenursing in healthcare corresponds to the provision of nursing facilities to the patients through telecommunication media. The term *telenursing* has been described as a subset of virtual consultation concerned with the nursing and care facilities. The service that is included is telenursing. The application of telemedicine may vary across different areas based on the patient needs and the demographics of the localities. The nurses serve as essential carers and perform a crucial part in determining the healthcare systems. According to the WHO, there is a massive dearth of primary healthcare workers across different countries. The availability of healthcare practitioners directly impacts the public health and community health of a geographical region. The public health care (PHC) is directly related to health status and the overall well-being of the community. Telephonic nursing or telenursing consultations, therefore, serve to be of essential importance in the management and availability of basic healthcare services. The role of nurses is not only to provide care facilities to the patients through consultations but also to educate them for the promotion of their health and well-being. This component of the telemedicine and virtual consultation is known as tele-education in the context of the healthcare system. The telenursing services that are often provided through telephonic consultations include assessment over call, advice, and referral along with the facilitation for the allied services required by the patient (Agboola et al., 2014). The telemedicine and telenursing consultations are popular as they allow for the treatment and management of the condition without on-site visits. The telenursing services are dependent upon three primary factors that include knowledge about tele-consultation, access and availability of resources, and nurse experience. The support systems are often used in the telemedicine and telenursing services to ease their access and the availability of the carers (Abrams, 2014). The use of these systems also safeguards that the safety and quality in the tele-consultation services are improved. This computerized assistance has been therefore reported to be of essential assistance and care for the nurses. Application of telemedicine is, therefore, also related to the availability and feasibility of the telemedicine-enabled equipment that is used in the care units of the hospitals.

11.1.1 TELEHEALTH ALGORITHMS AND MODELS

Many studies in the past have substantiated that the mortality rate of IHD (ischemic heart disease) can be strikingly decreased with the help of telemedicine. One of the approaches can be sending live real-time health information of a patient from an ambulance to the health center so that the doctors and their team can prepare timely, like readying the operation room and providing the treatment immediately once the patient arrives at the emergency unit. The reliability of this live health data being transmitted is critical in preventing misinterpretation and misdiagnosis by the doctors. Hence, it is likely with the 5G network, since it facilitates the best data speed and low latency. However, its network coverage is poor in comparison to 4G. It induces an elevated number of handovers when the ambulance navigates 5G networks at high speed, which might result in data corruption, resulting in degradation of telemedicine services. Hence, an effective and detailed handover decision-making algorithm was proposed that integrated the traveling time estimation (TTE), fuzzy analytic hierarchy process (FAHP), and also the technique for order of preference by similarity to ideal solution (TOPSIS), considerably reducing the number of unnecessary preventable handovers in 5G heterogeneous networks, and thereby improved the user satisfaction by choosing the lower cost network while retaining the quality of telemedicine service (Diong et al., 2021) (Figure 11.1).

The pandemic encouraged doctors to practice tele-consultations for some common otolaryngology illnesses. One of the issues faced was the judgment on the etiology of dizziness and if was otologic or non-otologic. The lateralization of symptoms cannot categorically differentiate one from the other, although it could be helpful. However, telemedicine could be used to speed up the supervision of dizzy patients during the times when a physical assessment is not possible. For an effective tele-consultation, a thorough and detailed history is the base. Examination of the shaky patient is compatible to an algorithmic solution that can be modified to a telemedicine setup for the general otolaryngologist (Chari et al., 2020) (Figure 11.2).

In tele-dermatology, the consulting doctor often requires high-quality images of the affected area for diagnosing the patient's condition and then uses the

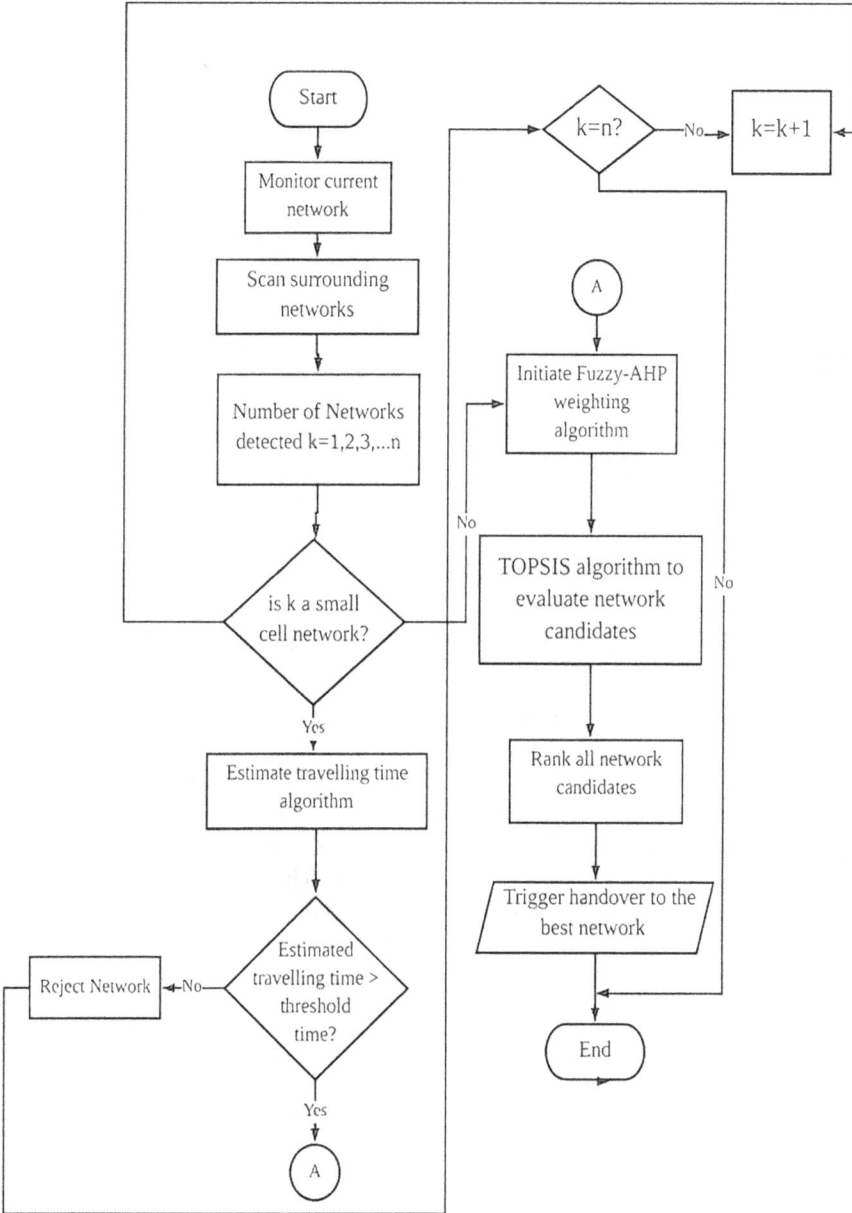

FIGURE 11.1 Proposed vertical handover algorithm (Diong et al., 2021).

tele-communication platform to converse with the patient instead for doing assessments. However, most times video-consult solutions do not have sufficient
resolution in imaging for analysis. Although patients are instructed on how to
click photos of the affected area, it is usual for patients to click images that
are blurred, in poor lighting conditions, or that do not properly show the lesion

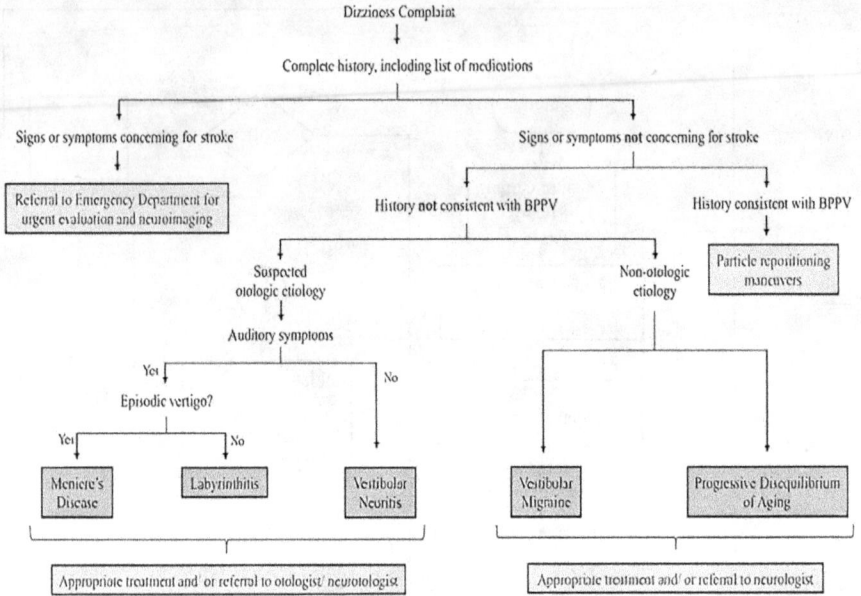

FIGURE 11.2   Algorithm for dizzy patients' management (Chari et al., 2020).

(Vyas, 2023). Hence, an automated machine learning technique was developed for evaluating the quality of the image and providing substantial feedback regarding how to get better photos when needed (like "turn on camera flash" in case of dim lighting). TrueImage provided a value score to compute how appropriate an image is for tele-dermatology. The dermatologist can establish a certain level of image resolution using the score. The algorithm is described in three stages: (1) semantic segmentation to identify skin regions, (2) feature generation, and (3) a quality classifier applied to these features (Vodrahalli et al., 2021) (Figure 11.3).

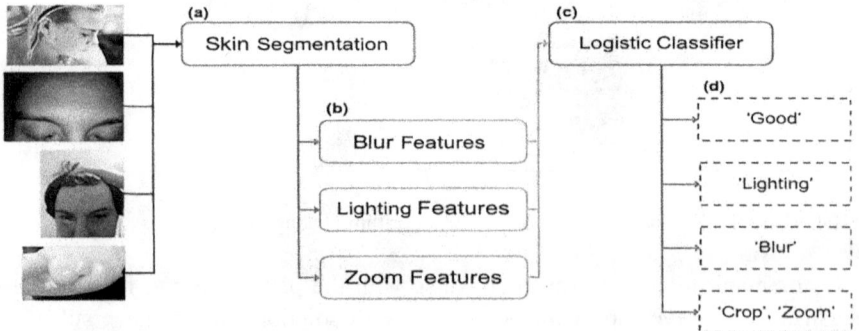

FIGURE 11.3   TrueImage dermatology image quality detection algorithm (Vodrahalli et al., 2021).

Five models have been identified in a telehealth report by McKinsey in 2021 for virtual non-acute care and assessed the total number and cost of healthcare that could be provided in this manner. They include:

a. On-request virtual emergency treatment as an alternative to ED trips and after-hours consultations. These are the telehealth use cases that insurers most frequently use right now, allowing a patient to remotely consult on request with an unknown consultant to address immediate concerns and avoid a trip to the ED.
b. Visits to a virtual hospital with a recognized healthcare giver for consultation that does not require bodily examination or any parallel procedure. Such a visit can be for primary care, behavioral or psychological health, and some specialty care. Along with virtual visits, this approach also gives doctors the tools they need to handle patients with chronic conditions more effectively. These tools include remote patient monitoring, digital therapeutics, and digital coaching.
c. By merging virtual access to physician consultation with "near home" sites for testing and immunizations, near-virtual hospital visits give patients the opportunity to conveniently access healthcare outside of a doctor's office.
d. Virtual home health services regulate virtual visits, observation from a distance, and digital patient engagement tools to enable some of these services to be delivered remotely. These services include physical therapy, occupational therapy, and speech therapy as well as education for patients and providers. Virtual home health services could improve the patient and carer experience, expand the reach of domiciliary health providers, and strengthen connections with the carer's team. Direct services, such as wound care and assistance with everyday activities, would still be provided in person.
e. Technology-enabled home-based medicine management permits patients will swap between getting some injectable and infusible medications at the clinic and at home. This change can be brought about by using remote monitoring to manage patients and track symptoms by offering self-service patient education tools, and by providing telehealth staff oversight.

Virtual healthcare models and business models are expanding and thriving, moving from purely "virtual urgent care" to a range of services enabling longitudinal virtual care, integration of telehealth with other virtual health solutions, and hybrid virtual/in-person care models, with the potential to improve consumer experience/convenience, access, outcomes, and affordability (Bestsennyy et al., 2021).

11.2 FUTURE DRIVERS OF TELEMEDICINE

Consumers belonging to different age ranges have been associated with the use of telemedicine as they may use it for consultations, management of chronic disease, and medication suggestions (Thaker et al., 2013). As the person-centered care models are being used and applied more effectively, the development of telemedicine must evolve with the changing needs and the demands of the people. The organized telemedicine services have assisted in the improvement of the virtual consultation care services and also in the improvement of the quality of medical care that is given to the patients. The use of these services in the future is assumed to reach a wider audience and different countries with customized applicability. The telemedicine care services become more readily accepted and can be designed to meet the care needs of a wider audience and to serve to the needs of people with varying care demands. With the advancement of artificial intelligence, the use of telemedicine has also been advanced and has become more accurate. Coupling of artificially intelligent technologies with telemedicine has helped in the improvement of the accuracy and the efficiency of the telemedical facilities (Udsen et al., 2014). The development and advancement of telemedicine have a huge dependency on the development of information communication technologies. It is important to consider the consumer-based development of telemedicine technology. Different platforms are preferred depending upon the specific clinical, technological, and accessibility to healthcare situations. The standards of care as well as practice evolve regularly based on the evidence and experience associated with telemedicine use. The development of telemedicine also relies upon the evidence and experience with the specificity of medical, technological, and accessibility to the care developments (Karthik et al., 2014).

11.3   NARRATIVE LITERATURE REVIEW REGARDING VC IN DEVELOPED AND DEVELOPING COUNTRIES

This chapter has synthesized and articulated literature from reliable databases like Elsevier, ResearchGate, and SCOPUS to retrieve and analyze the available literature using the PRISMA approach. The PRISMA checklist is used to help the authors find credible literature for the synthesis of information. The key terms and search strings that were used in this analysis included "Telemedicine in India" "telemedicine business in India," "Telemedicine business around the world," and "global impact of telemedicine." The PRISMA checklist provides information regarding the strengths and weaknesses of the paper by analyzing its recency, bias, title, rationale, and validation of the data. This allows for the screening of relevant literature. Using the search strings to retrieve several records identified through different databases was carefully assessed in this study. Only recent literature articles, not more than ten years old, were selected. A special exception was allowed for the inclusion of certain fundamental articles that were beyond the exclusion criteria to enhance the credibility of this research. The flow diagram for the PRISMA approach for literature synthesis is illustrated in Figure 11.4.

This evaluation allowed for the data collection and analysis that was used to develop a questionnaire to perform the SWOT and PESTLE to undertake market assessment for telemedicine in India for a comprehensive conclusion.

The literature review has been classified into three different sections for ease of comprehension and data summary. The first section of this literature review focuses on the global outlook of the telemedicine market. This has been compared and contrasted against the Indian telemedicine market in section two of this literature

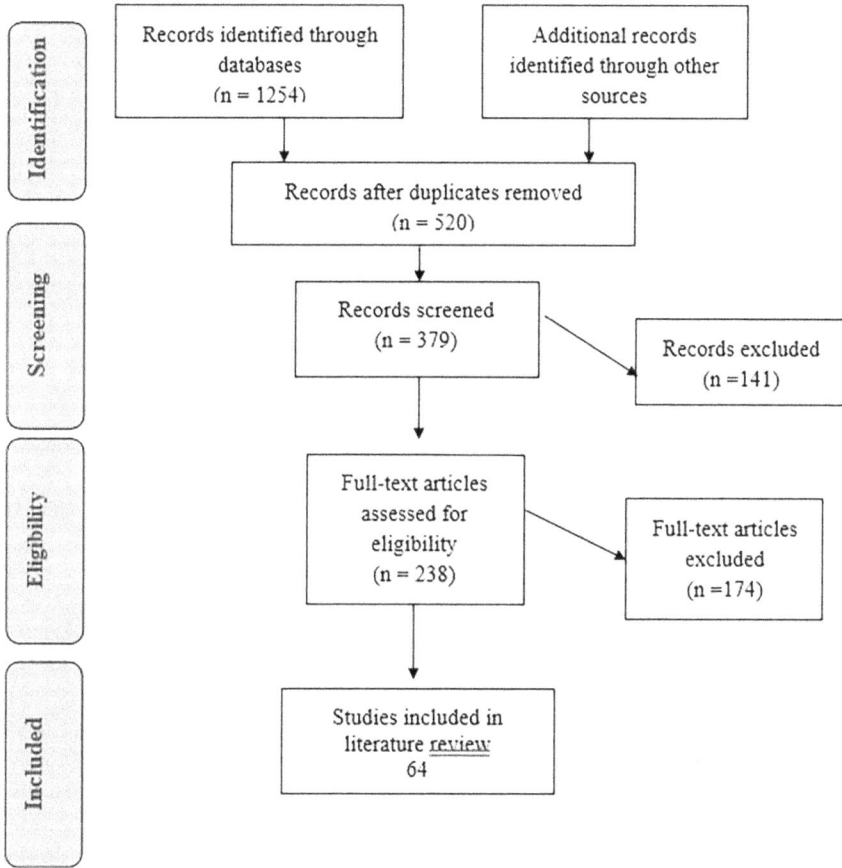

FIGURE 11.4 PRISMA flow chart for data extraction.

synthesis. The third section of this study is composed of literature synthesis and analysis of telemedicine application in the additional nations under development. Assessment of telemedicine application and implementation in the worldwide context has been done to determine the total spread and magnitude of business of telemedicine worldwide. This has been contrasted in section two of the review of literature that analyzes the applicability of telemedicine in the Indian market.

11.3.1 GLOBAL SCENARIO OF TELEMEDICINE

Healthcare costs are rising all around the globe and there is restricted time availability to get medical consultations. It has been studied by Doolittle and Spaulding (2017) that the telemedicine market in the United States of America had valued for about USD$21,446.33 million in the year 2018 and is further likely to grow up to USD$60,448.47 million by 2024. This ascend in the market has been credited to the increase in health facilities. Further, since the application of telemedicine promises to reduce costs and time invested in clinical

Telemedicine Market: Revenue Share (%), by Type,
Global, 2018

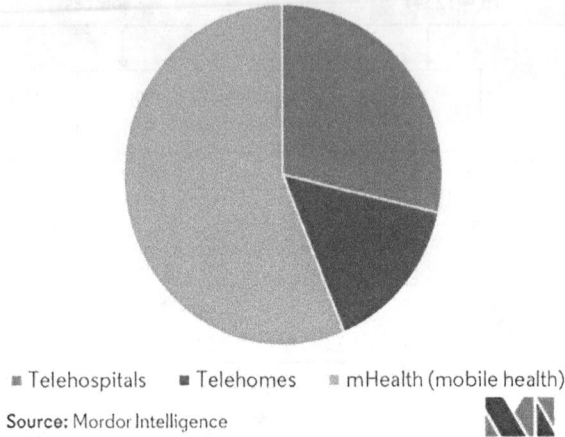

■ Telehospitals   ■ Telehomes   ■ mHealth (mobile health)

Source: Mordor Intelligence

FIGURE 11.5   Telemedicine market by type, global, 2018.

assessments, the mobile health inspectors that are highly accessible and feasible have gained consumer confidence and choice. The market expansion of telemedicine around the globe has also been supported by the indigenous programs that have promoted the application of telemedicine in healthcare industries in many countries. A comprehensive analysis by Mohr et al. (2019) asserts that tele-homecare provisions have been largely supported by the governments in the developed nations. The Whole System Demonstrator (WSD) system in the United Kingdom, Veterans Health Segmentation program (VHS) of the USA, and the TELEKART program in Denmark have played a fundamental role in starting and setting up the telemedicine business in these countries. In the global market of telemedicine, the revenue estimates indicate that North America possessed the maximum market share in telemedicine in the year 2018. A cohort analysis by Bashshur et al. (2016) indicated that in the global market share, the most widely used telemedicine facility was mobile health or "mHealth" followed by tele-hospitals and tele-homes (Figure 11.5). Ease of access and availability of clinical experts through mobile screens have served to be of extreme importance.

Another study by Eren and Webster (2018) estimated the telemedicine market growth rates across the world. This rate was centered on access to the Internet and digital literacy, providing scope for market occupancy in the global regions. The study evaluated that East Asia and Australia possess a high chance of developing the telemedicine market in the future. Latin America and Africa along with certain sections of the Middle East demonstrated a lack of potential growth for the telemedicine market in the near future, whereas North American and Western European countries are expected to demonstrate medium growth in the industry (Figure 11.6). Countries that reside in North America and Western Europe have well-developed healthcare systems and access to advanced

FIGURE 11.6   Telemedicine market scope across the globe.

technologies. Further, the telemedicine market has been largely accepted and is already flourishing in these nations and therefore only a medium-paced growth in the market niche of telemedicine is expected from these zones. On the contrary, countries in East Asia like India, Bangladesh, and Indonesia are rapidly developing with a tremendous increase in their information technology infrastructure (Klaassen et al., 2016). The Internet access in these regions has also upgraded and therefore they possess a true potential for market growth.

The mobile health market in telemedicine has presented itself to be highly promising. The WHO describes mHealth as "medical and public health practice supported by mobile devices such as mobile phones, patient monitoring devices, personal digital assistants (PDAs) and other wireless devices" (Vyas & Gupta, 2022). It has been reported by Trettel et al. (2018) that in the year 2017, there were 3 lakh and 25,000 mobile health apps and 84,000 mHealth app publishers. This paper also estimated that by 2022, this market is bound to grow by 32.5% CAGR. Europe is one of the leading nations in the mHealth market, with 30% of total market attainment, followed by the United States with 28% of the total market share. It has been expected that the global market for mHealth alone is bound to grow further at a tremendous rate with a 61.6% CAGR growth possibility for Europe alone. This expected increase has also been validated by the analysis of Avanesova and Shamliyan (2019), who have estimated that the mHealth application downloads have increased tremendously in the last seven years. It was calculated that 1.7 billion downloads for telemedicine applications had occurred in the year 2013 and that increased to 3.7 billion by the year 2017. A crucial insight about the telemedicine market was also given by Ting et al. (2020), who recognized that one in three telemedicine mobile apps were devoted to mental health. The distribution of disease specific telemedicine apps that have been studied worldwide by Lee and Lee (2018) indicate that until 2015, 29% of the total applications attributed to

Percentatge distribution of telemedicine apps

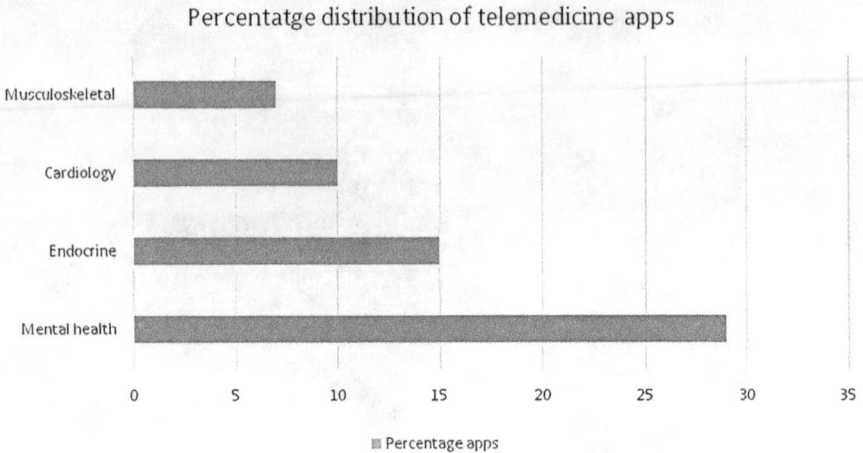

FIGURE 11.7   Global percentage distribution of telemedicine apps, 2015.

mental health, 15% for endocrine health, 10% for cardiology, and 7% for the musculoskeletal health (Figure 11.7).

As per the American Telemedicine Association (2017), more than 200 medical service providers in the USA provide telemedicinal services through video-based consulting across the world. It has also been stated that about 74% of the total patient population of the United States accesses virtual consultation services. Kamdar et al. (2018) have estimated that by 2024, the global market of telemedicine may grow up to 18.5%. The success and establishment of the telemedicine market globally have been possible due to two-way communication that has been immensely facilitated by technological advances. Factors affecting telemedicine market growth have been analyzed by Kim and Zuckerman. The paper argues that the evolution of telemedicine can be seen in parallel to the advancement of information technology and the obtainability of the resources. The telemedicine market has grown in conjunction with the demands and the diversion in the sector has been observed in tele-radiology; that is, the transmission of X-rays, CT, MRI, SPEC, and PET scans. The benefits of telemedicine include cost reduction, easy availability, and convenience. Ekanoye et al. (2017) argue that as the population increases, it leads to an amplified demand for the health practitioners that must be fulfilled on a large scale. Therefore, it depends on the availability of the infrastructure pertaining to information technology in the country. Another potential trait of the medical service is the potential threat of data breach and data security. However, even with its key limitations, virtual consultation has been regarded as a game changer in the health industry (Miller et al., 2018). As telemedicine brings healthcare facilities directly to the doorsteps of the consumer, it is bound to enhance the overall health statistics of any country it is applied to. As per the research report "Global telemedicine market outlook 2022," the market is trending towards an enormous growth in the coming years. Telemedicine deployment has remained subjective to the digitalization of the country to determine the success story of telemedicine solutions. Volterrani and Sposato (2019) have defined different levels of telemedicine solutions that are appropriate in the global market. These include product, platform, and database levels of software. Product solutions in the telemedicine market are patient-level solutions and largely serve by collection of the data. The platform solutions in telemedicine function as an intermediary interface between the patients/consumers and clinicians and functions through data sharing. The database solutions are highly advanced and function via data storage and analysis.

An extensive analysis by Okoroh et al. (2016) studied the telemedicine market worldwide and may be classified in two ways: solution service between healthcare professionals and patients and virtual consultation market for devices for self-care. The analysis reveals that out of all the products available in the virtual consultation domain around the world, 39% cater to the requirements of patients by directly connecting them to the medical professionals, with 5% of the services used only by the professionals and with 11% of apps catering the self-care services. The key players in the virtual consultation market sphere comprise of Aerotel Medical Systems, AMD Global Telemedicine INC., Resideo Technologies, in touch technologies INC. and All Scripts Healthcare Solutions Inc., etc. To analyze the global market trends, Lilly et al. (2017) carried out a rigorous investigation to identify the potential market expansion as well as telemedicine which are calculated as per several considerations. These considerations include that the United States has been a pioneer of telemedicine, with Canada being a forerunner along with Europe and limited involvement of Japan and other nations in the market niche.

The mounting costs of health treatments and limited accessibility and affordability of those solutions in the market niche are the core reasons for the booming telemedicine market. The global trends indicate that the telemedicine market is scheduled to advance globally due to an increase in digital literacy and improvement in information technology solutions worldwide. Further, since the technology is relatively new and unexplored in maximum of the countries, a huge population there can be catered to by the pioneers. This will enhance opportunities for the entrepreneurs in the telemedicine market and also allow for the expansion of the healthcare structure of the country that plans to adopt it. Telemedicine can, therefore, be termed a market niche that is waiting to get itself tapped for market utility and resource utilization. Successful application of telemedicine in the developed countries is indicative of the availability of mass digitalization and the availability of strong information communication and technological infrastructure (Young et al., 2019). As the developed nations are intensifying towards digitalization, they form a potential expansion destination for the telemedicine industry. Since the developing countries like India usually struggle with an underdeveloped healthcare system, the market potential of telemedicine introduction increases further. It has also been validated by Lupton and Maslen (2017) that developing nations are the most potential market niche for telemedicine expansion as it will not only aid in business development for the companies but will also assist in refining the healthcare system of the country by providing more feasible care to the patients and helping them improve the entire healthcare infrastructure of these nations.

11.3.2 TELEMEDICINE SITUATION IN OTHER COUNTRIES UNDER DEVELOPMENT

A staggering disparity is present in the acessibility and quality of healthcare available in the developed and the developing nations. A comprehensive study by Scott and Mars (2018) was conducted to identify the present and future prospects of telemedicine application in the developing nations. The study explores the existing virtual consultation projects that are running in the Caribbean, African, and Asian countries. The "potential" of virtual consultation is often discussed for the provision of the acess to the healthcare in the remote and rural regions. The need of improved medical care is a neccessity in the under developed and in the developing countries. The paper highlighted the plight of the healthcare status in the developing countries, for instance, Africa, houses the 14% people of the global inhabitants and bears 24% of the worldwide disease load due to limited healthcare services (Zolfo et al., 2011). The paper by Scott and Mars (2018) also identifies the health concerns that need to be catered to in the developing countries through the application of advanced telehealthcare services. These issues include aging population, poverty, high incidence of infectious diseases, poor woman and child health, increased aggression and conflicts, and shortage of healthcare workforce (Prieto-Egido et al., 2014). The other issues that need to be improved in the existing healthcare systems of the developing countries include poor healthcare systems and limited access, high costs of healthcare, limited literacy, and shortage of healthcare information systems.

11.3.2.1 Virtual Consultation Care in Latin America and Caribbean

The application of virtual consultation in the Latin American and Caribbean regions has grown significantly. The telemedicine and virtual consultation units are also associated with the three national networks to generate pilot projects that have been supplemented by the government support. This support has led to national-level telemedicine and telehealth program development of virtual consultation consultation and education support programs. A classical example of this successful project comes from the application of virtual consultation in "Minas Gerais: in the south of Brazil." The virtual consultation services have been applied for the 19 million people that are distributed in the 835 municipalities of the region. The state government had established a minor virtual consultation care network in the locality that focused primarily on the cardiology services. This network was established in association with the public universities in the year 2005. By the year 2012, this virtual consultation care network had expanded to 300 municipalities and had been functional with about 42% of the population of the state (De Melo et al., 2018). This project was successful as it utilized inexpensive instrumentation for the testing procedures and provided trusted consultations from the physicians, only excluding telenursing.

11.3.2.2 Telemedicine and Virtual Consultation Service Applications in Asian Countries

Application of the telemedicine services in Asian countries has remained largely unexplored. A crucial study was performed by Durani and Khoja that focused on the analysis of telemedicine application in Asian countries. The study showed that the most common methods of telemedicine and virtual consultation application in Asian countries were through store and forward technology comprising of about 43% of total consultations. This was followed by videoconferencing that constituted 355 of the total consultations followed by adoption of a hybrid approach. Asia is composed of a wide range of communities with extreme conditions. A successful model of virtual consultation application in the developing countries in Asia comes from Malaysia. In the year 1996, Malaysia had developed and implemented various virtual consultation care associated laws that allowed for the development of a "Multi-Media Supercorridor" initiative. The country had also developed a "Lifetime Health Plan" with "lifelong PHR" services. In China, "infer Vision" has developed partnership with several hospitals for providing AI diagnostics for several chest conditions with a partnership of more than 100 radiologists (Zhao, 2010).

11.3.2.3   Telemedicine in African Nations

The use of telemedicine in Africa has been focused on the use of virtual consultation, health education, and medicine-related research. There are about 51 mobile health programs that are being run in 26 developing states of Africa. Several mobile health and virtual consultation applications have advanced in the region that serve a promising future in virtual consultation in Africa (Kamsu et al., 2014). Mars studied the application of telemedicine in Africa and reported that significant developments have been made in the fields of tele-pathology, tele-cardiology, tele-radiology, tele-dermatological studies, tele-psychiatry, tele-obstetrics, and tele-ophthalmology. The advances have also been observed in the field of tele-education, which has assisted in the development of more localized trials and project activities that include promotion of HIV testing, help in maintenance of appointments and providing prompts, encouragement in medication adherence, and assisting patients with managing chronic health conditions.

Multiple challenges are related with the application of telemedicine and virtual consultation services in the developing countries. The common challenges are inclusive of availability of health human resources, populace demographics, responsiveness and prospect, dual burden of the disease, availability of technical support and communication mediums, and poverty (Puustjärvi & Puustjärvi, 2013). Several challenges associated with the health sector are also widespread. Successful and effective application of the telemedicine technologies in the developing countries is also constrained by the challenges that are postured by the healthcare systems of these countries. The reforms that are needed for this merger are larger and serve as one of the main hindrances in the usage of telemedicine technology in the developing nations. The primary care direction of the developing nations is to meet the health inequity, prevent spread, and maintain a healthy population with low health costs (Ryu, 2012). There has been persistent deficiency of evidence for the effective use of virtual consultation in the developing nations. Successful application of telehealth in the developing nations can involve a multitude of modern technologies that require transmission of the information through text, video, audio, and still images. The major constraint of the developing nations is the dearth of adequate healthcare infrastructure of the successful application of the telemedicine technologies (Ryu, 2012).

Khalifehsoltani and Germai (2010) studied that it is important to work with a collaborative approach towards health. The developing countries often face a shortfall of the skilled healthcare personnel that are related with the healthcare sector for the overall management of the healthcare sector and the disease burden of the countries. This also causes a massive health inequity due to lack of adequate servicepeople available to cater to the population.

The population demographics in the developing countries also play an important role. It was studied by Blaya et al. (2010) that there is a general trend in the last 50 years inclining towards the older population in the developing countries as well. Other than the increasing illnesses of the elderly, Syed-Abdul (2012) has studied that the developing nations are also under the disease burden of the communicable diseases and several non-communicable diseases impacting the healthcare system of the developing nations.

Another essential factor that serves as challenge to successful application of telemedicine in the developing nations is poverty. Khanal et al. (2015) assert that poverty exists significantly in many regions of the globe with non-uniform stages in various countries and regions. As per a report published in 2011, two-thirds of the worldwide poverty resides in only five countries with 1.2 individuals classified as "extremely poor." India has 33% of this share, China with 13%, Nigeria amounting for 9% of this population, and Bangladesh and the Republic of Congo estimating to about 5% of this population each. The low-income countries in the developing countries are generally poor. The reachability of the telemedicine solutions in these regions have therefore been largely limited. People with poor economic stability may not be able to have the primary basic essentials for the telemedicine operations. These include the accessibility of the technical services and of communication technology, mobile devices, and Internet access. The technology and communications availability are primary for the successful application of telemedicine.

11.4 TELEMEDICINE SITUATION IN INDIA

Research by Bajpai and Wadhwa (2019) has mentioned that the term *health care telemedicine* typically refers solely to the supply of clinical and non-clinical services like medical education, administration, and analysis to the people. They, in their systematic analysis, have asserted that telemedicine is marked as the start of a new age in the field of medical science; such a concept has been rising since it was offered to the public properly. According to Bali et al. (2016), India is a vast country with no national insurance or single health policy catering to the entire nation. The government of India has reinforced the healthcare delivery system that charts the three-tier structure and taken it as the principal concern of the individual state. Further, it has been observed that there is a great deal of inequality in the availability of healthcare facilities between rural and urban areas (Kakkar et al., 2017). This could be a median through telemedicine technology to use it like a tool to integrate with the ongoing healthcare delivery system. Both sectors of our country, private and government, have been vigorously participating in the telemedicine program to improve the virtual consultation infrastructure and outreach of our country. Therefore, telemedicine has gained and has witnessed an incredible growth during last two years in the Indian market; mainly the reason was timely merging in the fields of the technology, data, communication, and medical care laterally with the schemes of electronic health and e-medicine including telemedicine in the MoHFW (Ministry of Health and Family Welfare) of India (Bakshi & Tandon, 2019).

Telemedicine serves as a simple tool for the professionals as it allows for discussion of the matter over a telephone, or via satellite expertise and videoconferencing kit for conducting real-time consultations for the delivery of medical care using information technology. Telehealth has been gradually gaining base in the state in developing countries where the poor and rural neighborhoods have limited access to healthcare facilities. In India, public-private collaborative initiatives are affecting the lives of rural people by providing them with the necessary care and services. It is of high consideration that patients of rural areas travel large distances and also bear additional costs to get access to super specialist medicines due to limited healthcare service availability in underdeveloped regions of the country. According to Brindha (2013), "telemedicine could be a generic terminology that is employed to outline numerous aspects of health care at a distance." Thus, it can be asserted that via the long-term application of this technology, telemedicine can provide extreme help and aid the population by providing healthcare through telecommunication technologies. Research conducted by Pal et al. (2017) also argues that telemedicine can also be applied to supply the medical data and services. This assertion is centered on the fact that telecommunication could be a developing application in countries like India for clinic medication wherever clinical information is transmitted through the mobile phone or the net. Telemedicine could serve as a direct and straightforward tool for the use of medical professionals that can discuss the matter over the phone, or the satellite technology and teleconferencing arrangement to do a timely consultation. This can even allow collaboration between specialists in two completely different countries for more comprehensive care for the patient. According to Sim et al. (2015), a significant fraction of the population is suffering from chronic diseases, like diabetes, globally for years, and by 2030 over half of the population is at high risk for developing diabetic retinopathy. This may lead to the increase of an insurmountable burden for the professionals catering to the patients. Telecommunication is a process to provide the medicine faster and to any virtual location and address. Consequently, it has been readily accepted in countries like India that have a lack of accessibility to eye care services for the majority of the rural and remote population. In several programs, a heavy and specially trained retinal image grader has been used as a resource that is supplied worldwide by the patients.

As per the research conducted by Bali et al. (2016), the service of telemedicine application for several automatic retinal images analyzing systems has become commercially obtainable in India. This is comparable to other applications that are yet to make their impact in the Indian health market. This system has established itself as it has the potential to mend how diabetes eye care is taken care of and provided with the automated real-time evaluation to expedite diagnosis and for further consultation if required. According to Faruqi (2018), telemedicine has been regarded as a tool that can be used for remote diagnosis and treatment of patients through the means of telecommunication; it grips a large potential to improve the quality of care, affordability, and accessibility of healthcare and holds a large scope for development in the developing nations where the market niche of telemedicine is restricted. Faruqi (2018) emphasizes that the contribution of telemedicine has an influence on both sectors of society as it affects the public sectors as well as the private sectors for providing comprehensive care services that are immediate and accessible. In countries with less accessible medical facilities, telemedicine provides technology-based basic healthcare services at the lowest costs possible. "The name of the technology is store and forward." Digital cameras and mics are readily used in the application of telemedicine and the videos and images that are stored are forwarded to another site by a computer for obtaining expert opinion. Research published by Mathur et al. (2017) states that "Challenges like poverty, gender equality, illiteracy, and lack of healthcare facilities are faced by the world's most popular and largest democratic countries. Healthcare problem is mainly due to the shortage of absence of essential organization and qualified healthcare experts, particularly in remote parts of the country." Thus, telemedicine can serve as constructive means and a productive tool for the delivery of healthcare and other medical facilities to a large section of the Indian population and assist them in bettering the quality of life. Telemedicine has been successfully tested and has resulted in having everyone's promise (Devanbu et al., 2019).

This technology has been used in many healthcare centers and has resulted in an effective use to overcome and overpower the problems of healthcare facilities in unserved areas; this process and approach of delivering the healthcare facility can be seen as the replica of the world, mainly in developed countries. According to Pal and Mukhopadhyay (2017), telemedicine is a contributing component of the e-medicine and e-health industry that uses the information and communication technology (ICT) for delivery of healthcare services to cope with the distance and are able to connect the medical service provider with the patient. Telemedicine and e-health are just not about the patient and treatments on phone, it is also all about the knowledge and the equipment as viable and trustworthy (Devanbu et al., 2019). Telemedicine has not only helped in getting tele-consultations or medical healthcare, but it is also been recognized to help the Indian doctors who are practicing in local areas to get tele-education by facilitating efficient delivery of medical care to the remote area and vulnerable groups of the aging population. According to Maheshwari et al. (2016), telemedicine is prospering worldwide, whereas Indian healthcare is facing challenges in the remote areas for medical care delivery. Both sectors, public and private healthcare, have been trying and implementing various modes to reach the remote and rural population to provide healthcare and virtual consultation, in general, and have confronted many challenges because of the inadequate infrastructure of the country.

There have been only a few models that have been successful to fight against the untrained paramedical workforces and lack of a sustained economic and business model. An initiative has been taken by the private sector of the society to provide medical help through virtual consultation that can reach the rural areas of our country and also help the medical specialist to reach the people who are unable to get treatment and to know about their condition. It is a time and money-saving method. According to Bali et al. (2016), the purpose of virtual consultation is to manage the places that are substantially unequal in health services distribution. Telemedicine service facilities are to be provided in all the states and fast. All civil hospitals and medical schools in districts with nodes were visited by a three-member team. In nodes in Madhya Pradesh (ISRO), most of the centers failed to provide consultations in telemedicine. Further, they suggested that the system should be substituted by another system that can be available in a wide range of areas and networks (SWAN). Also, it has been recommended that the concept of virtual out-patient departments improve healthcare. Another crucial research by Kakkar et al. (2017) established that in these recent years people have moved towards the technology in healthcare and this has significantly reduced the human effort. Telemedicine has emerged as one of the most implemented concepts by people and has gained admiration among people by providing them easy accessibility to medicines and medical care. The very concept of virtual consultation is to provide health facilities from a distance and also it is highly convenient, which has resulted in its high rate of acceptance and an opportunity for further growth in the country (Singh et al., 2016).

Chelliyan et al. (2019) provide a detailed analysis of the telemedicine application in India. A "National Telemedicine Taskforce" was established in India in the year 2005 and helped in successful completion of several pilot projects in India's healthcare system. Some of the popular healthcare projects associated with these include "ICMR-AROGYASREE, NeHA and VRCs." The usage of telemedicine has also helped in the reachability of the healthcare services. In a country like India where the population is extensively large, application of telemedicine in the healthcare system has served to be a major boon. The Telemedicine Pilot Project in India was established by ISRO in 2001 and linked Apollo's Rural Hospital situated in a village called Aragondian in the Chittoor district (Andhra Pradesh) with the Apollo Hospital, Chennai. Other popular projects of the government of India that have been linked with the use of telemedicine include "National Rural Telemedicine Network, Integrated Disease Surveillance Project (IDSP), National Cancer Network (ONCONET), and National Medical College Network and the Digital Medical Library Network." The paper by Chelliyan et al. (2019) also discusses some individual pilot projects that have served to be of remarkable success in association with the Indian telemedicine healthcare structure. These are inclusive of "oncology at Regional cancer center, Trivandrum; mammography services at Sir Ganga Ram Hospital, Delhi" and "surgical services at Sanjay Gandhi Postgraduate Institute of Medical Sciences, School of Telemedicine and Biomedical Informatics." In the present scenario of the telemedicine application in Indian healthcare, the telehealth system in India is regulated by the MoHFW and the Department of Information Technology. "National Telemedicine Portal," a telemedicine division of MoHFW, for the telemedicine update by the Indian government has been established. The purpose of this portal has been identified as to intertwine the accessible medical colleges across the country for sharing electronic knowledge and education and delivery of e-healthcare in the remote and rural locations of the country. The high acceptability of telemedicine in India highlights its scope for further development.

Telemedicine in the Indian healthcare delivery system has been found to be operative in reducing the burden on the primary healthcare system. An interesting study by Chandwani and Dwivedi (2015) has found that both rural and urban healthcare systems of India are highly overworked with management of the several communicable, non-communicable, and the chronic illnesses of the population along with patient care management of the large population and the medical care follow-ups, making the need to develop a strong telemedical system essential for the country. Development of a uniform medical infrastructure for the upkeep of the records at the care centers and the nursing homes of India will aid in the supervision of health records and systemize the healthcare system. Several telemedicine systems have been developed in collaboration with state and central governments and the research institutes; for example, the government of West Bengal in coalition with IIT Kharagpur developed Webel ECS for multiple primary medical centers (Sapkal & Bairagi, 2011). The Onconet-Kerala or the Kerala Oncology Network is one of the most fruitful telemedicine networks in India. The project has been established with the purpose of identification of early stages of cancer, prognosis, and treatment. Acharya and Rai (2016) have found that these services were also used for management of pain and follow-up services for the patients. Another example of a successful telemedicine project of India is the TEJHAS, the Telemedicine Enabled Java-Based Hospital Automation System (an electronic database), and was developed to make the patient information obtainable to all the medical hubs involved in the venture, along with the maintenance of the patient medical records.

The telemedicine market is expected to grow significantly in India and argues that the telemedicine market is relatively nascent in India. The use of telemedicine has spiked under the ongoing pandemic of COVID-19. In the year 2019, it was predicted that the telemedicine market in the country India would reach $32 million by the year 2020. There is an upsurge of telemedicine of at least 178% under the current pandemic for remote consultations for flu or cold and its symptoms in the patients. In association with the same, MoHFW has issued fresh guidelines for the implementation of telemedicine projects and the groups that function in association with the Medical Council of India and the NITI Ayog. It is speculated that various telemedicine markets in India will see a rise. One of the clinical assistance tele-platforms in India, "mFine," has recorded a dramatic rise of four times in its patient volumes recently. A large number of individuals in India have also been highly active in purchasing their medication online via telemedical service providers and the online pharmacies.

The study by McKinsey Global estimated that successful implementation of the telemedicine technology in India could result in saving $4 billion to $5 billion annually in the country. The rise has also been estimated in conjunction with the 95% reduction of the data expenses in the country since 2013, making the Internet as well as telemedicine services highly accessible. The comprehensive report by the institution illustrates that the Internet subscriber base of India has grown to 560 million, placing second to China. The telemedicine market in India highly relies

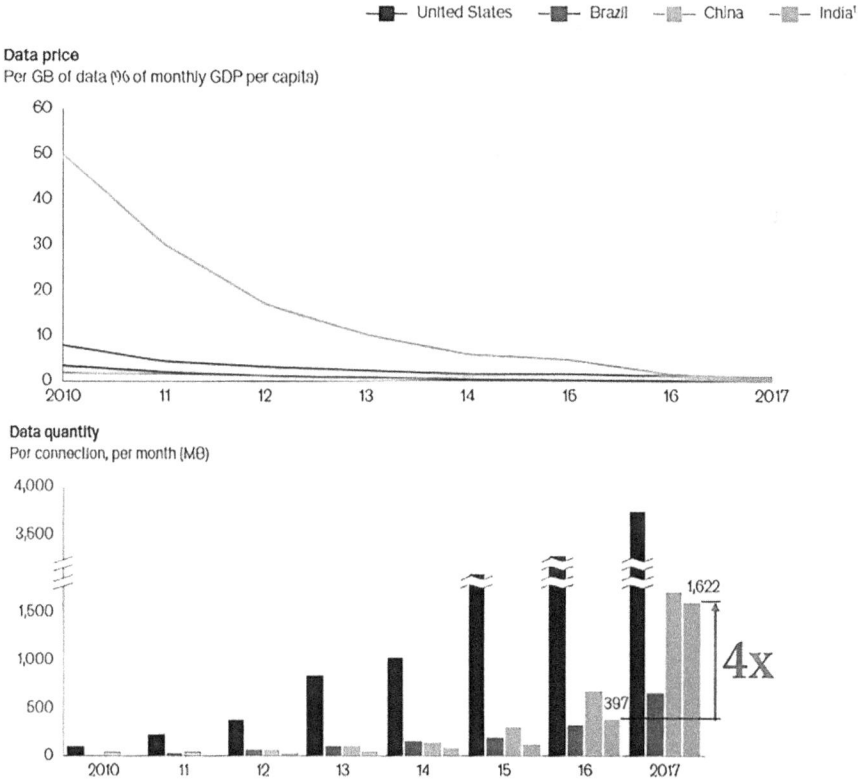

FIGURE 11.8 Internet data prices and data consumption per connection in the countries (McKinsey Global, 2019).

on the access and development of the technology in the common services centers that deliver primary aid to the seekers for the care over the mobile phones. The report asserts that "India could save up to $10 billion if telemedicine could replace 30 to 40% of in-person outpatient consultations, as well as improve care for the poor and residing in rural parts of the country. It is estimated that telemedicine could replace half of the total personal outpatient consultations, and with an accelerated execution plan the country could tap to 60 to 80% of this prospective by 2025." The growth at this stage will ultimately reach to saving of $4–5 billion annually in the India healthcare system. Major attributes of telemedicine, effective recordkeeping, and database maintenance are essential for the development of virtual consultation care services in India. The electronic health records (EHRs) development is therefore critical for the long-term medical record maintenance of the patients and the care seekers for the accurate diagnosis and further development of the technology. Successful application and execution of these structured records in the healthcare organization of India will also contribute to reduction in the administrative costs. Digital Information Security in Healthcare Act (DISHA) by the Indian government has also been a supportive initiative for telemedicine in India. The motive of this act is to safeguard the confidentiality and reliability of the health stats generated digitally and for the effective regulation of how this information is collected, stored, maintained, transmitted, and applied in the healthcare systems (McKinsey Global, 2019). The report also asserts that successful application of telemedicine could lead to the creation of new jobs, helping the successful development of the telemedicine market in the country as a healthcare reform (Figure 11.8).

The telemedicine health services in India therefore could be progressive by connecting, automating, and analyzing. A model for the same has been presented by McKinsey Global. The objective of this study was to assert how the digital solutions in the healthcare system can help in the alleviation of the demand and supply mismatch. It has been studied that application of telemedicine services successfully will result in saving up to 15% of the time of the medical professionals that can too result in improving the quality (Figure 11.9).

Telehealth services allow the doctors to consult their patients digitally. The research by Pratibha and Rema (2011) indicates that this can help in the reduction of about 30% of health costs. The telemedicine models possess a technical capability to manage about 50% of the doctor and patient consultations. The global trends indicate that successful implementation can result in about 30% of cost savings in the medical consultations and therefore can be of high utility in

Retail in the future: data-enriched client experiences, online and in stores.[1]

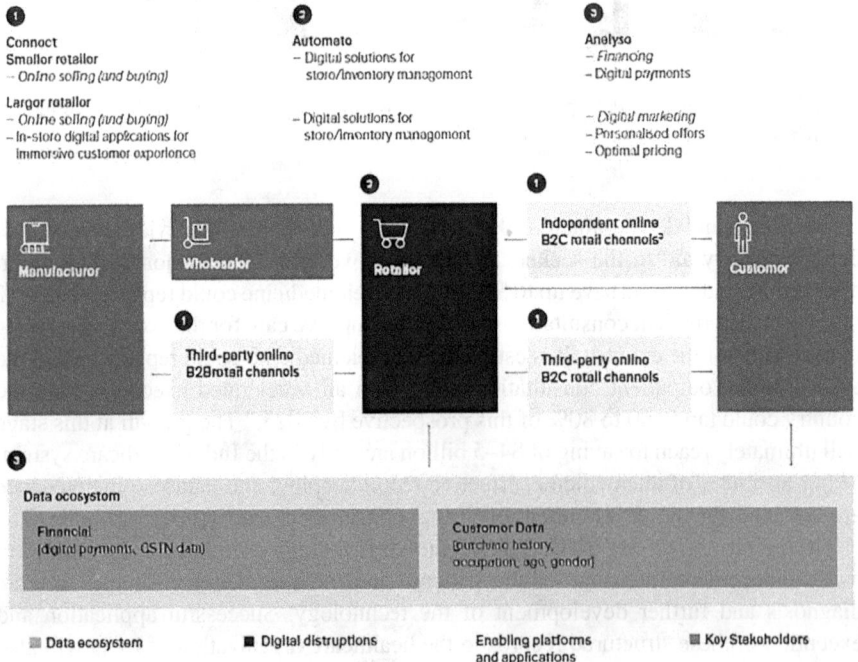

FIGURE 11.9 Functioning of the telemedicine markets, a hypothetical chain (McKinsey Global, 2019).

the management of the healthcare services by reducing the costs and assisting in the management of the healthcare systems. An inclusive study was conducted in rural Maharashtra by Fatehi et al. (2017), wherein less than 50% of the patients suffering from diabetes and hypertension followed the treatment plans with about 50% who failed to adhere to the medication services. Obtainability of inexpensive smartphones has led to a rise in the engagement of medical services with the patients. Telemedicine services also assist the medical care service providers in India from improving the quality and the medical care standards in India. The Manipal hospitals in India have installed a cognitive computing platform, "IBM Watson," for the oncology services for the analysis of the patients (McKinsey Global, 2019).

11.5 SWOT ANALYSIS

The business development and expansion of telemedicine in the Indian market can be evaluated by performing a rigorous analysis of several factors that can be assessed by performing a SWOT analysis of telemedicine in the Indian market. A SWOT analysis provides insights into the strengths (S), provides details about the potential weaknesses (W), available opportunities (O), and the potential threats (T) (Gurel & Tat, 2017). Other than the strengths, limitations, opportunities, and threats as reflected by the data collected through the questionnaire, the following can also be derived after rigorously reviewing the literature for telemedicine market in India:

11.5.1 STRENGTHS

- Increased political focus on healthcare improvement and digitalization: The government of India has recently launched several programs that focus on "mass digitalization" of the Indian population by providing the population Internet access. Further, the government has also launched several health benefiting schemes. Therefore, it is likely that introduction and expansion of telemedicine in India will be largely welcomed.
- Support towards medical technologies: India has greatly advanced itself with heavy investments and a focus towards healthy improvement and towards inclusion of several technically advanced models in the healthcare system.
- Enhanced collaboration with USA for healthcare: USA is the leading giant in telemedicine. As the strong ties between India and USA are further strengthening, it is a pronounced opportunity for the telemedicine market to venture into India and establish a strong position.
- Increased intellectual property protection: As the availability and access to Internet enhances in India, there is also increased stress on data security, making telemedicine a safe alternative for the users.
- Huge patient-doctor service gap: The fact that the total available doctors and healthcare professionals is not enough to cater to the large population of India provides a major scope for the growth of telemedicine.

11.5.2 WEAKNESSES

- Need of support towards quality: There have been limited numbers of large-scale application-based evidences that have proven the success of telemedicine in the developing nations. Therefore, it may get difficult to assert the beneficence of telemedicine in India.
- Limited digital literacy: One of the key limitations of the telemedicine market in India is lack of adequate digital literacy among the population. Telemedicine is an interactive model and, therefore, demands digital literacy for its successful application in the country.
- Indebted healthcare systems: The healthcare systems of India are not equally well funded and therefore the crunch can affect the healthcare spending, affecting the application and expansion of telemedicine in India.
- Reduction in individual disposable income: Availing of telemedicine may emerge as "expensive" for the small setting hospitals and the rural healthcare centers of India. The population may fail to lift the costs of technological implementation.
- Stringent regulations: India is divided into states with local governments and, therefore, there is stringency in application of any technology as it must abide by both state and central rules that may hinder interoperability and implementation and expansion of telemedicine in India.
- Weak consumer base: Even though there is an increase in the availability of the Internet, the majority of the population may lack digital literacy. Further, it will be difficult for the telemedicine market to gain and build trust in the community as a new reliable exposure in the medical system of country. Therefore, even though telemedicine market could be encouraged by the professionals, the same cannot be expected from the local population.

11.5.3 OPPORTUNITIES

- Advancement in technology: The country is pacing and advancing in the area of Information communication and technology. This provides a true potential for telemedicine market growth with a readily available infrastructure in the country.
- Limited accessibility to healthcare amenities in rural parts: The healthcare system of India is highly uneven, with limited healthcare services available to the rural and remote segments of the country. Therefore, introduction of telemedicine may promise a better healthcare structure in the country.
- Potential growth of market as per global trends: The telemedicine market trends have predicted the growth of the market, affirming the likelihood of expansion and growth in telemedicine.
- Rising urbanization: As more and more of the population of India is acquainted with urban life and digital literacy increases, the chances of success of the telemedicine market expansion in India also increase.

- Shortage of healthcare practitioners and doctors: The doctor-patient ratio of India is 1:1,800 and therefore there is an excessive dearth of doctors in our country. Availability of telemedicine will allow for a compensation for this shortage, guaranteeing industrial growth for telemedicine.
- Lack of competition: The concept of telemedicine is new in the Indian market with limited operational organizations. This eliminates business threats and provides a gigantic opportunity for the firms that may wish to invest and work in the telemedicine market of India.

11.5.4 THREATS

- Cyber security risks posed by digital platforms: With an increase in the availability of the Internet and expansion of data sciences, various phishing platforms have also emerged that may be involved in data leakage and may pose data and cyber security risks. Since the telemedicine platform is entirely digital, it is highly vulnerable to these information technology risks.
- Bargaining power of the suppliers: As the telemedicine market of India will expand, there will be a surplus of the suppliers that may affect the price of goods and services, marginalizing the overall profits.
- Bargaining power of the consumers: The Indian markets in general are consumer-driven markets and, therefore, the cost availed by the services will be reduced compared to other countries, affecting the overall profit ratio.

11.6 PESTLE ANALYSIS

The PESTLE analysis is done to identify the role of macro-environment factors that can affect the performance of an industry. It is inclusive of political factors, economic factors, sociocultural factors, technological factors, legislative factors, and environmental factors (Christodoulou & Cullinane, 2019).

11.7 TELEMEDICINE PRACTICE GUIDELINES

The following rules have been issued by the Ministry of Health and Family Welfare in collaboration with Board of Governors, Niti Ayog, and Medical Council of India in 2020. Under these guidelines, telemedicine has been defined as "The delivery of healthcare services, where distance is an important factor, by all health care professionals using communication and information technologies for the exchange of legal information for diagnosis, treatment and prevention of injuries and disease, research and evaluation, and for the continuing education of health care providers, all in the interests of advancing the health of individuals and their communities."

Now it is perfectly legal to provide telemedicine services by registered medical practitioners. These guidelines are a modification to the regulations by the Indian

Medical Council that came in the year 2002 and give regulatory support and a base to practice teleconsultations in India. According to these guidelines:

- For chronic ailments like hypertension, diabetes, or asthma, medicines should not be prescribed through telehealth, unless it is a refill of an old prescription attained via an in-person consultation not less than six months ago. However, one way of prescribing a medicine for chronic disease in teleconsultation is by using video only.
- Teleconsultants have a right to choose the medium of telemedicine service.
- Teleclinicians shall have the same level of care during teleconsultation as during in-person consultation.
- Patients will be held accountable for providing accurate information.
- Patient's personal information cannot be revealed or transferred to anyone without taking the patient's consent.
- Doctors are not to deny telemedicine service in the case of emergency, but should restrict it to immediate relief or first aid.
- In order to prevent drug abuse, every teleconsultant can "prescribe medications via telemedicine only when he/she is content to have collected satisfactory and appropriate information relating to the medical condition of the patient and the medicines which are prescribed are in the best interest of the patient."
- "A professional misconduct can be filed for prescribing medicines without an appropriate diagnosis. If the tele-consultation was not done over video, then drugs cannot be prescribed except over the counter medications."

Until these guidelines, the virtual consultation sector had a regulatory anonymity that was seen as a barrier to its acceptance and adoption by all patients, care providers, and hospitals.

11.8   CONCLUSION

Telemedicine has come out as a truly convenient option in the pandemic times, where hundreds of medical and healthcare professionals have lost their lives due to uncertain exposure to the COVID-19 virus. A teleconsultation can be conducted without exposing the teleconsultant to any infection during such humanitarian crises. Digital health platforms have witnessed the outpouring since the lockdown began in March 2020. The teleconsultations have been growing every week. Queries with respect to fever, cough, sore throat, and body pain have multiplied. Also, there is approximately a 50% rise in the number of consultants joining a digital health platform. Also, the new practice guidelines have given authenticity and confidence to teleconsultants to get on the platform.

Moreover, the new telemedicine guidelines and COVID-19 pandemic have acted as a booster shot for the sector. It would not be wrong to state that the unforeseen limitations during the COVID-19 pandemic acted as a blessing in disguise and also paved the way for future acceptance of new telemedicine technologies. Digital technologies have now been embraced by clinicians and healthcare providers as a usual means of functioning. Policymakers and healthcare professionals should be careful to accept the benefits of providing treatment virtually and should support the rapid creation of relevant policies.

REFERENCES

Abrams, H.B. (2014). What is a health systems innovation centre and why does everyone want one? *General Medicine*, *15*(1), 5–13.

Acharya, R.V., & Rai, J.J. (2016). Evaluation of patient and doctor perception toward the use of telemedicine in Apollo Tele Health Services, India. *Journal of Family Medicine and Primary Care*, *5*(4), 798.

Agboola, S., Hale, T.M., Masters, C., Kvedar, J., & Jethwani, K. (2014). "Real-world" practical evaluation strategies: A review of telehealth evaluation. *JMIR Research Protocols*, *3*(4), 75.

American Telemedicine Association. (2017). What is telemedicine? Retrieved from http://www.americantelemed.org/about-telemedicine/what-is-telemedicine.

Avanesova, A.A., & Shamliyan, T.A. (2019). Worldwide implementation of telemedicine programs in association with research performance and health policy. *Health Policy and Technology*, *8*(2), 179–191.

Bajpai, N., & Wadhwa, M. (2019). Accessing specialist services via telemedicine in India. *Diabetes Technology & Therapeutics*, *21*(S2), S2-1.

Bakshi, S., & Tandon, U. (2019). Drivers and barriers of telemedicine in India: Seeking a new paradigm. *Journal of Computational and Theoretical Nanoscience*, *16*(10), 4367–4373.

Bali, S., Gupta, A., Khan, A., & Pakhare, A. (2016). Evaluation of telemedicine centres in Madhya Pradesh, Central India. *Journal of Telemedicine and Telecare*, *22*(3), 183–188.

Bashshur, R.L., Shannon, G.W., Bashshur, N., & Yellowlees, P.M. (2016). The empirical evidence for telemedicine interventions in mental disorders. *Telemedicine and e-Health*, *22*(2), 87–113.

Bestsennyy O., Gilbert, G., Harris, A., & Rost, J. (2021). Telehealth: A quarter-trillion-dollar post-COVID-19 reality. *Mckinsey Report*.

Blaya, J.A., Fraser, H.S., & Holt, B. (2010). E-health technologies show promise in developing countries. *Health Affairs*, *29*(2), 244–251.

Brindha, G. (2013). Emerging trends of telemedicine in India. *Indian Journal of Science and Technology*, *6*, 5.

Chandwani, R.K., & Dwivedi, Y.K. (2015). Telemedicine in India: Current state, challenges and opportunities. *Transforming Government: People, Process and Policy*.

Chari, D.A., Wu M.J., Crowson, M.G., Kozin, E.D., & Rauch, S.D. (2020). Telemedicine algorithm for the management of dizzy patients. *Otolaryngology–Head and Neck Surgery*, 163(5), 857–859.

Chellaiyan, V.G., Nirupama, A.Y., & Taneja, N. (2019). Telemedicine in India: Where do we stand?. *Journal of Family Medicine and Primary Care*, *8*(6), 1872–1876.

Christodoulou, A., & Cullinane, K. (2019). Identifying the main opportunities and challenges from the implementation of a port energy management system: A SWOT/PESTLE analysis. *Sustainability*, *11*(21), 6046.

De Melo, M.D.C.B., Nunes, M.V., Resende, R.F., Figueiredo, R.R., Ruas, S.S.M., dos Santos, A.D.F., ... & de Aguiar, R.A.T. (2018). Belo horizontetelehealth: incorporation of teleconsultations in a health primary care system. *Telemedicine and e-Health*, *24*(8), 631–638.

Devanbu, V.G.C., Nirupama, A.Y., & Taneja, N. (2019). Telemedicine: New technology, new promises? *Indian Journal of Community Health*, *31*(4), 437–441.

Diong, B.W., Goh, M.I., Chung, S.K., Chekima, A., & Yew, H.T. (2021). Vertical handover algorithm for telemedicine application in 5G heterogenous wireless networks. *International Journal of Advanced Computer Science and Applications*, *12*(8), 611–617.

Doolittle, G.C., & Spaulding, R.J. (2017). Defining the needs of a telemedicine service. In *Introduction to telemedicine, second edition* (pp. 79–92). United Kingdom: CRC Press.

Ekanoye, F., Ayeni, F., Olokunde, T., Nina, V., Donalds, C., & Mbarika, V. (2017). Telemedicine diffusion in a developing country: A case of Nigeria. *Science Journal of Public Health*, *5*(4), 341–346.

Eren, H., & Webster, J.G. (Eds.). (2018). *Telemedicine and electronic medicine*. United Kingdom: CRC Press.

Faruqi, A. (2018). Role of telemedicine in Indian healthcare system. *International Journal of Management*, *9*(1), 42–49.

Fatehi, F., Smith, A.C., Maeder, A., Wade, V., & Gray, L.C. (2017). How to formulate research questions and design studies for telehealth assessment and evaluation. *Journal of Telemedicine andTelecare*, *23*(9), 759–763.

Gürel, E., & Tat, M. (2017). SWOT analysis: A theoretical review. *Journal of International Social Research*, *10*(51), 994–1006.

Kakkar, A., Naushad, S., & Khatri, S.K. (2017). Telemedicine and EHR integrated approach for an effective e-governance healthcare framework. *International Journal of Medical Research & Health Sciences*, *6*(5), 108–114.

Kamdar, N., Huverserian, A., Regev, A., Beck, L., & Mahajan, A. (2018). Telemedicine for anesthesiologists: Preoperative evaluation and beyond. *ASA Newsletter*, *82*(10), 16–19.

Kamsu-Foguem, B., & Foguem, C. (2014). Could telemedicine enhance traditional medicine practices? *European Research in Telemedicine/La RechercheEuropéenne en Télémédecine*, *3*(3), 117–123.

Khalifehsoltani, S.N., & Gerami, M.R. (2010). E-health Challenges, Opportunities and Experiences of Developing Countries. In *2010 International Conference on e-Education, e-Business, e-Management and e-Learning* (pp. 264–268). IEEE.

Khanal, S., Burgon, J., Leonard, S., Griffiths, M., & Eddowes, L.A. (2015). Recommendations for the improved effectiveness and reporting of telemedicine programs in developing countries: Results of a systematic literature review. *Telemedicine and e-Health*, *21*(11), 903–915.

Klaassen, B., van Beijnum, B.J., & Hermens, H.J. (2016). Usability in telemedicine systems—A literature survey. *International Journal of Medical Informatics*, *93*, 57–69.

Lee, J.Y., & Lee, S.W.H. (2018). Telemedicine cost–effectiveness for diabetes management: A systematic review. *Diabetes Technology & Therapeutics*, *20*(7), 492–500.

Lilly, C.M., Motzkus, C., Rincon, T., Cody, S.E., Landry, K., Irwin, R.S., & Group, U.M.C.C.O. (2017). ICU telemedicine program financial outcomes. *Chest*, *151*(2), 286–297.

Lupton, D., & Maslen, S. (2017). Telemedicine and the senses: A review. *Sociology of Health & Illness*, *39*(8), 1557–1571.

Maheshwari, S., Kalyanpur, A., Mehra, C., Seshadri, S., & Thomas, S. (2016). Telemedicine for Indian primary health centres: Is there a need for super specialist consultation. *International Journal of Telemedicine and Clinical Practices*, *1*(3), 238–243.

Mathur, P., Srivastava, S., Lalchandani, A., & Mehta, J.L. (2017). Evolving role of tele-medicine in health care delivery in India. *Prim Health Care*, *7*(260), 2167–1079.

Miller, A., Rhee, E., Gettman, M., & Spitz, A. (2018). The current state of telemedicine in urology. *The Medical Clinics of North America*, *102*(2), 387–398.

McKinsey Global (2019). Digital India: Technology to transform a connected nation. Retrieved from: https://www.google.com/url?sa=t&rct=j&q=&esrc=s&source=web& cd=&ved=2ahUKEwj6wLmEwNzpAhWVb30KHS8nBbEQFjAAegQIAhAB&url= https%3A%2F%2Fwww.mckinsey.com%2Fbusiness-functions%2Fmckinsey-digital %2Four-insights%2Fdigital-india-technology-to-transform-a-connected-nation&usg= AOvVaw1gGVmq_mmuq_HH-CJr9hxZ

Mohr, N.M., Hurst, E.K., MacKinney, A.C., Nash, E.C., Carr, B.G., & Skow, B. (2019). Telemedicine for early treatment of sepsis. In *Telemedicine in the ICU* (pp. 255–280). USA: Springer.

Okoroh, E.M., Kroelinger, C.D., Smith, A.M., Goodman, D.A., & Barfield, W.D. (2016). US and territory telemedicine policies: Identifying gaps in perinatal care. *American Journal of Obstetrics and Gynecology, 215*(6), 772-e1.

Pal, S., & Mukhopadhyay, A. (2017). A machine learning approach for telemedicine governance. *Journal of Marine Medical Society, 15*(3), 14.

Pratibha, V., & Rema, M. (2011). Teleophthalmology: A model for eye care delivery in rural and underserved areas of India. *International Journal of Family Medicine, 2011.*

Prieto-Egido, I., Simó-Reigadas, J., Liñán-Benítez, L., García-Giganto, V., & Martínez-Fernández, A. (2014). Telemedicine networks of EHAS foundation in Latin America. *Frontiers in Public Health, 2,* 188.

Puustjärvi, J., & Puustjärvi, L. (2013). Practising cloud–based telemedicine in developing countries. *International Journal of Electronic Healthcare, 7*(3), 181–204.

Ryu, S. (2012). Telemedicine: Opportunities and developments in member states: report on the second global survey on eHealth 2009 (global observatory for eHealth series, volume 2). *Healthcare Informatics Research, 18*(2), 153–155.

Sapkal, A.M., & Bairagi, V.K. (2011). Telemedicine in India: A review challenges and role of image compression. *Journal of Medical Imaging and Health Informatics, 1*(4), 300–306.

Scott Kruse, C., Karem, P., Shifflett, K., Vegi, L., Ravi, K., & Brooks, M. (2018). Evaluating barriers to adopting telemedicine worldwide: A systematic review. *Journal of Telemedicine and Telecare, 24*(1), 4–12.

Sim, D.A., Keane, P.A., Tufail, A., Egan, C.A., Aiello, L.P., & Silva, P.S. (2015). Automated retinal image analysis for diabetic retinopathy in telemedicine. *Current Diabetes Reports, 15*(3), 14.

Singh, A., Roy, A., & Goyal, P. (2016). Telemedicine and telehealth-The Indian scenario. *Journal of Integrated Health Sciences, 4*(1), 3.

Syed-Abdul, S., Scholl, J., Chen, C.C., Santos, M.D., Jian, W.S., Liou, D.M., & Li, Y.C. (2012). Telemedicine utilization to support the management of the burns treatment involving patient pathways in both developed and developing countries: A case study. *Journal of Burn Care & Research, 33*(4), e207–e212.

Thaker, D., Monypenny, R., Olver, I., & Sabesan, S. (2013). Cost savings from a telemedicine model of care in northern Queensland, Australia. *The Medical Journal of Australia, 199*(6), 414–417.

Ting, D.S., Gunasekeran, D.V., Wickham, L., & Wong, T.Y. (2020). Next generation telemedicine platforms to screen and triage. *Health Policy and Technology, 9*(2), 179–191.

Trettel, A., Eissing, L., & Augustin, M. (2018). Telemedicine in dermatology: Findings and experiences worldwide–a systematic literature review. *Journal of the European Academy of Dermatology and Venereology, 32*(2), 215–224.

Udsen, F.W., Hejlesen, O., & Ehlers, L.H. (2014). A systematic review of the cost and cost-effectiveness of telehealth for patients suffering from chronic obstructive pulmonary disease. *Journal of Telemedicine and Telecare, 20*(4), 212–220.

Vodrahalli, K., Daneshjou, R., Novoa, R.A., Chiou, A., Ko, J.M., & Zou, J. (2021). TrueImage: A machine learning algorithm to improve the quality of telehealth photos. *Pacific Symposium on Biocomputing*, 26, 220–231.

Volterrani, M., & Sposato, B. (2019). Remote monitoring and telemedicine. *European Heart Journal Supplements*, *21*(Supplement_M), 54–56.

Vyas, S. (2023). Extended reality and edge AI for healthcare 4.0: Systematic study. In *Extended reality for healthcare systems* (pp. 229–240). Academic Press.

Vyas, S., & Gupta, S. (2022). Case study on state-of-the-art wellness and health tracker devices. In *Handbook of research on lifestyle sustainability and management solutions using AI, big data analytics, and visualization* (pp. 325–337). IGI Global.

Young, K., Gupta, A., & Palacios, R. (2019). Impact of telemedicine in pediatric post-operativecare. *Telemedicine and e-Health*, 25(11), 1083–1089.

Zhao, J., Zhang, Z., Guo, H., Ren, L., & Chen, S. (2010). Development and recent achievements of telemedicine in China. *Telemedicine and e-Health*, *16*(5), 634–638.

Zolfo, M., Bateganya, M.H., Adetifa, I.M., Colebunders, R., & Lynen, L. (2011). A telemedicine service for HIV/AIDS physicians working in developing countries. *Journal of Telemedicine and Telecare*, *17*(2), 65–70.

12 Advance and Smart Healthcare System
A Case Study Calo – An AI-Based Health Utility Mobile Application

Gaytri Bakshi, Priyanka Yadav, Harsh Joshi, and Ashutosh Negi

12.1 INTRODUCTION

Health has seven dimensions, and we are healthy if we cover all these dimensions (Bruce & Fries, 2003). They are physical, mental, emotional, financial, spiritual, nutritional, and environmental. The insurance technology sector, also known as "insurtech," is making healthcare more accessible to a larger population (Aggarwal et al., 2013; Dutta, 2020). Hospital staffing levels are more effectively managed and patient wait times are decreasing as a result of artificial intelligence and predictive analytics (Carew et al., 2021; Krittanawong, 2018; Li et al., 2022). Ultra-precise robots that aid in operations and make some treatments less invasive are reducing surgical procedures and recovery times. Our current healthcare system could be made leaner with the help of healthtech (Benfredj, 2021; Chakraborty et al., 2021; Singh et al., 2021). Through tech-infused care, soaring costs, intolerable wait times, inefficiencies in drug research, and restricted access to insurance and healthcare providers are all improved (or at least addressed). There could be many notions of how to take care of oneself, but through this work, a health application, Calo, is demonstrated and that aims to achieve almost all of the dimensions of health. Calo is a personal, customizable, portable Android application that takes care of health from all dimensions.

The functionalities of the app include the following:

- Calorie detection in unpacked food just by clicking a picture of it using an object recognition algorithm with accuracy ranging from 90–100%.
- Monitoring heart rate (pulse rate) without needing any extra sensors (based on monochromatic red light sensing using LED).
- Tracking the physical activity of users based on step count detection and body movement.

- Intelligent medicine information and drug detection using optical character recognition and drug API.
- Offers smart suggestions and choices based on mental health and mood.
- Robust security mechanism implemented using Google firebase authentication.
- Removes fraud and spam user threats by two-way email verification.
- Scalable, multiple Android API supporting user interface with material UI implementation (works for 100% android devices > API 21).
- High-performance application because the machine learning implementation is carried on a cloud console.
- Intelligent warnings and suggestions based on the user's health status (BMI and goals on weight).
- Facilitates medicine reminders for keeping track of what medicine is to be taken and when should it be taken.
- Habit tracker, which helps to set goals and monitor progress resulting in a strong will toward goals affecting emotional well-being.
- Complete and thorough health analysis, which is as large as 30 days of analysis on nutrition intake, heartbeat trends, mood trends, and lots more.
- Quick search for nearby pharmacies was made easy with 100% accurate location searches and directions (integrated using robust Google Places API).

12.2   LITERATURE REVIEW

Denis Laure and Ilya Paramonov review current algorithms for heart rate measurement using mobile phones in their article "Improved Method for Heart Rate Measurement Using Mobile Phone Camera," and they then suggest an improved algorithm that is more effective than the reviewed ones (Paramonov, 2017). The article, "The use of mobile apps to improve nutrition outcomes," was printed in volume 1 of the 22nd Portuguese Conference on Pattern Recognition, RECPAD 2016, held in Aveiro, Portugal. To find out if the use of nutrition apps led to better outcomes, including knowledge and behavior, among healthy individuals, Wen-Hao Huang conducted a systematic review (DiFilippo et al., 2015). António J. R. Neves and Daniel Lopes present a practical study on the use of the library to study some of the Google Vision features in order to test their usability in practical applications in their paper, "A practical study about the Google Vision API," published in the 22nd Portuguese Conference on Pattern Recognition, RECPAD 2016 (Lopes, 2016). In their study, "A Study of Calorie Estimation in Pictures of Food," that was published in the *Interactive Journal of Medical Research*, Jun Zhou and Dane Bell outline techniques to precisely estimate food calories from still images and more effectively identify dietary patterns and food selections linked to health and health risks (Bell, 2016). In their study, "A Novel Method to Detect Heart Beat Rate Using a Mobile Phone," Tuvi Orbach and Kostas Marias suggest a method that can calculate the heart rate using only a mobile phone's camera (Marias, 2010).

Health is a primary matter of concern. Although with progressing technologies, health science has reached the epitome of success and proved to cure what was once called impossible, these sciences work under the roofs of large hospitals, under expert doctors, and with high-tech machines (Vyas & Gupta, 2023). The use of technologies created with the intention of enhancing any component of the healthcare system is known as "health tech," or "healthcare technology." There is a need to build a simple, user-friendly mobile application that can integrate basic healthcare utilities and could basically save us a plethora of resources.

It's crucial to look for ways to keep vulnerable people involved, connected, and safe. The logical method for doing this is through digital technology. There are a variety of options on the market; some will be more suitable for specific demands than others. India must prioritize the design and development of technology that considers regional limitations, including price. The Indian healthcare industry faces numerous regional and behavioral issues, but cost is still a significant factor. New technology needs to be created to address the country's limitations and priced appropriately for it to flourish and make an impact on a large scale. The good news is that AI delivers on its promises exactly. AI essentially amounts to transferring limited expert information to numerous beneficiaries by teaching algorithms robots to copy this knowledge, if it is implemented properly (Vyas, 2023).

The main objective of this work is to integrate the basic healthcare utilities in form of a simple, user-friendly mobile application. The solution aims to provide a complete health monitoring package with a lot of functionalities. It serves to provide users with an application to detect calories and nutrition in any packed or unpacked food just by clicking a picture of it. The app also serves to provide users with complete health and fitness tracking to monitor their health.

12.3 METHODOLOGY

The application is potentially a portable health gadget that can do more than conventional applications do. The intended users include people of all ages who wish to keep a track of their health status, use real-time monitoring services, detect nutrition content in both packed and unpacked food, look up to the application in case of fatal moods, and keep track of their body's physical activities, too. It eases daily hospital visits and brings a revolution in the healthcare industry with the intelligent solutions it offers. The application is user-friendly, with registration pages to create one's own account to store the details. The application has a main activity page from where it is divided into tracking various activities done by a human and performing analytics on it. The full block diagram of interfacing is shown in Figure 12.1. The system requirements for developing this application are the following:

- 500 GB Hard Disk (7200 RPM) + 512 GB SSD. GPU – (12 GB VRAM).
- Android Studio.
- Android SDK and Emulator.
- Cloud Platform – Hosting the model.
- 5th generation or above processors.
- Supporting operating system (Ubuntu 18.04).

The background working of the application is represented in Figure 12.2. In Firebase, authentication and validation are specified as declarative rules in the

Interfacing:

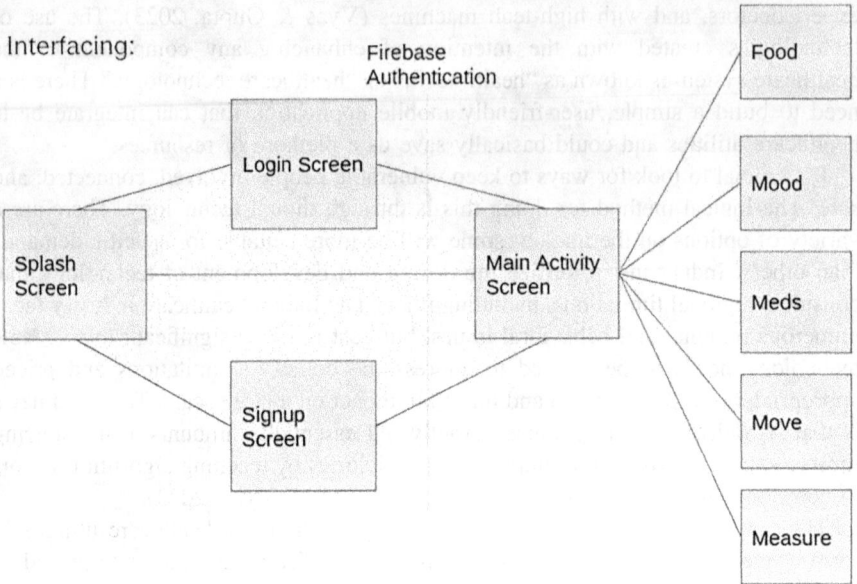

FIGURE 12.1 Block diagram of interfacing.

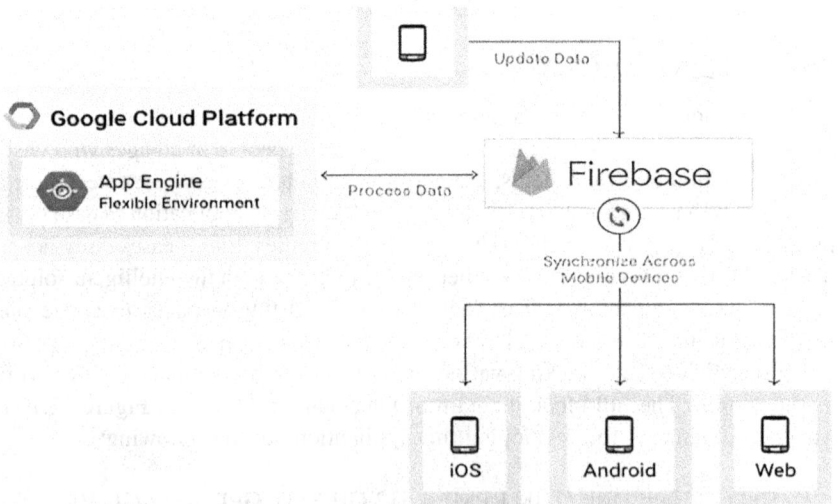

FIGURE 12.2 Backend working of the application Calo.

Firebase web user interface (UI), eliminating the necessity for writing imperative code. Additionally, these design patterns switch from Firebase, a fully managed platform, to Compute Engine, an entirely uncontrolled platform. Google manages updates and autoscaling on managed platforms but places some restrictions on the configuration.

12.4 IMPLEMENTATION

The major goal of this work is to incorporate essential medical services into a straightforward, user-friendly mobile application. The solution intends to offer an extensive set of features for health monitoring.

The functional project objectives are the following:

- Implementing robust security for authentication, authorization, and identity access management using Firebase.
- Implement modern UI for a smooth user experience using Google-powered material UI tools.
- Design and algorithm for implementing calorific value detection using unpacked food images.
- Using Google Cloud Platform to deploy the AI model to predict the calorific value.
- Use GCP SDK to connect the Android application with the GCP instance running the identification algorithm.
- Implement features to keep track of habits and goals and give a complete analysis of the progress.
- Implement smart mood suggestions using recommendation algorithms to suggest users with choices as per their mood (happy, sad, angry, etc.).
- Use RGB color component variation algorithm to detect the pulse rate of an individual with the rear camera of the phone and flashlight.
- Implement real-time physical activity capture (step counter) using Google's location API and Android's native motion sensors.
- Implement a feature to search all the available pharmacies nearby with directions using Google Maps API.
- Implement medicine reminders for people on regular medicine using Android's notification manager.
- Implement real-time medicine's information retrieval using web search with the help of optical character recognition and Google Search API.

12.4.1 QUALITY AND PERFORMANCE ATTRIBUTES

Key features relating to quality and performance:

- Robust Security: Uses Firebase authentication and two-way authentication to ensure high security.
- High Accuracy: Algorithms implemented have shown high results with scores on accuracy on ML algorithms ranging from the 90–100 percentile.
- High Performance: Since the app is cloud integrated, the implementation and processing of all the heavy algorithms take place on the cloud console. This allows even a 2 GB RAM user to swiftly use the application.
- Scalable Interface: Material UI implementation makes it scalable for almost 100% of Android devices.
- Intelligent Sensing: With optical character recognition, the app handles the medicine module smartly and efficiently.

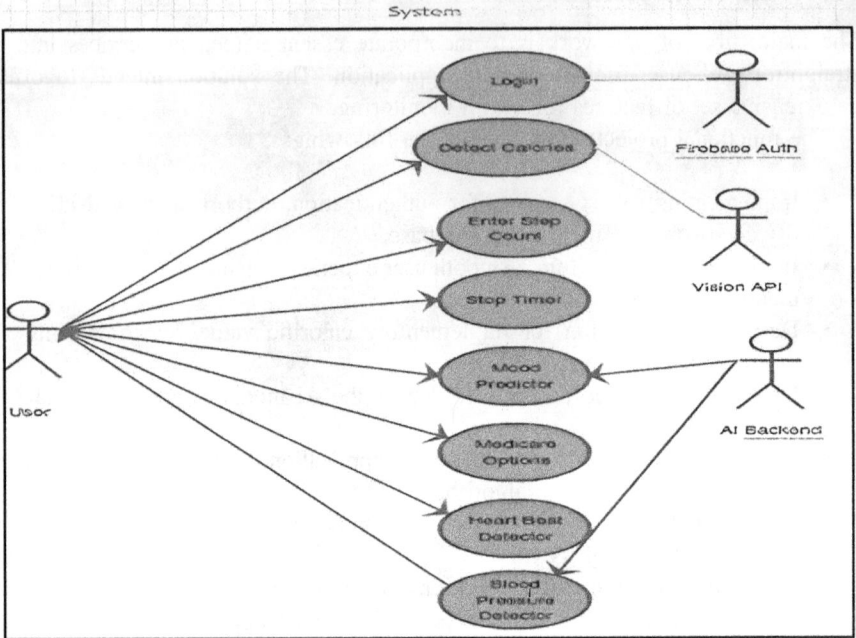

FIGURE 12.3   Use case diagram for application – Calo.

- Low Storage: Thanks to cloud integration, despite algorithms like image classification, OCR, and heart rate detection, the application is as small as 16 megabytes in size.
- Low Network Usage: Using the RESTful API rather than SOAP API optimizes the network performance of the application.
- Low Power Consumption: The app has been customized to seek less battery life as it does not require to work much during sleeping hours.

The working specification use case diagram of the application is shown in Figure 12.3. Many menu-driven features of this application are as follows.

12.4.2   ROBUST SECURITY USING FIREBASE AUTHENTICATION

Data in Cloud Firestore, Realtime Database, and Cloud Storage are protected by Firebase Security Rules, which offer strong, totally customizable security. Numerous popular authentication techniques are supported by Firebase authentication, which also combines with Firebase Security Rules to offer complete verification capabilities (Figure 12.4).

An open-source project called Material-UI includes React components that use Google's Material Design. One of the top user interface libraries is Material-UI, which has received over 35,000 GitHub stars.

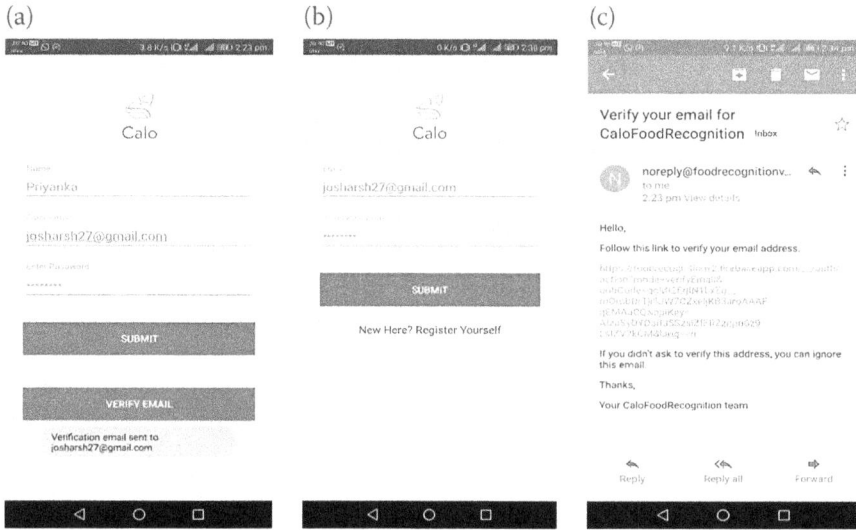

FIGURE 12.4 (a)(b)(c) Account registration using Firebase authentication.

12.4.3 Food Calorie Detection (Unpacked Food)

This app proposed to bring about a revolution by detecting calorie content from unpacked food. Using a robust self-learning object detection and classification algorithm, the application smartly categorizes the food based on the potential consumption content and, later integrated with one of the largest food calorie and nutrition databases, provides the actual calorie content in the food (Figure 12.5).

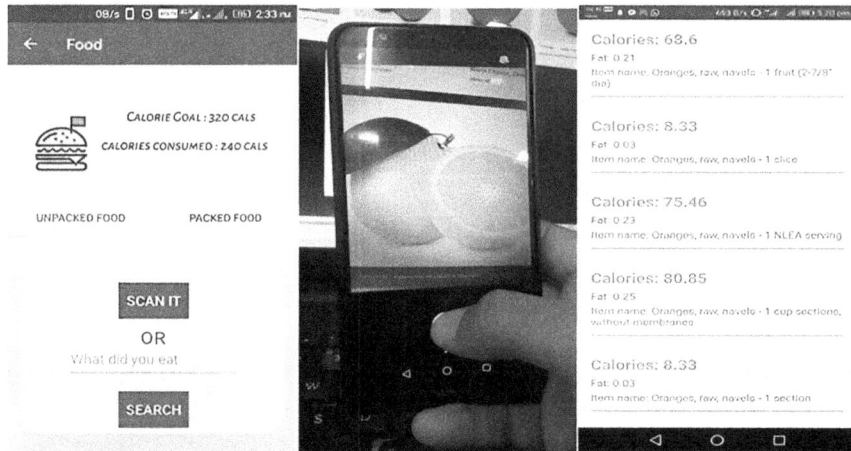

FIGURE 12.5 Estimation of food calories.

As the user is already being tracked on their physical routine, based on the health goal set for them (e.g., losing, maintaining, or gaining weight), the application intelligently tells the user if they should go with this food or not (Bell, 2016). The calorie formula is as follows:

$$\text{Calorie required for the day} = (\text{Calorie intake Set based on choice} - \text{calories consumed}) + \text{Calorie burnt.}$$

12.4.4 SMART MOOD AND SUGGESTIONS

The user is given a choice to select their mood based on the selection of smart suggestions made to better the mental state and provide the user with options to handle their mood (Figure 12.6).

12.4.5 REAL-TIME STEP COUNTER TO TRACK PHYSICAL ACTIVITY

This pedometer uses the built-in sensor to count the user's steps, integrated with GPS when needed for accuracy. It smartly senses the movement and detects the physical exercise done in a day.

Key Features: (Figure 12.7)

- Less battery consumption (not Internet based).
- 100% accuracy.
- Optimized results.
- Keeps a log.
- Helps to detect calories burned based on the number of steps walked.

12.4.6 REAL-TIME HEARTBEAT MONITOR

A fingertip is placed on the camera and the flash is turned on so that the change in blood density can be observed and an equivalent heartbeat can be obtained.

FIGURE 12.6 Mood expressions with suitable suggestions to follow.

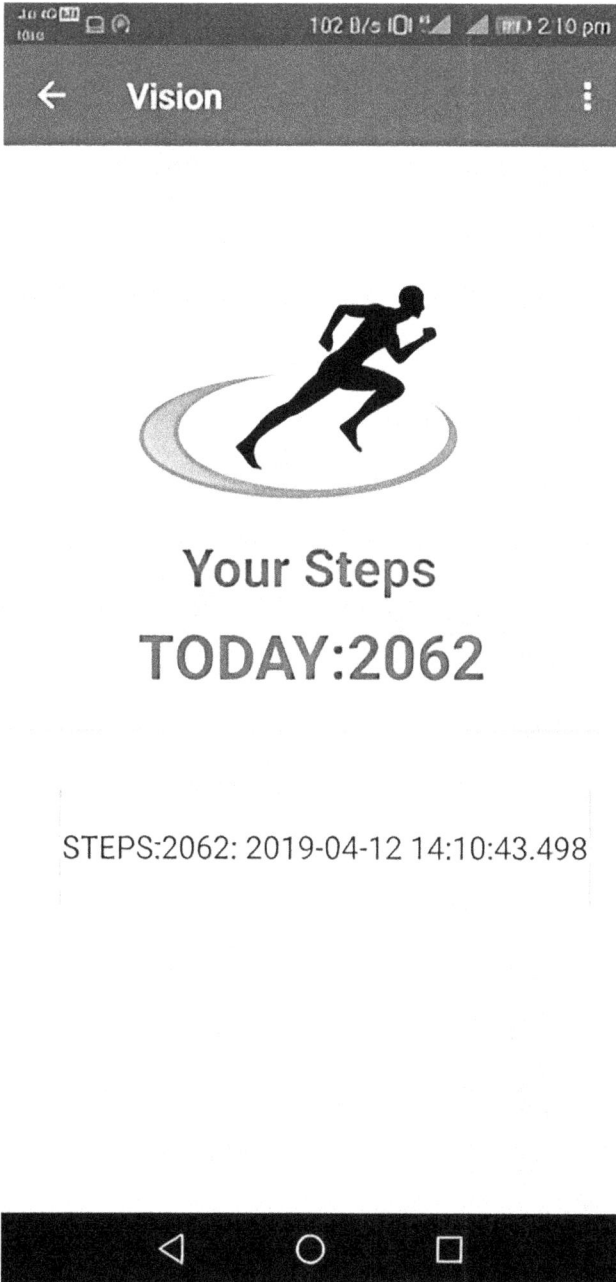

FIGURE 12.7 The calculated steps taken by a person in a day.

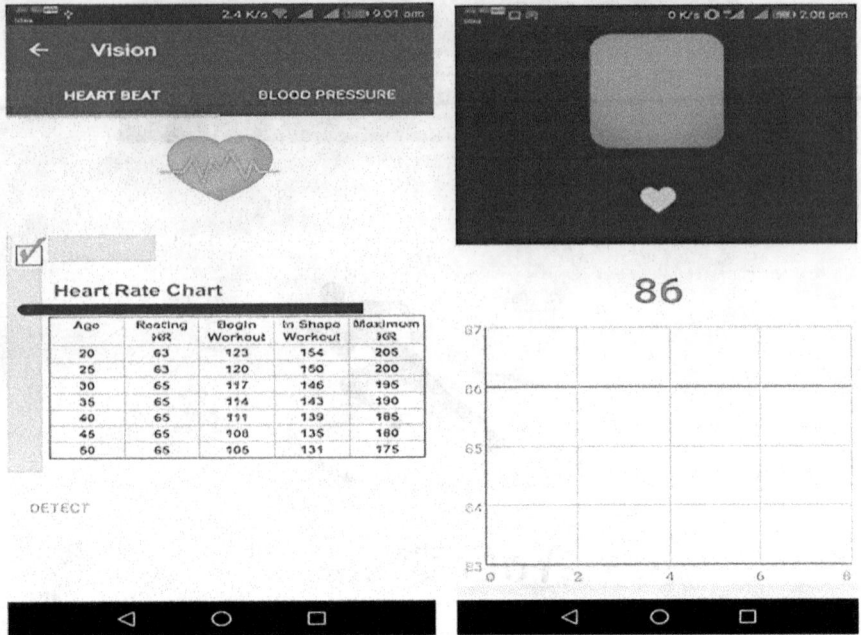

FIGURE 12.8 Heartbeat analyzer.

The real-time graph is also generated (Paramonov, 2017). A fingertip has been placed at the camera and the flash is turned on to capture the vessel's softness using red and green color intensities (Figure 12.8).

12.4.7 TRACKING NEARBY PHARMACIES

Calo provides users with the opportunity to search all the available pharmacies nearby with directions (Figure 12.9).

12.4.8 MEDICINE REMINDER

This module enables setting reminders for medicine intake. It uses Android's notification manager class.

Different types of notifications are as follows:

- A persistent icon that resides in the status bar and can be accessed through the launcher (when the user clicks it, a specific intent can be launched).
- The device turning on or flashing LEDs.

Notifying the user by flashing the backlight, playing a sound, or vibrating (Figure 12.10).

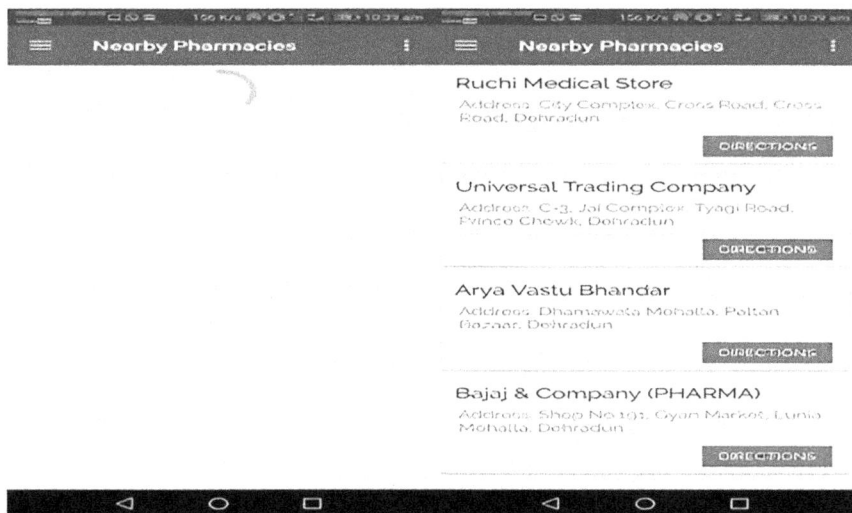

FIGURE 12.9   Location-wise, nearby pharmacies.

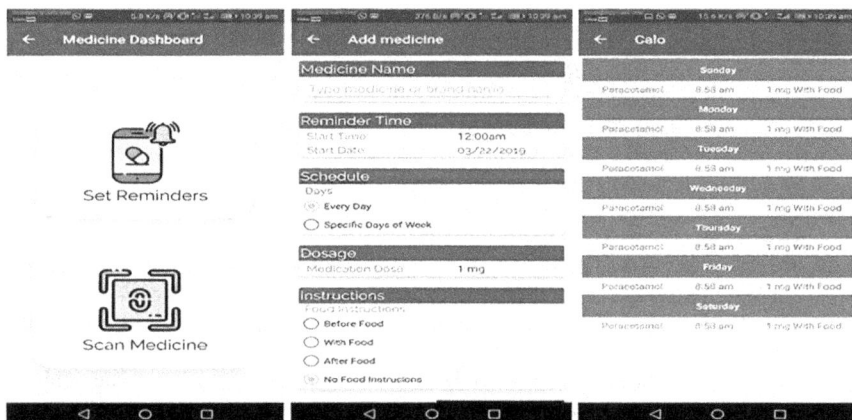

FIGURE 12.10   Medicine scanner and reminder.

12.5   CONCLUSION AND FUTURE PROSPECTUS

- Designing, developing, deploying, and maintaining a large database (India based) along with API to facilitate robust medicine information retrieval.
- Shifting the optical character recognition used in medicine module to handwriting recognition so that it gets easier to seek medicine information from the prescription slip.

REFERENCES

Aggarwal, A., Kapoor, N., & Gupta, A. (2013). Health Insurance: Innovation and challenges ahead. *Global Journal of Management and Business Studies*, *3*(5), 475–780.

Bell, J. Z. (2016). A study of calorie estimation in pictures of food. *Interactive Journal of Medical Research*.

Benfredj, R. (2021). How healthtech is transforming the future of rare disease diagnosis. *Future Rare Diseases*, *1*(1), FRD3.

Bruce, B., & Fries, J. F. (2003). The Stanford health assessment questionnaire: Dimensions and practical applications. *Health and Quality of Life Outcomes*, *1*, 1–6.

Carew, S., Nagarajan, M., Shechter, S., Arneja, J., & Skarsgard, E. (2021). Dynamic capacity allocation for elective surgeries: Reducing urgency-weighted wait times. *Manufacturing & Service Operations Management*, *23*(2), 407–424.

Chakraborty, I., Ilavarasan, P. V., & Edirippulige, S. (2021). Health-tech startups in healthcare service delivery: A scoping review. *Social Science & Medicine*, *278*, 113949.

DiFilippo, K. N., Huang, W.-H., Andrade, J. E., & Chapman-Novakofski, K. M. (2015). The use of mobile apps to improve nutrition outcomes: A systematic literature review. *Journal of Telemedicine and Telecare*, *21*(5), 243–253.

Dutta, M. M. (2020). Health insurance sector in India: An analysis of its performance. *Vilakshan-XIMB Journal of Management*, *17*(1/2), 97–109.

Krittanawong, C. (2018). The rise of artificial intelligence and the uncertain future for physicians. *European Journal of Internal Medicine*, *48*, e13–e14.

Li, L., Diouf, F., & Gorkhali, A. (2022). Managing outpatient flow via an artificial intelligence enabled solution. *Systems Research and Behavioral Science*, *39*(3), 415–427.

Lopes, A. J. (2016). A Practical Study about the Google Vision API. *22nd Portuguese Conference on Pattern Recognition*. outcomes, T. u. (2016). Aveiro, Portugal: 22nd Portuguese Conference on Pattern Recognition, RECPAD.

Marias, T. O. (2010). A Novel Method to Detect Heart Beat Rate Using a Mobile Phone. *Annual International Conference of the IEEE Engineering in Medicine and Biology Society*. IEEE Engineering in Medicine and Biology Society.

Paramonov, D. L. (2017). *Improved algorithm for heart rate measurement using mobile phone camera*. Atlanta, GA.

Singh, K., Misra, M., & Yadav, J. (2021, July). Artificial Intelligence and Machine Learning as a Tool for Combating COVID-19: A Case Study on Health-Tech Start-Ups. In *2021 12th International Conference on Computing Communication and Networking Technologies (ICCCNT)* (pp. 1–5). IEEE.

Vyas, S. (2023). Extended reality and edge AI for healthcare 4.0: Systematic study. In *Extended reality for healthcare systems* (pp. 229–240). Academic Press.

Vyas, S., & Gupta, S. (2023). WBAN-based remote monitoring system utilising machine learning for healthcare services. *International Journal of System of Systems Engineering*, *13*(1), 100–108.

13 Remote Patient Monitoring
An Overview of Technologies, Applications, and Challenges

Darpan Anand, Gurpreet Singh, and Vishan Kumar Gupta

13.1 INTRODUCTION

Remote patient monitoring (RPM) is a rapidly growing field in healthcare that uses digital technologies to monitor patients remotely. The use of RPM has the potential to improve patient outcomes, reduce healthcare costs, and increase access to care. RPM can be used to monitor patients with chronic diseases, post-operative patients, and patients with acute illnesses. RPM devices can be wearable, implantable, or ambient, and can measure a range of physiological and environmental parameters. RPM has many applications, including chronic disease management, post-operative care, and telemedicine. However, there are also several challenges associated with RPM, including data security and privacy concerns, interoperability, and the need for standardization (Hayes et al., 2023). RPM refers to the use of technology to monitor patients' health parameters outside of traditional clinical settings, including homes, long-term care facilities, and remote areas. RPM technology can be used to monitor a variety of health parameters, such as vital signs, activity levels, medication adherence, and disease-specific symptoms. RPM enables healthcare providers to remotely monitor patients' health and intervene in a timely manner, when necessary, thus reducing hospital readmissions, improving patient outcomes, and reducing healthcare costs (Chowdhury & Haque, 2023).

"Remote patient monitoring: A review of current technologies and practices" by Bashi et al. (2015). This literature review provides an overview of the current state of RPM technology and practices, including the various types of RPM devices, their applications, and the challenges associated with implementing RPM (Bashi et al., 2017).

"Remote Patient Monitoring: A Comprehensive Review," published in the *Journal of Biomedical Informatics* in 2018, provides a comprehensive review of RPM technologies, including their history, current state of the art, and future directions. The authors conclude that RPM has the potential to improve patient outcomes, reduce healthcare costs, and enhance patient satisfaction (Farias et al., 2020). "Remote Patient Monitoring in Chronic Disease Management: A Systematic Review of Reviews," published in the *Journal of Medical Internet Research* in 2017, reviewed 38 systematic reviews of RPM in chronic disease management. The authors found that RPM can lead to improvements in clinical outcomes, patient satisfaction, and healthcare utilization (Farias et al., 2020).

"Remote Patient Monitoring for Chronic Disease: A Landscape Assessment of Devices and Applications," published in the *Journal of Diabetes Science and Technology* in 2018, assessed the current state of RPM devices and applications for chronic disease management. The authors found that RPM is becoming increasingly common and that devices are becoming more sophisticated and user-friendly (Anand, 2021).

"Remote Patient Monitoring in Heart Failure: A Systematic Review," published in the *European Journal of Heart Failure* in 2019, reviewed 37 studies of RPM in heart failure management. The authors found that RPM can lead to improvements in clinical outcomes, reduce hospitalizations, and enhance patient satisfaction (Anand, 2021).

"Remote Patient Monitoring for the Management of Chronic Pain: A Systematic Review," published in the *Journal of Telemedicine and Telecare* in 2019, reviewed 15 studies of RPM in chronic pain management. The authors found that RPM can lead to reductions in pain severity, improved function, and enhanced patient satisfaction. "Remote patient monitoring: A comprehensive study" by Malasinghe et al. (2019) provides an in-depth analysis of RPM, including its history, current state, and future prospects. The authors also discuss the various challenges associated with implementing RPM, including regulatory and reimbursement issues, and the need for improved data security and privacy protections (Ali et al., 2023).

"Remote patient monitoring in chronic disease management: A systematic review of current evidence" by Inglis et al. (2018) review focuses specifically on RPM in the context of chronic disease management and provides an analysis of the current evidence for the effectiveness of RPM in improving patient outcomes and reducing healthcare costs (Ali et al., 2023).

"Remote patient monitoring for heart failure: A systematic review" (2019) focuses specifically on RPM for heart failure and provides an analysis of the current evidence for the effectiveness of RPM in reducing hospital readmissions, improving patient outcomes, and reducing healthcare costs (Bashi et al., 2017).

"Remote patient monitoring and telemedicine in neonatal and pediatric settings: Scoping literature review" by Sasangohar et al. (2021) focuses on the use of RPM and telemedicine in neonatal and pediatric settings and provides an analysis of the current state of the literature in this area, including the various types of RPM devices and their applications (Malasinghe et al., 2019).

"A Literature Review of Remote Patient Monitoring: A Case for Chronic Heart Failure Interventions" (2021) examines the use of RPM for chronic heart failure management. The authors found that RPM can help to improve patient outcomes and reduce healthcare costs, but that more research is needed to fully evaluate its effectiveness (Anand, 2021).

"Remote patient monitoring for chronic disease: A landscape assessment of current state of the art and future opportunities" (2020) provides an overview of current RPM technologies and their applications for chronic disease management. The authors identify opportunities for future research and development, including the integration of RPM with other healthcare technologies and the use of data analytics for population health management.

"A systematic review of remote patient monitoring interventions for chronic heart failure patients, with or without implantable devices" (2021) evaluates the effectiveness of RPM interventions for chronic heart failure patients. The authors found that RPM can help to reduce hospitalizations and improve patient outcomes, but that further research is needed to determine the optimal RPM approach for different patient populations (Sasangohar et al., 2018).

"Remote patient monitoring technologies for managing chronic obstructive pulmonary disease (COPD): a systematic review" (2021) examines the use of RPM for managing COPD. The authors found that RPM can help to improve patient outcomes and reduce healthcare costs, but that more research is needed to determine the optimal RPM approach for different patient populations.

"The effectiveness of remote patient monitoring for self-management of chronic obstructive pulmonary disease: A systematic review" (2021) evaluates the effectiveness of RPM for self-management of COPD. The authors found that RPM can help to improve patient outcomes and reduce healthcare costs, but that more research is needed to determine the optimal RPM approach for different patient populations (Wood & Laffel, 2007).

A review of remote patient monitoring technologies: In this article, the authors provide a comprehensive overview of the different types of RPM technologies currently available, including sensors, wearables, and mobile health applications. They also discuss the benefits and challenges of RPM, as well as potential future directions.

Remote patient monitoring in cardiology: This study examines the use of RPM in the management of patients with heart disease, including the use of implanted devices such as pacemakers and defibrillators. The authors discuss the benefits of RPM in terms of improved clinical outcomes and reduced healthcare costs, as well as potential challenges such as patient privacy concerns (Makary et al., 2020).

Remote patient monitoring in diabetes care: This review article focuses on the use of RPM in the management of diabetes, including the use of continuous glucose monitoring (CGM) systems and other monitoring devices. The authors discuss the benefits of RPM in terms of improved glycemia control and reduced healthcare costs, as well as potential challenges such as device accuracy and patient adherence.

A systematic review of RPM for chronic obstructive pulmonary disease (COPD): This study examines the use of RPM in the management of COPD, a chronic lung disease that affects millions of people worldwide. The authors review the literature on RPM for COPD, including studies on the use of telemonitoring, self-monitoring, and other technologies. They discuss the benefits of RPM in terms of improved clinical outcomes and reduced healthcare costs, as well as potential challenges such as patient adherence and device usability (Zhang et al., 2019).

Remote patient monitoring for mental health: This review article discusses the use of RPM in the management of mental health conditions, including depression, anxiety, and bipolar disorder. The authors review the literature on RPM for mental health, including studies on the use of mobile health applications, tele-monitoring, and other technologies. They discuss the benefits of RPM in terms of improved patient outcomes and reduced healthcare costs, as well as potential challenges such as patient privacy concerns and the need for more research in this area (Farias et al., 2020).

13.2   TYPES OF RPM DEVICES

RPM devices are used in various clinical domains, including chronic disease management, postoperative care, and mental health. They are designed to enable remote monitoring of patients and early detection of health problems. RPM devices have the potential to improve patient outcomes while reducing healthcare costs by enabling early detection of health problems, reducing hospital read-missions, and improving medication adherence. The data collected by remote patient monitoring devices is transmitted to healthcare providers through communication networks such as Wi-Fi, cellular networks, or Bluetooth. The healthcare providers can then analyze the data and provide feedback or alerts to the patient as needed. Remote patient monitoring devices can help improve patient outcomes by enabling early detection of health problems, reducing hospital readmissions, and improving medication adherence. They can also help reduce healthcare costs by decreasing the need for in-person visits and hospitalizations (Farias et al., 2020).

RPM devices can be classified into three categories:

- Wearable devices: These devices are worn on the body and can collect data such as heart rate, blood pressure, and activity levels. Examples of wearable devices include smartwatches, fitness trackers, and patch sensors.
- Implantable devices: These devices are implanted in the body and can collect data such as heart function and blood glucose levels. Examples of implantable devices include pacemakers, implantable cardioverter-defibrillators, and continuous glucose monitors.
- Ambient devices/home monitoring kits: These devices are used at home to collect data such as blood glucose levels, blood pressure, and oxygen saturation levels. Examples of home monitoring kits include blood glucose monitors, blood pressure monitors, and pulse oximeters.

Wearable devices are the most common type of RPM device and include devices such as fitness trackers and smartwatches. These devices can measure a range of physiological parameters, including heart rate, blood pressure, and blood glucose levels. Implantable devices, such as pacemakers and defibrillators, are used to monitor and treat patients with heart conditions. Ambient devices, such as smart home devices, can monitor a patient's environment and provide information on factors such as temperature, humidity, and air quality (Zhang et al., 2019).

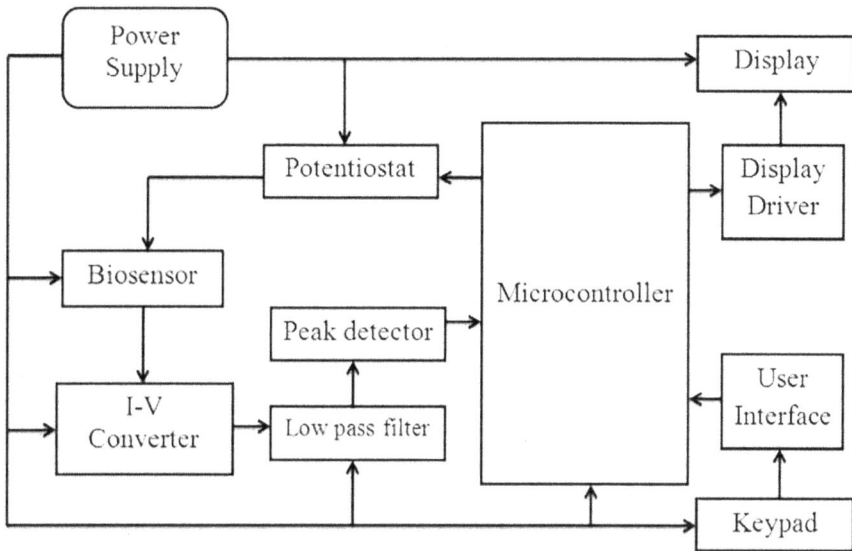

FIGURE 13.1   Block diagram of blood glucose monitors.

Remote patient monitoring (RPM) devices are technological tools used for collecting, transmitting, and analyzing patient health data from outside of traditional healthcare settings. These devices typically include sensors and communication technologies to transmit health data to healthcare providers for analysis and monitoring. Here are some common types of remote patient monitoring devices:

1. **Blood Glucose Monitors:** These devices are used to measure blood glucose levels in patients with diabetes. The device consists of a small sensor that is placed on the skin and a transmitter that sends data to a remote receiver. The block diagram for a digital glucose monitor is shown in Figure 13.1.

Blood glucose monitors are remote patient monitoring devices used to measure blood glucose levels in individuals with diabetes. These devices are also known as glucometers or blood glucose meters. Blood glucose monitoring is an essential part of diabetes management, as it allows individuals with diabetes to monitor their blood glucose levels and adjust their insulin or medication doses accordingly.

Blood glucose monitors typically consist of a meter, lancets, and test strips. The individual pricks their finger with a lancet to obtain a drop of blood, which is then placed on a test strip. The test strip is inserted into the meter, which then measures the blood glucose level and displays it on the screen (Sasangohar et al., 2018).

In recent years, there have been advances in blood glucose monitoring technology, including continuous glucose monitoring (CGM) systems. CGM systems use a sensor that is inserted under the skin to continuously monitor glucose levels in the interstitial fluid. The data is then transmitted to a receiver or a smartphone app, which allows individuals with diabetes to monitor their glucose levels in real time and adjust their insulin or medication doses as needed.

Blood glucose monitors and CGM systems are important tools for individuals with diabetes to manage their blood glucose levels and prevent complications of diabetes such as heart disease, kidney disease, and nerve damage. They also help healthcare providers monitor the effectiveness of diabetes management plans and adjust treatment plans as needed.

Blood glucose monitors are remote patient monitoring devices used to measure the level of glucose (sugar) in the blood. These devices are commonly used by people with diabetes to monitor their blood sugar levels and to manage their diabetes (Wood & Laffel, 2007).

Blood glucose monitors typically consist of a lancet, which is used to prick the finger and collect a small drop of blood, and a glucose meter, which is used to analyze the blood sample and measure the glucose level. Some glucose meters are also equipped with wireless communication capabilities, which allow the data to be transmitted to healthcare providers for remote monitoring.

There are several types of blood glucose monitors available, including traditional glucose meters, continuous glucose monitoring (CGM) systems, and flash glucose monitoring systems. Traditional glucose meters require a blood sample to be collected and analyzed manually, while CGM systems continuously monitor glucose levels using a sensor inserted under the skin. Flash glucose monitoring systems, on the other hand, use a sensor that is worn on the skin and can be scanned to obtain glucose readings.

Blood glucose monitors can help people with diabetes monitor their blood sugar levels and make informed decisions about their diet, medication, and lifestyle choices. By tracking glucose levels over time, people with diabetes can also identify trends and patterns in their blood sugar levels and adjust their management strategies accordingly.

In addition to helping people with diabetes manage their condition, blood glucose monitors can also be used in hospitals and healthcare settings to monitor patients with critical illnesses or to manage patients undergoing surgery.

2. **Blood Pressure Monitors:** These devices are used to measure blood pressure in patients with hypertension or other cardiovascular conditions. The device consists of an inflatable cuff that is placed around the patient's arm and a digital display that shows the readings.

Blood pressure monitors are remote patient monitoring devices used to measure the force of blood against the walls of the arteries. These devices are commonly used to monitor and manage hypertension, a condition characterized by high blood pressure that can increase the risk of heart disease and stroke.

Blood pressure monitors typically consist of a cuff, a pump, and a digital display or monitor. The cuff is placed around the upper arm and inflated by the pump to restrict blood flow. The monitor then measures the pressure exerted by the blood against the cuff, which is displayed as two numbers: systolic pressure (the pressure when the heart beats) and diastolic pressure (the pressure when the heart is at rest).

There are two main types of blood pressure monitors: manual and automatic. Manual blood pressure monitors require the user to inflate the cuff and listen to a stethoscope to detect the pulse and measure the blood pressure. Automatic blood pressure monitors, on the other hand, use an electronic sensor to detect the pulse and measure the blood pressure automatically.

Blood pressure monitors can help people with hypertension monitor their blood pressure levels and make informed decisions about their diet, medication, and lifestyle choices. By tracking blood pressure readings over time, people with hypertension can also identify trends and patterns in their blood pressure levels and adjust their management strategies accordingly (Sasangohar et al., 2018).

In addition to helping people with hypertension manage their condition, blood pressure monitors can also be used in hospitals and healthcare settings to monitor patients with critical illnesses or to manage patients undergoing surgery.

Blood pressure monitors are remote patient monitoring devices used to measure the force of blood against the walls of arteries as it circulates through the body. Blood pressure is an important indicator of cardiovascular health and can help identify conditions such as hypertension, which can increase the risk of heart disease, stroke, and other health problems.

Blood pressure monitors come in two main types: manual and automatic. Manual blood pressure monitors typically consist of an inflatable cuff, a stethoscope, and a gauge. The cuff is wrapped around the upper arm and inflated with a rubber bulb to temporarily stop blood flow through the artery. The stethoscope is used to listen for the sound of blood flowing through the artery as the cuff is slowly deflated, and the gauge is used to measure the pressure of the blood flow at different stages of the process (Burrell et al., 2022).

Automatic blood pressure monitors, on the other hand, use electronic sensors to detect the pulse and calculate the blood pressure automatically. These devices typically consist of an inflatable cuff that is attached to a small machine with a digital display. The cuff is wrapped around the upper arm and automatically inflates and deflates to measure blood pressure (Kour et al., 2021).

Both manual and automatic blood pressure monitors can be used at home or in healthcare settings to monitor blood pressure levels. Some blood pressure monitors are also equipped with wireless communication capabilities, which allow the data to be transmitted to healthcare providers for remote monitoring. A block diagram for the BPM is provided in Figure 13.2.

Blood pressure monitors can help people with hypertension or other cardiovascular conditions monitor their blood pressure levels and make informed decisions about their diet, medication, and lifestyle choices. By tracking blood pressure over time, people can identify trends and patterns and adjust their management strategies accordingly. In addition, blood pressure monitors can help healthcare providers identify and manage hypertension in patients and reduce the risk of cardiovascular disease (Gupta, 2022).

3. **Pulse Oximeters:** These devices are used to measure the oxygen saturation level in a patient's blood. The device consists of a small sensor that is placed on the finger or earlobe and a digital display that shows the readings.

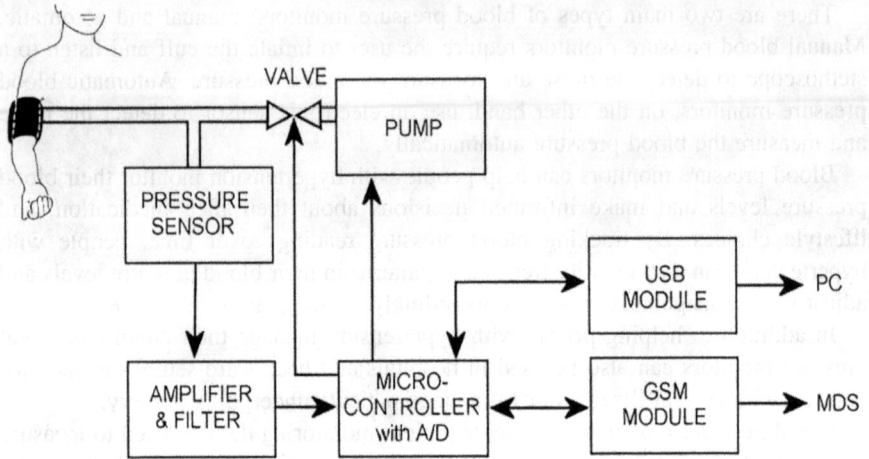

FIGURE 13.2 Blood pressure monitor system.

Pulse oximeters are remote patient monitoring devices used to measure the oxygen saturation level in the blood. These devices are commonly used in healthcare settings and by people with respiratory conditions, such as asthma or chronic obstructive pulmonary disease (COPD) (Kumar et al., 2022).

Pulse oximeters typically consist of a small device with a clip or probe that is attached to a finger, toe, or earlobe. The device uses light sensors to measure the amount of oxygen carried in the blood by detecting changes in the absorption of light. The readings are then displayed on a digital screen. Pulse oximeters can help people with respiratory conditions monitor their oxygen levels and adjust their oxygen therapy as needed. They can also be used by healthcare providers to monitor patients with critical illnesses or during surgical procedures.

In addition, pulse oximeters can be useful for people who participate in high-altitude activities, such as hiking or climbing, as well as for athletes who want to monitor their oxygen levels during exercise. Pulse oximeters are non-invasive and can provide quick and accurate readings of oxygen saturation levels in the blood. They are easy to use and can be used at home or on-the-go. Some pulse oximeters are also equipped with wireless communication capabilities, which allow the data to be transmitted to healthcare providers for remote monitoring. Pulse oximeters are remote patient monitoring devices used to measure the oxygen saturation level in the blood. The device works by shining a light through a person's skin and measuring the amount of light that is absorbed by oxygenated versus deoxygenated blood (Tomar & Agarwal, 2013).

Pulse oximeters typically consist of a small clip or probe that is attached to a person's fingertip or earlobe. The device may have a digital display that shows the oxygen saturation level as a percentage, as well as a graph that shows the person's pulse rate over time. Some pulse oximeters may also have alarms that sound if the oxygen saturation level falls below a certain threshold.

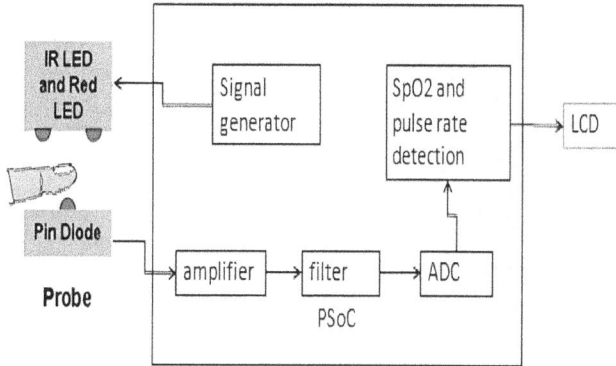

FIGURE 13.3 Pulse oximeter system.

Pulse oximeters are commonly used in healthcare settings, such as hospitals and clinics, to monitor the oxygen saturation level in patients with respiratory or cardiovascular conditions. They are also used by people with conditions such as sleep apnea or chronic obstructive pulmonary disease (COPD) to monitor their oxygen levels at home.

Pulse oximeters (as depicted in Figure 13.3) can help people and healthcare providers monitor oxygen saturation levels and make informed decisions about oxygen therapy or other treatments. They can also help identify potential complications, such as hypoxia (low oxygen levels), and allow for early intervention to prevent serious health problems (Kumar Gupta & Singh Rana, 2021).

Overall, pulse oximeters are a simple and non-invasive way to monitor oxygen saturation levels and can provide valuable information for people with respiratory or cardiovascular conditions, as well as healthcare providers (Babič et al., 2017).

4. **ECG Monitors:** These devices are used to measure a patient's heart rate and rhythm. The device consists of electrodes that are attached to the patient's chest and a small monitor that displays the readings. ECG (electrocardiogram) monitors are remote patient monitoring devices used to measure the electrical activity of the heart. The device works by detecting the electrical impulses that are generated by the heart as it beats and translating them into a graphical representation of the heart's activity.

ECG monitors typically consist of a set of electrodes that are attached to a person's chest, arms, and legs. The electrodes are connected to a small machine with a screen that displays the ECG waveform in real time. The machine may also have the capability to store the ECG readings for later analysis. ECG monitors can be used in a variety of settings, including hospitals, clinics, and at home. They are commonly used to diagnose and monitor heart conditions such as arrhythmias, heart attacks, and heart failure. ECG monitors can also be used to monitor the effects of medications or treatments on the heart, as well as to detect potential heart problems before they become more serious (Gupta & Rana, 2019).

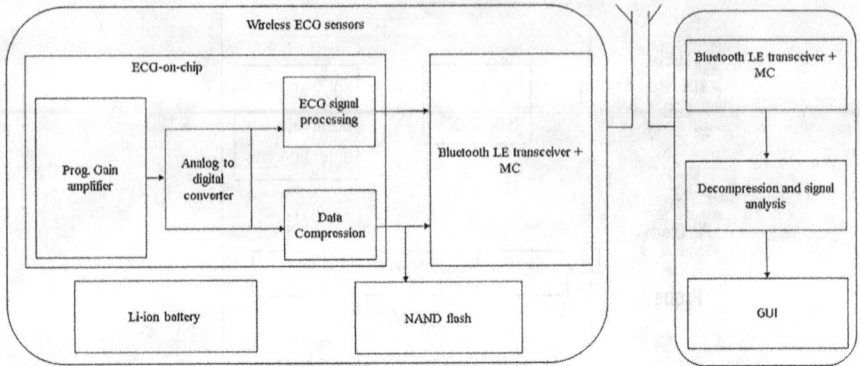

FIGURE 13.4   ECG monitor.

In addition to traditional ECG monitors, there are also portable ECG monitors that can be worn on the body, such as on the wrist or chest. These devices can be especially useful for people who need to monitor their heart activity on a regular basis, such as those with arrhythmias or other cardiovascular conditions. Overall, ECG monitors are a valuable tool for diagnosing and monitoring heart conditions and can provide important information to help people and healthcare providers make informed decisions about treatment and management.

ECG monitors, also known as electrocardiogram monitors, are remote patient monitoring devices used to record the electrical activity of the heart (block diagram in Figure 13.4). The device works by attaching small electrodes to a person's chest, arms, and legs, which detect and transmit the electrical signals generated by the heart (Srivastava & Choubey, 2020). ECG monitors typically consist of a small device with a screen that displays the electrical activity of the heart as a waveform pattern. The device may also have alarms that sound if the heart rate or rhythm falls outside of a certain range. Some ECG monitors may be portable, allowing people to wear them for extended periods of time to monitor their heart activity throughout the day. ECG monitors are commonly used in healthcare settings to diagnose and monitor conditions such as arrhythmias, heart attacks, and heart failure. They can also be used by people with heart conditions to monitor their heart activity at home and provide valuable information for healthcare providers.

ECG monitors can help healthcare providers identify and manage heart conditions, monitor the effectiveness of treatments, and make informed decisions about the need for further testing or intervention. They can also help people with heart conditions better understand their heart activity and take proactive steps to manage their health (Anand & Khemchandani, 2020). Overall, ECG monitors are a valuable tool for diagnosing and monitoring heart conditions and can provide important information for people and healthcare providers.

5. **Wearable Activity Trackers:** These devices are used to monitor a patient's physical activity and exercise habits. The device consists of a small sensor that is worn on the wrist or clipped onto clothing, and it tracks

steps, calories burned, and other physical activity metrics. Wearable activity trackers are remote patient monitoring devices used to track a person's physical activity and monitor their overall health and fitness. These devices are typically worn on the wrist like a watch and can track a variety of metrics such as steps taken, distance traveled, calories burned, and heart rate. Wearable activity trackers typically consist of a small device with sensors that track movement and other biometric data. The device may have a screen that displays the data in real time, as well as the ability to sync with a smartphone app or computer for more detailed analysis and tracking.

Wearable activity trackers are commonly used by people to monitor their fitness levels and set goals for physical activity. They can also be used to track sleep patterns, which can provide valuable information about a person's overall health and well-being. Some activity trackers may also have additional features such as GPS tracking, music controls, and smartphone notifications (Gupta et al., 2022). Wearable activity trackers can help people stay motivated and track their progress towards fitness goals. They can also provide valuable information for healthcare providers to monitor a person's physical activity and make recommendations for lifestyle changes that can improve overall health. Overall, wearable activity trackers are a convenient and useful tool for tracking physical activity and promoting overall health and fitness.

6. **Smart Pill Dispensers:** These devices are used to help patients manage their medication regimens. The device dispenses pills at predetermined times and sends reminders to patients to take their medication. Smart pill dispensers are remote patient monitoring devices used to help people manage their medication schedules and improve medication adherence. The device works by dispensing medication at pre-programmed times and reminding the person when it is time to take their medication.

Smart pill dispensers typically consist of a container for medication with an electronic system that dispenses the medication at the programmed time. The device may also have a screen or audio alerts to remind the person when it is time to take their medication. Some smart pill dispensers may also have sensors that detect when the medication has been dispensed, providing additional feedback and monitoring.

Smart pill dispensers can be particularly useful for people with chronic conditions who need to take multiple medications throughout the day or who may have memory or cognitive impairments that make it difficult to remember their medication schedule. They can also be helpful for caregivers who may be responsible for managing the medication schedules of multiple people (Ghiasi et al., 2020). Smart pill dispensers can help people adhere to their medication schedule and improve overall health outcomes. By reducing the risk of missed or incorrect doses, they can also help prevent complications and hospitalizations related to medication errors. Overall, smart pill dispensers are a useful tool for medication management and can provide peace of mind for both people and their caregivers.

13.3 APPLICATIONS OF RPM

RPM has many applications in healthcare, including chronic disease management, post-operative care, and telemedicine. Chronic disease management is one of the most common applications of RPM. RPM devices can be used to monitor patients with conditions such as diabetes, hypertension, and heart disease. These devices can provide patients with real-time feedback on their health status and help healthcare providers make more informed treatment decisions. Post-operative care is another application of RPM. RPM devices can be used to monitor patients after surgery and detect any complications or adverse events. Telemedicine is another application of RPM. RPM devices can be used to remotely monitor patients and provide virtual consultations with healthcare providers (Ali et al., 2023; Farias et al., 2020). RPM has a wide range of applications in healthcare. Here are some of the key applications of RPM:

- Chronic disease management: RPM can be used to monitor people with chronic diseases such as diabetes, heart disease, and hypertension. The technology can help healthcare providers monitor vital signs, track medication adherence, and detect early warning signs of complications.
- Post-acute care: RPM can be used to monitor people who are recovering from surgery or who have been discharged from the hospital after an acute illness. The technology can help healthcare providers monitor vital signs, track medication adherence, and detect early warning signs of complications.
- Aging in place: RPM can be used to help seniors age in place by monitoring their health status and providing alerts when intervention is needed. The technology can help seniors remain independent while also providing peace of mind for family members and caregivers.
- Mental health: RPM can be used to monitor people with mental health conditions such as depression, anxiety, and bipolar disorder. The technology can help healthcare providers monitor mood changes, track medication adherence, and provide early intervention when needed.
- Pregnancy monitoring: RPM can be used to monitor pregnant women with high-risk pregnancies. The technology can help healthcare providers monitor vital signs, track fetal development, and detect early warning signs of complications.
- Rehabilitation: RPM can be used to monitor people undergoing rehabilitation after an injury or surgery. The technology can help healthcare providers monitor progress, track adherence to exercise programs, and provide early intervention when needed.
- Chronic disease management: RPM can help people with chronic conditions such as diabetes, heart disease, and chronic obstructive pulmonary disease (COPD) to manage their conditions more effectively by monitoring vital signs, symptoms, and medication adherence.
- Postoperative care: RPM can be used to monitor people after surgery to ensure they are recovering properly and to detect any complications that may arise.

- Elderly care: RPM can be particularly useful for elderly people who may have multiple chronic conditions and may require frequent monitoring and medication adjustments.
- Mental health: RPM can be used to monitor people with mental health conditions such as depression and anxiety, providing valuable information about symptoms and treatment effectiveness.
- Maternal and fetal health: RPM can be used to monitor the health of pregnant women and their unborn babies, providing early detection and intervention for any potential complications.
- Sports medicine: RPM can be used to monitor the health and performance of athletes, providing valuable information about vital signs, oxygen levels, and other metrics.

Overall, RPM has the potential to improve healthcare outcomes and reduce healthcare costs by providing more timely and efficient care to people with a wide range of healthcare needs (Gupta et al., 2021).

13.4 CHALLENGES OF RPM

While RPM has many potential benefits, there are also several challenges associated with the technology. One of the main challenges is data security and privacy concerns. RPM devices collect sensitive health information, and there is a risk that this information could be accessed by unauthorized parties. Interoperability is another challenge of RPM. RPM devices from different manufacturers may not be compatible with each other, making it difficult to share data between devices. The lack of standardization is another challenge of RPM. There is currently no standardized approach to RPM, which can make it difficult to compare data across devices and systems (Gupta et al., 2023; Vyas & Gupta, 2022).

Despite the potential benefits of RPM, there are several challenges associated with its implementation. Data security and privacy concerns are significant barriers to the adoption of RPM technology, as patient health data is transmitted over the Internet or other communication networks. Additionally, regulatory issues, such as reimbursement policies and licensing requirements, can limit the widespread adoption of RPM. Finally, patient acceptance of RPM technology can be a challenge, as patients may be reluctant to use new technology or may not feel comfortable with continuous monitoring of their health parameters (Han et al., 2022; Sooch & Anand, 2021).

While RPM has the potential to improve health outcomes and increase patient satisfaction, there are several challenges associated with its implementation and use. Some of these challenges include:

- **Cost:** RPM devices and software can be expensive, and the cost may not be covered by insurance. This can make it difficult for some people to access RPM services.
- **Data security and privacy:** The transmission and storage of patient data raises concerns about data security and privacy. It is important for RPM systems to have adequate security measures in place to protect patient information.

- **Technical issues:** RPM devices and software may experience technical issues, such as connectivity problems or software glitches. These issues can lead to inaccurate data or interruptions in care.
- **User compliance:** For RPM to be effective, patients must be willing and able to use the devices and software correctly. This can be challenging for some people, particularly elderly or technologically inexperienced patients.
- **Integration with existing healthcare systems:** Integrating RPM with existing healthcare systems can be difficult, particularly in large healthcare organizations with complex IT systems.
- **Reimbursement:** RPM is not always reimbursed by insurance providers, which can create financial challenges for healthcare organizations and limit access to RPM for patients.
- **Data management:** The volume of data generated by RPM devices can be overwhelming, and healthcare providers may struggle to manage and interpret the data effectively.
- **Privacy and security:** The use of RPM devices raises concerns about data privacy and security, particularly in light of the growing prevalence of cyber threats and data breaches.
- **Patient acceptance and engagement:** Patients may be hesitant to use RPM devices or may struggle to engage with the technology, particularly if they are not comfortable with digital technology or have concerns about data privacy.
- **Regulatory compliance:** The use of RPM devices is subject to a range of regulatory requirements, and healthcare providers must ensure that they are in compliance with relevant laws and regulations.

Overall, while RPM has the potential to improve healthcare delivery and patient outcomes, it is important to address these challenges in order to ensure the effective implementation and use of RPM systems.

13.5   ADVANCEMENTS IN RPM

RPM technology has advanced significantly in recent years, with the development of wearable devices and sensors that can continuously monitor patients' vital signs, activity levels, and medication adherence (Han et al., 2022). These devices can be integrated with smartphones or other mobile devices to transmit data to healthcare providers for review and intervention. RPM technology has been used in a variety of clinical settings, including post-operative care, chronic disease management, and geriatric care. RPM technology is continually advancing, driven by a growing need for more personalized, proactive healthcare solutions. Some of the recent advancements in RPM technology include:

- Wearable devices: Wearable devices such as smartwatches and fitness trackers are increasingly being used for RPM, providing continuous monitoring of vital signs such as heart rate, blood pressure, and oxygen saturation.

- Artificial intelligence (AI) and machine learning: AI and machine learning algorithms can be used to analyze RPM data, providing valuable insights into patient health and helping to identify potential issues before they become serious.
- Telemedicine: Telemedicine platforms allow healthcare providers to remotely consult with patients and monitor their health status, reducing the need for in-person visits and providing more efficient and cost-effective care.
- Cloud-based data storage: Cloud-based storage solutions can be used to securely store and manage RPM data, making it easily accessible to healthcare providers and enabling real-time monitoring and intervention.
- Remote diagnosis and treatment: RPM technology is increasingly being used to diagnose and treat a range of conditions, including sleep apnea and chronic pain, providing more effective and accessible healthcare solutions for patients.

Overall, these advancements in RPM technology are helping to transform the way healthcare is delivered, providing more personalized, proactive, and accessible care for patients. As the technology continues to evolve, it has the potential to drive significant improvements in health outcomes and patient satisfaction.

13.6 BENEFITS OF RPM

The use of RPM has been associated with a number of benefits, including reduced hospital readmissions, improved patient outcomes, and reduced healthcare costs. RPM technology can also improve patient engagement and self-management, by providing patients with real-time feedback on their health parameters and encouraging healthy behaviors. RPM offers a range of benefits for patients, healthcare providers, and the healthcare system as a whole. Some of the key benefits of RPM include:

- **Improved patient outcomes:** RPM allows for more proactive and personalized healthcare, enabling early detection and intervention for potential health issues before they become serious.
- **Enhanced patient engagement:** RPM empowers patients to take a more active role in their own healthcare, allowing them to track their health status, monitor symptoms, and make informed decisions about their care.
- **Increased efficiency:** RPM can reduce the need for in-person visits and hospitalizations, saving time and resources for healthcare providers and reducing healthcare costs for patients and insurers.
- **Better access to care:** RPM technology can help to overcome barriers to care, particularly for patients who live in remote or underserved areas, by enabling remote consultation and monitoring.
- **Reduced healthcare costs:** RPM can help to reduce healthcare costs by reducing the need for in-person visits, hospitalizations, and emergency room visits. This can be particularly beneficial for patients with chronic conditions who require frequent monitoring and management.

- **Improved care coordination:** RPM facilitates better communication and coordination between healthcare providers, allowing for more effective care management and reducing the risk of medical errors.
- **Enhanced patient experience:** RPM can improve the patient experience by providing more personalized and convenient care. Patients can receive care in their own homes, without the need for travel or disruption to their daily routine.
- **Increased patient engagement:** RPM can increase patient engagement and empowerment by providing patients with greater visibility into their health status and enabling them to take an active role in their care.
- **Better use of healthcare resources:** RPM can help to optimize the use of healthcare resources by enabling healthcare providers to focus their attention on patients who require the most urgent care.
- **Reduced healthcare disparities**: RPM has the potential to reduce healthcare disparities by providing more equitable access to care for patients from diverse backgrounds and with varying levels of healthcare access.

Overall, the benefits of RPM are significant, offering a range of opportunities for improving health outcomes, enhancing patient satisfaction, and reducing healthcare costs. As the technology continues to evolve, it is likely that the benefits of RPM will become even more pronounced, providing more personalized and proactive healthcare solutions for patients and healthcare providers alike.

13.7   FUTURE DIRECTIONS OF RPM

Potential future directions for RPM include the integration of AI and ML algorithms to improve clinical decision making and personalized healthcare. RPM data can be used to develop predictive models that can identify patients at risk for adverse health events, enabling healthcare providers to intervene before a serious event occurs. Additionally, RPM data can be used to personalize treatment plans and improve patient outcomes (Yadav et al., 2019).

The future of RPM is very promising, with ongoing advancements in technology and increasing adoption by healthcare providers. Some of the key directions in which RPM is likely to evolve include:

- Greater integration with electronic health records (EHRs): RPM devices are increasingly being integrated with EHRs, allowing healthcare providers to access RPM data alongside other patient health information.
- Expansion of RPM applications: RPM technology is likely to be increasingly used to monitor a wider range of health conditions, including mental health conditions and respiratory diseases.
- Increased use of AI and machine learning: AI and machine learning algorithms will play an increasingly important role in RPM, helping healthcare providers to analyze and interpret large volumes of data and provide more personalized care (Tomar et al., 2022).

- Adoption of 5G and other advanced connectivity technologies: The rollout of 5G networks and other advanced connectivity technologies will enable more seamless and reliable RPM data transmission, enabling healthcare providers to monitor patients in real time.
- Focus on patient-centric design: Future RPM devices will be designed with a greater focus on patient needs and preferences, making them more intuitive and user-friendly.
- Expansion of RPM to new areas: RPM is already being used to monitor a range of conditions, including chronic diseases such as diabetes, hypertension, and heart disease. In the future, RPM is likely to be used to monitor an even wider range of conditions, including mental health conditions and neurological disorders.
- Integration with other healthcare technologies: RPM is increasingly being integrated with other healthcare technologies, such as telemedicine platforms, electronic health records (EHRs), and AI-powered diagnostic tools. This integration is likely to accelerate in the future, as healthcare providers seek more comprehensive and streamlined solutions for patient care.
- Increased use of wearables: Wearable devices such as smartwatches, fitness trackers, and biosensors are becoming increasingly sophisticated, and are likely to play an even greater role in RPM in the future. These devices can provide continuous monitoring of vital signs, as well as other health metrics such as activity levels and sleep patterns.
- Adoption of personalized medicine: RPM can help to enable personalized medicine by providing more detailed and individualized data on patient health status. This can help healthcare providers develop more targeted and effective treatment plans that take into account each patient's unique needs and circumstances (Sharma et al., 2022).
- Emphasis on data analytics and population health management: As the volume of RPM data continues to grow, healthcare providers are likely to increasingly focus on data analytics and population health management. By analyzing RPM data at a population level, healthcare providers can identify trends, patterns, and risk factors, and develop more effective strategies for disease prevention and management.

Overall, the future of RPM is likely to be characterized by increasing integration, personalization, and sophistication, with the potential to transform the way healthcare is delivered and improve health outcomes for patients. As RPM technology continues to evolve, it will be important for healthcare providers and policymakers to ensure that it is deployed in a way that maximizes its benefits while addressing any associated challenges or risks (Sharma et al., 2022).

13.8 CONCLUSION

RPM is a rapidly growing field in healthcare that has the potential to improve patient outcomes, reduce healthcare costs, and increase access to care. RPM devices can be wearable, implantable, or ambient and can measure a range of physiological and environmental parameters. RPM has many applications, including chronic disease management, post-operative care, and telemedicine. However, there are also several challenges associated with RPM, including data security and privacy concerns, interoperability, and the need for standardization. RPM has emerged as a promising approach to providing healthcare services outside of traditional clinical settings. Advances in RPM technology have enabled continuous monitoring of patient health parameters, which can be transmitted to healthcare providers for review and intervention when necessary. Despite the potential benefits of RPM, there are several challenges associated with its implementation, including data security and privacy concerns, regulatory issues, and patient acceptance. However, the potential future directions of RPM, such as the integration of AI and ML algorithms, offer promising opportunities to improve clinical decision making and personalized healthcare.

REFERENCES

Ali, H., Qadir, J., & Shah, Z. (2023). ChatGPT and large language models (LLMs) in Healthcare: Opportunities and risks. *Institute of Electrical and Electronics Engineers (IEEE)*.

Anand, D. (2021). Role of artificial intelligence in the field of medical and healthcare: A systematic review. *Annals of the Romanian Society for Cell Biology*, 3729–3737.

Anand, D., & Khemchandani, V. (2020). Data security and privacy functions in fog computing for healthcare 4.0. *Fog Data Analytics for IoT Applications: Next Generation Process Model with State of the Art Technologies*, 387–420.

Babič, F., Olejár, J., Vantová, Z., & Paralič, J. (2017, September). Predictive and Descriptive Analysis for Heart Disease Diagnosis. In *2017 Federated Conference on Computer Science and Information Systems (Fedcsis)* (pp. 155–163). IEEE.

Bashi, N., Karunanithi, M., Fatehi, F., Ding, H., & Walters, D. (2017, Jan 20). Remote monitoring of patients with heart failure: An overview of systematic reviews. *Journal of Medical Internet Research*, *19*(1), e18.

Burrell, A., Zrubka, Z., Champion, A., Zah, V., Vinuesa, L., Holtorf, A. P., & Paul, S. (2022). How useful are digital health terms for outcomes research? An ISPOR special interest group report. *Value in Health*, *25*(9), 1469–1479.

Chowdhury, M.N.U.R., & Haque, A. (2023). Chatbots: A game changer in mHealth. *Institute of Electrical and Electronics Engineers (IEEE)*.

Farias, F.A.C., Dagostini, C.M., Bicca, Y.A., Falavigna, V.F., & Falavigna, A. (2020 May). Remote patient monitoring: A systematic review. *Telemedicine and e-Health*, *26*(5), 576–583. 10.1089/tmj.2019.0066. Epub 2019 Jul 17. PMID: 31314689.

Ghiasi, M. M., Zendehboudi, S., & Mohsenipour, A. A. (2020). Decision tree-based diagnosis of coronary artery disease: CART model. *Computer Methods and Programs in Biomedicine*, *192*, 105400.

Gupta, S., Sharma, H.K., & Kapoor, M. (2023). Digital medical records (DMR) security and privacy challenges in smart healthcare system. In *Blockchain for secure healthcare using internet of medical things (IoMT)*. Cham: Springer. 10.1007/978-3-031-18896-1_6

Gupta, V. K. (2022). Toxicity detection of small drug molecules of the mitochondrial membrane potential signalling pathway using bagging-based ensemble learning. *International Journal of Data Mining and Bioinformatics*, *27*(1-3), 201–220.

Gupta, V. K., & Rana, P. S. (2019). Toxicity prediction of small drug molecules of aryl hydrocarbon receptor using a proposed ensemble model. *Turkish Journal of Electrical Engineering and Computer Sciences*, 27(4), 2833–2849.

Gupta, V. K., Gupta, A., Kumar, D., & Sardana, A. (2021). Prediction of COVID-19 confirmed, death, and cured cases in India using random forest model. *Big Data Mining and Analytics*, 4(2), 116–123.

Gupta, V. K., Gupta, A., Jain, P., & Kumar, P. (2022). Linear B-cell epitopes prediction using bagging based proposed ensemble model. *International Journal of Information Technology*, 1–10.

Han, J., Pei, J., & Tong, H. (2022). *Data mining: concepts and techniques*. Morgan Kaufmann Publishers, Elsevier, Second Edition, 2009.

Hayes, C.J., Dawson, L., McCoy, H., Hernandez, M., Andersen, J., Ali, M.M., Bogulski, C.A., & Eswaran, H. (2023). Utilization of remote patient monitoring within the united states health care system: A scoping review. *Telemedicine and e-Health*.

Kour, H., Sabharwal, M., Suvanov, S., & Anand, D. (2021). An Assessment of Type-2 Diabetes Risk Prediction Using Machine Learning Techniques. In *Proceedings of International Conference on Big Data, Machine Learning and their Applications: ICBMA 2019* (pp. 113–122). Springer Singapore.

Kumar, P., Gupta, V.K., & Singh, D.P. (2022, November). Face Mask Detection Using Convolution Neural Network. In *2022 3rd International Conference on Issues and Challenges in Intelligent Computing Techniques (ICICT)* (pp. 1–6). IEEE.

Kumar Gupta, V., & Singh Rana, P. (2021). Ensemble technique for toxicity prediction of small drug molecules of the antioxidant response element signalling pathway. *The Computer Journal*, 64(12), 1861–1875.

Makary, M.S., da Silva, A., Kingsbury, J., Bozer, J., Dowell, J.D., Nguyen, X.V. (2020). Noninvasive approaches for anxiety reduction during interventional radiology procedures. *Topics in Magnetic Resonance Imaging*.

Malasinghe, L.P., Ramzan, N. & Dahal, K. (2019). Remote patient monitoring: a comprehensive study. *Journal of Ambient Intelligence and Humanized Computing 10*, 57–76.

Sasangohar, F., Davis, E., Kash, B.A., & Shah, S.R. (2018 Dec 20). Remote patient monitoring and telemedicine in neonatal and pediatric settings: scoping literature review. *Journal of Medical Internet Research*, 20(12), e295. 10.2196/jmir.9403. PMID: 30573451; PMCID: PMC6320401.

Sharma, K., Anand, D., Mishra, K. K., & Harit, S. (2022). Progressive study and investigation of machine learning techniques to enhance the efficiency and effectiveness of industry 4.0. *International Journal of Software Science and Computational Intelligence (IJSSCI)*, 14(1), 1–14.

Sooch, S. K., & Anand, D. (2021, June). Smart Health Monitoring during Pandemic Using Internet of Things. In *2021 10th IEEE International Conference on Communication Systems and Network Technologies (CSNT)* (pp. 489–493). IEEE.

Srivastava, K., & Choubey, D. K. (2020). Heart disease prediction using machine learning and data mining. *International Journal of Recent Technology and Engineering*, 9(1), 212–219.

Tomar, A., Kumar, S., & Pant, B. (2022, March). Crowd Analysis in Video Surveillance: A Review. In *2022 International Conference on Decision Aid Sciences and Applications (DASA)* (pp. 162–168). IEEE.

Tomar, D., & Agarwal, S. (2013). A survey on data mining approaches for healthcare. *International Journal of Bio-Science and Bio-Technology*, 5(5), 241–266.

Vyas, S., & Gupta, S. (2022). Case study on state-of-the-art wellness and health tracker devices. In *Handbook of research on lifestyle sustainability and management solutions using AI, big data analytics, and visualization* (pp. 325–337). IGI Global.

Wood, J. R., & Laffel, L.M.B. (2007). Technology and intensive management in youth with type 1 diabetes: State of the art. *Current Diabetes Reports*.

Yadav, V., Goel, U., & Kumar, T. (2019). Drug activity prediction of small drug molecules using Random Forest model. *International Journal for Advance Research and Development*, *4*(4), 83–86. International Journal of Information Sciences and Application, 11(1), 1-5, 2019.

Zhang, S., Zhang, X., Ma, S., Purmann, C., Davis, K., Wong, W. H., & Urban, A. E. (2019). Network effects of the neuropsychiatric 15q13. 3 microdeletion on the transcriptome and epigenome in human induced neurons. *bioRxiv*, 772541.

14 Artificial Intelligence and Augmented Reality
Legal and Ethical Issues in the Telemedicine/Telehealth Sphere

Ridoan Karim and Sonali Vyas

14.1 INTRODUCTION: BACKGROUND AND DRIVING FORCES

The healthcare industry is currently facing significant challenges due to the rising patient demand, budgetary constraints, and increasing prevalence of chronic diseases. Along with the growing data volumes, the utilization of digital health technologies is also on the rise in all healthcare settings. Innovative digital technologies have the potential to enable healthcare providers to focus on the underlying causes of diseases, as well as monitor the effectiveness of preventive measures and treatments. Clinical entrepreneurs and computer and data scientists predict that artificial intelligence (AI), specifically machine learning, will be a vital component of healthcare reform (Pereira et al., 2021). Additionally, augmented reality (AR) will complement AI for such reforms, as it encompasses interactions, sensory comprehension, and adaptability mechanisms (Ara et al., 2021). Healthcare providers can use emerging technologies like artificial intelligence and augmented reality to boost operations and improve and expand healthcare by incorporating critical data from these technologies into decisions and conclusions (Davenport & Kalakota, 2019).

The need for a remote tool in the healthcare industry has increased significantly due to COVID. And, as such, AI and AR can usher disruptive services in diagnostics, drug research, epidemiology, treatment, and administrative effectiveness (Aggarwal et al., 2020). Healthcare professionals and patients are using telemedicine and telehealth technology more often because of COVID-19 (El-Sherif et al., 2022). AI gathers, sorts, and analyses billions of pieces of information quickly. It can identify trends and patterns that a human may miss, forecast people's future habits, comprehend people's personalities, and act on their behalf. At the intersection of telemedicine, AI and AR are a perfect storm with fresh perspectives that may lead the future healthcare industry. Nevertheless, to incorporate these technologies into the changing landscape of healthcare delivery systems and attain

DOI: 10.1201/9781003346289-14

buy-in from providers and patients alike, healthcare organizations must promptly adjust their IT budgets, plans, and scopes.

While telehealth is essential for managing and effectively delivering patient care, the digital footprint also has certain disadvantages. New technology's use raises concerns about its potential insecurity, errors, and data leaks (Coventry & Branley, 2018). These potential insecurities, errors, and data leaks have certain legal implications (Karim et al., 2018). Hence, the aim of this chapter is to provide readers with a comprehensive understanding of how artificial intelligence (AI) and augmented reality (AR) work together to improve healthcare security and process improvement plans, as well as the legal implications of digitalization. Due to the potential for catastrophic legal consequences, particularly in high-risk industries like healthcare, due to data breaches or errors, it is essential to identify the ethical and legal applications of AI and AR deployment in the telemedicine/telehealth arena. Therefore, a chapter like this is indispensable for making informed decisions.

14.2   APPLICATIONS OF AI AND AR IN HEALTHCARE

The emergence of the fifth Industrial Revolution has resulted in the replacement of the fourth Industrial Revolution, which involved the integration of technology with the physical, digital, and biological realms. This new era, powered by AI and AR, is rapidly transforming the technological landscape. A pandemic like COVID-19 broke out amid this change and accelerated the AR and AI adoption process. And healthcare is one of the prominent industries that have already adopted AI and AR. Even before the COVID-19 pandemic, healthcare has struggled due to limited resources, rationing, funding, accessibility, efficiency, and other significant issues (Williams & Bryan, 2007). COVID-19 just reemphasized these hurdles and called for greater reform. AI and AR, together as a means of interactions, sensory comprehension, and adaptability mechanisms, call for extended healthcare applications.

The healthcare industry may leverage artificial intelligence as a powerful resource to become a more data-driven medical service provider (Majeed & Oun Hwang, 2022). The latest breakthroughs in computer technology, storage capacity, sensors, machine learning, and deep learning, along with cross-disciplinary research, have opened up new opportunities for the extensive development and implementation of AI techniques in healthcare and medicine. Technological advancements like 5G, edge computing, and robots have all helped to achieve the ultimate goal of digitizing healthcare through AI (Dawson & Szakonyi, 2020). For example, 5G technology offers increased mobile broadband capacity, a more consistent user experience, and lower latency, enabling the enhancement of display and equipment quality for AR and VR devices (Alshammari et al., 2022). Virtually created workplaces that are completely interactive and replicate in-office work settings might be some of the biggest consequences of 5G across sectors (Painuly et al., 2021), and it equally promises to change healthcare drastically.

Medical and healthcare professionals may utilize both virtual and physical AI technologies. Where virtual features of AI can identify trends in surgical operations to increase best practices, physical elements like surgical robots can improve the control accuracy with submillimeter precision (Thai et al., 2020). Prior to COVID-19, telemedicine and telehealth equipment known as bidirectional webcams often produced audio and pictures in two opposing directions or a bidirectional flow (Talal et al., 2020). Nevertheless, today's web camera systems with more advanced features offer additional capabilities such as camera control and digital picture processing (Venkatesan et al., 2021). Mobile device platforms (such as physical inspection and image review) and advanced digital interaction approaches are also emerging (Trenfield et al., 2022). Hence, with the above-mentioned recent updates, AR telemedicine solutions are revolutionizing how doctors and patients interact, giving medical professionals precise information in real time to choose the best course of treatment for patients. AR technology enables a virtual interactive presence that facilitates communication between surgeons and patients, improving verbal, visual, and physical collaboration. With the development of AR technology, surgeons now have a way to practice operations that are thought to be more trustworthy, rapid, and affordable.

Let's look more at how the AI- and AR-based telemedicine/telehealth sphere can usher in changes in the healthcare industry.

14.2.1 Patient Monitoring

One of the earliest and most popular applications of telemedicine was patient monitoring. This approach enables faster and more cost-effective doctor-patient consultations while still evaluating the patient's current state and clinical outcomes. Through the use of videoconferencing and digital medical equipment to capture and record the patient's clinical data, it has evolved to resemble in-person counseling. The goal is to provide increased accessibility, convenience, efficiency, and cost savings compared to traditional physical patient monitoring methods.

The latest designs of telepresence robots are intended to be operated remotely using a software interface that connects the user to the robot via a Wi-Fi connection, allowing it to move freely through rooms and hallways (Koceski & Koceska, 2016). The use of vision systems and artificial intelligence in conjunction with navigation and obstacle detection has recently been added to this notion. One instance of this is the Dr. Rho Medical Telepresence Robot, which is equipped with a mobile body and a screen that facilitates communication between patients and doctors (Pacis et al., 2018). It also includes a micro-projector for group exams and procedures and a straightforward visual system that directs the cameras to imitate the physician's movements and gestures (Pacis et al., 2018). Telepresence robot models such as the Dr. Rho Medical Telepresence Robot have already been implemented in several hospitals (Pacis et al., 2018).

With the integration of artificial intelligence and machine learning, a robot's path can be planned using simultaneous mapping and navigation sensors. Additionally, the physical configuration of the robot, including the height and angle of the screen or camera lens, can be adjusted for the patient's comfort. 3D images can also be captured using wireless sensors that transmit data to devices and continuously monitor the development of a patient (Sathyanarayana et al., 2018). A technique to gather sickle cell disease patient data faster and in real time is by utilizing a web-based management system that incorporates machine learning algorithms (Abd et al., 2017). The system's primary functions include management, patient accommodations, and monitoring, but the AI element allows it to forecast how much medication will be needed based on historical data (Abd et al., 2017). Advancements in patient monitoring have paved the way for the use of self-diagnosing techniques and telemedicine not only in hospitals but also in homes (Casariego-Vales et al., 2021). To fully utilize telemedicine for devices beyond hospitals, each device should have a standardized data architecture (Casariego-Vales et al., 2021). Most patient monitoring equipment and technologies used outside of hospitals are frequently used to keep an eye on geriatric patients or patients with long-term illnesses (El-Rashidy et al., 2021).

14.2.2 HEALTHCARE INFORMATION TECHNOLOGY

With the increasing use of self-diagnosing devices and manual data collection in hospitals, the volume of healthcare data being generated is becoming overwhelming, making it challenging to access and manage medical information. Additionally, a unified record-keeping system must be developed for all participating institutions, taking into account that telemedicine's main objective is to link medical experts and patients from all over the world (Attaran, 2022). Hence, an AI-based blockchain management system can and should play a big role in combining big data analytics and neural networks, which is a popular technology, to maintain and retrieve electronic health records efficiently (Attaran, 2022).

AI is increasingly being used to systematize data collection and analysis, which also tackles challenges associated with medical operations. According to a research conducted by the American College of Physicians, physicians spend half of their time on administrative work and electronic health records (Erickson et al., 2020), which is why AI-based management is developed. One approach to streamlining the time-consuming process of checking patients' vital signs is through a system that uses a chat interface questionnaire, saves the data, and provides the results to the doctor. This approach successfully manages patient data, allowing for quicker prescription and information delivery, as patients can upload photos or videos for immediate evaluation. Another advancement in patient monitoring involves the use of cloud computing to address issues with low bandwidth and implementation difficulties in some locations (Sodhro & Zahid, 2021). The objective of this initiative is to enhance the speed and effectiveness of gathering and distributing patient data by deploying the service in data centers and utilizing remote servers for information storage and management. A study titled "A Cloud Computing Based Telemedicine Service" showcased how cloud computing was utilized to hasten medical analysis by transmitting ECG wave data to multiple locations (Matlani & Londhe, 2013). To integrate established healthcare information system architecture with emerging eHealth consumer electronics technology, cloud computing has been employed. This approach is exemplified by modern wearables, which are equipped with blood pressure sensors and a corresponding, well-defined application programming interface

(Mandel et al., 2016). The capacity for transactional processing and data retrieval speed will be increased by using cloud computing in this situation. This may be a step toward regional medical information and record standardization, which will be crucial for data processing, cooperation, and the general implementation of an automated system in regional hospitals (Mandel et al., 2016). As a result, telemedicine processes will be completed more quickly, and new initiatives will be implemented consistently.

Since telemedicine relies on wireless connections and continuous transfer of patient information, there is a concern regarding the security and privacy of patients' data. Wavelet-based watermarking has been used in several projects, including the use of AI and image processing to safeguard data (Soni & Kumar, 2020). To protect the anonymity of the diagnosis, various scholars suggest using an algorithm for the digital watermarking of medical photographs (Soni & Kumar, 2020).

14.2.3 INTELLIGENT ASSISTANCE AND DIAGNOSIS

The incorporation of assisting intelligent diagnostics is another emerging trend for AI and AR in healthcare. It is designed to help patients, either directly or by studying the patient's previous data. AI and AR technology can work with neural networks and machine learning to advance continually by utilizing the data and findings that are added to the system (Li et al., 2021). The use of intelligent diagnostics has become increasingly prevalent in various telehealth self-diagnosis software and mobile applications (Ur Rehman et al., 2021). The trend in medical electronics has led to the emergence of self-diagnosing software and hardware that can quickly evaluate vital signs such as breathing, heart rate, and pulse (Pacis et al., 2018).

After the initial assessment, the patient will be classified according to the severity of their condition. Physicians will then assess the situation accurately and may conduct consultations over the phone or through AR technology. This procedure will compile the patient's data and, if necessary, set up the consultation. The tool will also keep track of individuals who are taking medication or exhibiting reoccurring symptoms. Wireless telemedicine is also used in emergency circumstances, such as when ambulances share medical information with adjacent hospitals or medical facilities to work together and provide the arriving patient with prompt, high-quality care (Napi et al., 2019).

14.2.4 INFORMATION ANALYSIS AND COLLABORATION

The incorporation of AI and AR in telemedicine can enhance academic teaching, medical research, and consultations. Connecting medical professionals across different regions through technology can facilitate the collaboration on medical data and views, leading to more effective diagnoses. Additionally, integrating the results of clinical trials with big data analytics and using genetic neural networks to identify trends and analyse data will transform pharmaceutical research (Mehta et al., 2019).

By relying solely on results, computer analysis can eliminate biases and personal perspectives. In the study "A Predictive Model for Assistive Technology Adoption for People with Dementia," neural networks were used to assess a patient's inclination to adopt technology by analyzing their characteristics, behaviors, and other factors through an algorithm (Zhang et al., 2014). These models aim to develop prediction tools for home-based care that may be used for improved comprehension and interpretation (Zhang et al., 2014). One of the key uses of neural networks is to establish links between patient data or medical diagnoses in electronic health systems (Zhang et al., 2014). To use and analyze larger amounts of data, AI and AR might combine with other existing information technologies of numerous hospitals or institutions. If this deployment is successful, it may provide new information on connecting patients' data individually and collectively.

14.3 ETHICAL CHALLENGES

AI applications have been criticized in the past (Guan, 2019). Despite the widespread use of AI, more is needed about how to manage these programs effectively. Unlike AI, augmented reality (AR) adds computer-generated stuff (such as images, sounds, and texts) to users' real-world perceptions. The use of augmented reality requires a response to the quick adoption of new technologies and an in-depth understanding of cybersecurity challenges. In addition to the competence formation, a real-time cybersecurity tracker is required for robust AR apps to function (Burton, 2021).

Networked devices that interact with the physical world (cyber-physical systems) are relevant to evaluate within the ethical frameworks of AR and AI, since cybersecurity is essential to their operation. The demand for technology adoption in healthcare is rising along with the ICTs' astronomical expansion (Burton, 2021). Despite the benefits highlighted, there are drawbacks. AI and AR raise questions about whether they can and should alter the healthcare service process. The question of whether AI belongs to the current legal categories or whether a new category with its features and implications should be established is an ongoing debate. The use of AI and AR in clinical settings has the potential to enhance healthcare significantly, but it also raises ethical concerns that must be addressed.

In order to harness the full potential of AI and AR in healthcare, there are four key ethical concerns that need to be addressed: obtaining consent for data usage, ensuring safety and transparency, addressing algorithmic fairness and biases, and protecting data privacy. The legal status of AI systems remains a contentious issue in both political and legal spheres. The aim is to assist policymakers in proactively addressing the ethical challenges associated with the use of AI and AR in healthcare settings.

The constraints of algorithmic transparency have had a significant impact on most legal discussions surrounding AI (Gerke et al., 2020). Given the prevalence of AI in high-risk situations, it is crucial to prioritize responsible, fair, and transparent AI design and governance. The pillars of accessibility and comprehensibility of information are paramount. Machines that can learn and follow arbitrary rules or develop new patterns of behavior are perceived as a potential threat. Some argue that autonomous AI violates the moral and legal foundations of society, raising concerns about its use (Giuffrida, 2019). The accountability for any potential harm resulting from the use of AI needs to be established. The utilization of AI and AR technologies may limit our ability to attribute responsibility and assign culpability, posing an unknown risk.

A modern computer technology known as an artificial intelligent system (AIS) has the potential to conceal the reasoning behind its actions (Naik et al., 2022). An AIS system can be highly complex, making it difficult for a clinical user with technical training to use it effectively, while being relatively easy for a computer science expert with competence in that particular area (Naik et al., 2022). AISs such as IBM's Watson for Cancer are designed to aid clinical users and have a direct influence on clinical decision-making (Smith & Fotheringham, 2020). The AIS would then assess the data and propose a treatment strategy for the patient, which could lead to new dynamics in clinical decision making if employed. Medical professionals, including doctors, nurses, and other personnel, are interested in the secure use of new technology in the clinical environment (Kamal et al., 2020).

On the other hand, it would be unfair to overlook the benefits of AI. In fact, there are various goals, approaches, and potential uses of emerging machine learning healthcare applications (ML-HCAs). The range of ML-HCAs available for primary care settings is broad, from non-autonomous death predictions to fully autonomous synthetic intelligence diabetic retinopathy diagnosis and resource allocation (Char et al., 2020). To ensure the integration of outcomes into research, researchers must explain their methods and projections. These outcomes should also inform the cost of clinical trials and guide future research. However, ethical considerations must be a top priority for AI in healthcare to safeguard patient well-being in an ever-changing environment with frequent disruptions. Nevertheless, the ability of healthcare software to assess plans and identify potential vulnerabilities relies on various factors.

14.4 LEGAL CHALLENGES

With every dramatic transition of new technologies, AI and AR also pose a range of legal concerns in the healthcare industry. Legal concerns vary from data privacy to liability, consumer protection to intellectual property, based on the nature and scope of AI and AR applications. Some of the legal concerns of AI and AR relating to healthcare are highlighted here:

14.4.1 Data Protection

"Data is the lifeline of AI." The importance of data in the field of artificial intelligence is best reflected in this assertion. All algorithmic technologies—and machine learning in particular—are data-driven and necessitate a well-functioning data ecosystem underpinned by a solid legislative framework that fosters trust and data accessibility while preserving privacy rights. By enacting personal data protection laws in so many jurisdictions, policymakers have played a crucial role in establishing the desired perspective on data privacy as a fundamental human right. General data protection regulation (GDPR) is such an example, and it does serve not only Europe but also the whole world at large. With a focus on individual liberties, European ideals, and trust, it has set a new global norm (Rustad & Koenig, 2019).

Most privacy laws worldwide consider six major principles. First and foremost, personal data must be handled honestly, legally, and transparently. It must be gathered for clear, stated, and legal reasons. The "data minimization" principle requires that personal data be sufficient, relevant, and restricted to what is required for the processing purpose. Personal information must be relevant and correct. In theory, it must only be stored for the time that is necessary. Last but not least, the processing of personal data should be conducted with the utmost integrity and secrecy, allowing for the necessary level of security and protection against improper or unlawful processing. This notion has to be implemented through appropriate organizational and technological methods (Gupta et al., 2023).

Since AI and machine learning require substantial data volumes, among which personal data occupy an essential role, the application of data protection laws on algorithmic systems is unquestionable. However, it could be particularly challenging to satisfy the standards of specific data protection laws in an AI-driven healthcare system. First, using skewed health data raises questions about data security and accuracy. It must be emphasized that developing AI will be a multi-disciplinary effort. To lessen the negative effects of AI technology, designers must work closely with committees made up of specialists from many areas (legal, ethics, sociology). Such a strategy would support the operation of participatory risk assessment of health data quality. A restriction on the purpose of the data may provide another difficulty (Xiang & Cai, 2021). AI systems frequently make use of data that has been gathered incidentally for their primary goal. In this situation, the data collector must get extra consent from the data subject in order to comply with data protection regulations. Then again, the data minimization rule poses another challenge. Who would decide how much data is necessary to analyze a health situation? When will the data be stored? If not stored for a long time, how will the AI understand the health pattern of a data subject?

Finally, transparency and the right to information are the two issues that provide the greatest challenge to data protection compliance. Transparency dictates data controllers to inform data subjects on how data is the collected data is processed, its purpose and reason, as well as the timeframe of data retention. This is also regarded as the right to be informed/right to receive an explanation of data processing. This idea might not make sense in the context of the explainability criterion in the area of AI. It is unclear between the legal academics how to interpret the right to an explanation of automated judgments (Asan et al., 2020).

Many scholars believe the solution is simple as they state that in the involvement of automated data process, the data controller must give the data subject meaningful information about the logic involved, as well as the significance and the anticipated consequences of the processing of the personal data (Murdoch, 2021). However, the right to explanation, in principle, is not that simple to comprehend. The question here is how to convey the message to the data subject and what amount of "relevant" information regarding the underlying logic should be disclosed to the data subject? Will it be not a threat to the overall process? Will the information about the AI processing not be categorized as sensitive information? What will happen if the hackers understand the system and the patterns of AI? Are there more threats to data security if the information of AI processing is disclosed with the data subject? These questions should be evaluated from the viewpoint of the person requesting the information as it is somewhat subjective in nature. Hence, when striving to adhere to international data protection standards, healthcare practitioners who use personal information for AI may experience difficulties (Murdoch, 2021).

Several nations even have stringent data protection regulations that prohibit the use of AI and automated decision making when making decisions concerning personal data (Wachter et al., 2017). Healthcare providers trying to adhere to the fairness principle may face difficulties due to the intricacy of the AI approaches being deployed. For instance, machine learning algorithms may contain the prejudices of the people who created them. Reliance on historical criteria might replicate previous biases in situations where machine learning predicts behavioral consequences. Additionally, skewed results may be caused by inadequate data, data oddities, and algorithmic mistakes. An algorithm that uses information from one region of the world, for instance, might not work well elsewhere.

As a result, the GDPR advises organizations using automated data processing, such as AI, to take certain precautions, such as using the right mathematical or statistical techniques for profiling, putting in place organizational and technical measures to address causes of inaccurate personal information, and reducing the likelihood of errors (Casey et al., 2019). The Federal Trade Commission (FTC) Guidance mandates that healthcare industries using AI must ensure that their decisions are fair, requiring them to: give patients/consumers access to and control over information used in decision making and implement access controls and other safeguards to protect their algorithms from unauthorized use (Wagner, 2020).

14.4.2 LIABILITY REGIME

The liability and the legal obligations of people involved in developing AI and AR technologies remain as one the fundamental questions when analyzing the legal framework. There is a general belief among legal practitioners that investors, or more generally, producers acting on behalf of investors, should be held firmly accountable for any flaws in AI technology or even in AR (Čerka et al., 2015). That should also be taken to mean that they are responsible even if the flaws are discovered after the product has been used, provided they are in charge of technical software updates (Čerka et al., 2015). Hence, in most cases, strict liability will be imposed on the producer of the AR and AI products, regardless of whether those defects are based on hardware or software. The reasoning behind it is that responsibility associated with a product's flaws remains constant regardless of its fundamental or functional properties, even if the product or only one of its components is in digital/software form.

AI and AR innovations and products will also be subject to the general rule of product liability. The general principles related to product liability dictate a fair distribution of risks and benefits resulting from commercial production, which calculates the costs of individual harm and an attributable responsibility for the prevention of harm (Schönberger, 2019). This means that if it is established that a specific AI and AR product has caused harm, the burden of proving that the producer/investors have done everything at par lies with them.

In any case, the investor is still liable if the flaw stems from software, hardware, or service offered with the product (Bathaee, 2020). Hence, timely updates, both software and hardware, lie with the responsibility of the producer/investor even after the release of the AI and AR system to maintain the required level of safety and efficacy. In the case of AI, it is always challenging to forecast a product's performance due to the interconnectedness of appliances and cybersecurity concerns. As the AI system connects different devices, hence, the system will have to accommodate different devices manufactured by different companies/manufacturers. Hence, applying product or strict liability provides a broad framework but does not specify specific legal/practical concerns. A flaw in the AI system may happen for a specific product; however, applying liability to all manufacturers might increase the legal risks of all stakeholders concerned. Nevertheless, it is essential to note that in circumstances where the flaw could be anticipated, the risk defense, which shields the producer from liability for unforeseen faults, should not be accessible.

It is also hard to perceive the level of safety a user is entitled to anticipate due to the unpredictable nature of AI systems. The same holds true for determining anything that may be categorized as a failure to achieve the desired degree of safety in the system. These traits may lead to a circumstance in which it is simpler for the producer to substantiate important data. The imbalance between the investor, the producer acting in his capacity, and the user (healthcare providers), supports shifting the burden of proof to the user to prove that the system failed to perform with the standard criteria. Additionally, the user's/healthcare provider's responsibility to decide on the AI system outcome relies on the probable accurate function of the system. Hence, if the system fails, the medical decision taken by the user will also fail. Hence, it will be easier for the aggrieved party (patient) to bring a legal suit against the healthcare provider rather than bringing a lawsuit against the producer/manufacturer/investor. Generally, it is harder for the plaintiff/aggrieved party to develop the liability claim against the producer because of the increasing complexity of this technology (Reed, 2018).

Other than the producers and users, operators are equally accountable for the malfunction of the system (Cabral, 2020). Producers must thus ensure that the design, description, and marketing of AI and AR devices enable operators to carry out their responsibilities. The rules requiring operators to perform their duties comprehensively have been implemented in several jurisdictions under tort law, mainly in common law countries (Sullivan & Schweikart, 2019). Software developers should be held accountable for the harm caused by flaws in their services, much like manufacturers of digital technology goods and services (Schönberger, 2019).

Hence, there are several aspects of liability in this AI and AR environment that ought to be taken into account for implementation. First, there should be a mandatory liability to ensure that victims may obtain compensation and that pro-spective tortfeasors are protected from the danger of liability in situations where there is a possibility that third parties may be exposed to an elevated risk of injury.

Second, as healthcare is a sensitive industry, a more refined duty of care should be imposed on the manufacturer, user, operator, and developer, which will even-tually help the victims with the claim of damages. Thirdly, the burden of proof should initially lie with the manufacturer, user, operator, and developer that they have taken standard measures to ensure the safety of the system and the patient, and if they fail as such to prove, the victim/aggrieved party should receive compensable damage.

The appropriate and just response of law and regulatory requirements to the liability questions should be worked out regardless of the complications imposed by the AI or AR in healthcare. Nevertheless, the policymakers should not hesitate to make any essential modifications and alterations to the current liability regimes in order to introduce a more comprehensive legal and regulatory structure of AI and AR. It may be necessary to develop a number of solutions rather than a single, all-encompassing one, taking into account a wide range of dangers linked with the AR-based AI and its numerous intricacies. However, notwithstanding the diversity of available options, identical risks should be covered by equivalent liability structures. These rules should eventually make it apparent whether losses are recoverable and to what amount. Both product and strict liability should continue to exist, as with other available remedies. That ought to make it possible for a victim to file claims for damages against many defendants. Rule on numerous tortfeasors should thus apply. Additionally, contractual lia-bility or other compensation regimes may apply alongside or in place of tortious responsibility.

14.4.3 CONSUMER PROTECTION

In AI and AR-driven healthcare, the patients as customers/consumers need special consideration. In many scholarly discussions on patient empowerment and involvement, the terms *patient* and *consumer* are sometimes used interchangeably (Goldstein & Bowers, 2015). However, the question arises whether two phrases represent the same thing, or is a "consumer" an informed, deliberate decision maker and a "patient" a passive player in the healthcare system? However, the recent commercialization of healthcare dictates that two words correspond to each other. Hence, when we talk about AI- and AR-driven impact on patients, we should also understand how patients as consumers would fit in the technological transition. Generally, there are several concerns about the consumer's position in the algo-rithmic environment, which include the increased vulnerability in biased decision making, invasion of privacy, deception, commercialization, and exploitation.

Consumer protection through EU legislation primarily aims to protect consumers by taking control of the decision-making process with adequate information (Busch et al., 2016). The law also enshrines the principles of transparency, freedom of choice, equality, etc. (Busch et al., 2016). Nevertheless, not only in Europe but also in other jurisdictions, the existing regulatory environment relating to consumer protection is complicated and often sector-specific. Most of the consumer protection laws address specific consumer protection issues in banking, real estate, tourism, or other financial institutions. Very few see healthcare as a separate and important sector for the consumer protection regime. There are, however, additional regulations that are more general in nature and that, regardless of how particular they are, must be implemented in all AI-enabled systems, even the healthcare industry. For example, in the EU, directive 2019/216/EU on better enforcement of consumer protection, directive 2011/83/EU on consumer rights, and directive 2005/29/EC on unfair commercial practices (UCPD) may be similarly applicable for AI- and AR-driven healthcare system.

Our goal is not to analyze these legal acts in depth but rather to pinpoint the pertinent concerns from the consumer or user's perspective of AI- and AR-based healthcare services or products and place them within the context of the current legal framework, which is meant to offer some guidance on the moral and legal use of algorithmic systems. The following broad principles of consumer law should be advantageous to patients: protection of the weaker party, controlled autonomy, non-discrimination (equal treatment), and privacy. Autonomy and privacy/data protection seem the most crucial ones when we talk about AI and AR in healthcare. We have discussed data protection in the previous sub-section; hence, let us focus on the issues relating to "autonomous consumer."

The notion of regulated autonomy extends beyond the purview of consumer law. The term *autonomous consumer* refers to a person who makes decisions independently and considers all available information and possibilities (Jabłonowska & Pałka, 2020). Consumers should receive clear information on the usage, features, and qualities of acquired items when it comes to AI- and AR-enabled healthcare services, just like they would in any other type of transaction. That means that even an AI-AR healthcare system should provide a patient with the option to choose whether they want to be admitted to a hospital or, if yes, which one they want to choose. It is not always that the physician will be notified by the AI system that the patient is in a vulnerable condition, and the physician can take autocratic medical decisions through AR on behalf of the patient. It is also important that the patient provide consent to be the subject to algorithmic analysis, which is based on the data provided by the patient. If the researchers are required to save the data and reuse it for research purposes, it is equally important that they have taken the consent from the patient.

14.4.4 INTELLECTUAL PROPERTY

There are several ways in which intellectual property law, AI, AR, and healthcare are closely related. AI and AR technologies provide managing solutions that confirm intellectual property rights, as they are in many other sectors today, making it easier to initiate patent, trademark, and design searches. Nevertheless, there are certain challenges as well. AI- and AR-related products' patentability may significantly affect how businesses in the healthcare industry safeguard their intellectual property, particularly their patents. Recent healthcare-related choices have made the problem even more complicated.

The various intricacies involved in patenting artificial intelligence (AI) and ideas produced by AI are highlighted by a recent Western District of Texas in U.S. rulings regarding machine learning technology (Ben-Ari et al., 2017). A Texas court ruled in *Health Discovery Corp. v. Intel Corp.* that two patents covering the use of extremely sophisticated machine learning systems to intelligently and automatically aid humans in analyzing and finding patterns in genes and other biological systems were directed to non-patentable subject matter under 35 U.S.C. 101. The Court determined that the claims only "increase" or "better" a mathematical analysis, which is an abstract concept, excluding them from patent protection.

In *Alice Corp. Pty. v. CLS Bank Int'l,* 573 U.S. 208 (2014), the Supreme Court outlined multiple processes for deciding whether a patent application asserts an unpatentable idea (Hsiao, 2019). The Court must look at whether a claim of the patent is classified under the natural laws or it is just an abstract concept; if it is an abstract idea, then the Court must analyse the claim's character enough and its usefulness to qualify as a patent-eligible application (Hsiao, 2019). The second phase is often met when a patent claim is made for the healthcare industry (Hsiao, 2019).

Due to the difficulty in determining whether a patent actually covers invalid subject matter or not, Alice's application by courts remains unpredictable and inconsistent, especially in the case of the healthcare industry (Hsiao, 2019). Health Discovery Corp. contributes to further guidance on Alice's application to AI, despite the fact that patenting in healthcare industry is still challenging (Hsiao, 2019). When it comes to AI, a claim must be made that "improves an existing technical or computer feature" in order to be patentable (Hsiao, 2019). For instance, just enhancing a process with computer capabilities to make it faster or more effective does not make an otherwise abstract notion patentable. Furthermore, a claimed invention does not automatically become patentable only because it produces better data quality or accurate information as compared to traditional techniques. In reality, a judge is more likely to agree that a machine learning invention qualifies for patent protection when the claims are expressed as an advancement of a novel idea, such as a new piece of hardware or software (Hsiao, 2019). However, where the claims are made in terms of improvements to a mathematical algorithm rather than improvements to a technology that uses a mathematical algorithm, as was the case in Health Discovery Corp., the invention is more likely to be deemed to be just an improvement of an abstract concept.

According to conventional wisdom, patents promote inventive and creative labor by granting inventors monopoly (or exclusive) rights over their ideas for pre-determined amounts of time. The patent system plays a significant role in fostering technical advancement. There is also a different viewpoint, which states that the societal costs connected to patent rights have largely been overlooked while the incentives they give have been greatly overstated (Roin, 2014). Nevertheless, whatever it is, we cannot deny that the major technology players will not jump to the AI in healthcare and develop the overall technology if they do not see the commercial benefits. Patents act as a means of maximizing the value of an idea by restricting access to it. Given the fast-paced, knowledge-driven, and interconnected nature of today's society, novel ideas and technologies have a relatively brief life span. Hence, it will need the combined efforts of a sizable network of participants to access one another's knowledge and ideas to realize commercial value from these discoveries in a reasonable amount of time. Effectively, patenting provides such a cooperative endeavor.

Software engineers have long recognized the need for widespread cooperation across like-minded communities of practice and interest as necessary to create highly complex applications software that will power today's healthcare industry. Hence, it is important that the engineers may patent their developments relating to AI and AR, without having many legal challenges. Such an easy patent measure might promote innovations in the AI and AR in healthcare and bring all the industry players abroad for commercial benefit. The era of innovation requires an easy patent process to safeguard the information-driven healthcare industry. However, the challenges to dream of such an industry are quite large.

14.4.5 ANTITRUST ISSUES

The U.S. Department of Justice's (DOJ) Antitrust Division has issued observations about algorithmic collusion that may have an influence on how AI is used in the healthcare industry. The DOJ has acknowledged that algorithmic pricing may be extremely competitive and coordinated action to fix prices may happen when companies agree to employ similar software to understand the market. As a result, the antitrust concerns may outweigh the efficiency gains from utilizing AI with price information.

14.5 RECOMMENDATION AND CONCLUSION

Generally, the existing legal provisions worldwide ignore historical disparities that have impeded the development of AI technologies and led to uneven access to digital resources (Mohamed et al., 2020). There are more ethical and social questions, in general, relating to AI, in addition to the legal ones. Increased educational objectives and strategies for successful human-AI collaboration are urgently needed, in addition to the new institutional enforcement that the existing legal regimes mandates. An AI system is an all-compassing system that incorporates humans, devices, and organizations. It is necessary to expand the value-based approach to building and regulating AI systems to accommodate those who will utilize this technology and those who choose not to.

Legal implementation of AI and AR also has further technical challenges; for example, it is important to understand how the law would work when healthcare providers over-rely on or over-interpret AI; how patients perceive the utilization of AR; how data is collected, analyzed, and updated to achieve the best achievable performance, etc. Healthcare providers utilizing AI and AR should be aware of an algorithm's limitations and avoid asking questions whose responses can be influenced by algorithmic bias in order to adhere to ethical fairness. Healthcare providers must guarantee datasets are accurate and representative of the populations to which the health practitioners will use the algorithms, and they must use impartial data produced through a regulated procedure to construct algorithms. Additionally, ethical and legal principles of AI mandate that healthcare providers inform the patients directly about automated data processing. For instance, Articles 13 and 14 of the GDPR mandate that healthcare providers notify people when automated decision-making, such as profiling, is being used and where such decision making may have substantial implications for the data subject's rights or other interests. This material needs to contain pertinent details on the reasoning for, the relevance of, and the anticipated effects of the processing on the individual personal data.

To comply with data privacy regulations, healthcare providers are obligated to only gather personal information for specific, lawful, and transparent purposes, and not use it in any way that deviates from those objectives, except in rare cases. However, defining the purpose of processing personal information may pose a challenge for healthcare providers when implementing AI, as it is impossible to predict what insights algorithms may uncover or how the generated data will be used. AI algorithms may identify previously unknown correlations that enable new applications of the data. As a result, healthcare organizations must routinely evaluate if they are using personal data in AI for the same or similar objectives as those communicated to the data subject at the time of data collection. If the processing of personal data in AI is not aligned with the original purpose, healthcare providers must take certain actions, such as seeking renewed consent and establishing a new legal basis for the processing.

Additionally, although the number of scholarly articles reporting on AI and AR in healthcare is growing fast, there is little study on the legal implications of AI, and essentially none that we are aware of at the time of this writing that deals with the cutting edge of AI and AR in healthcare. Even while the ethical disparities brought on by AI are receiving attention, such as the potential for AI to either eliminate or raise inequalities or racial prejudices in healthcare with new paradigms, the legal implications are not yet examined or addressed.

However, it is vital to remember that the legal challenges of AI or AR in healthcare are not entirely different from the legal challenges of AI or AR implementation in other industries. Although the scholarly publications in many sectors of AI application place more emphasis on accuracy and resilience on particular tasks, such studies provide less emphasis on efficiency, privacy, fairness, wider model relevance, and interpretability. These factors frequently require more time and effort and are less well supported by established testing resources or benchmarks.

REFERENCES

Abd, D., Alwan, J.K., Ibrahim, M., & Naeem, M.B. (2017). The Utilisation of Machine Learning Approaches for Medical Data Classification and Personal Care System Management for Sickle Cell Disease. In *2017 Annual Conference on New Trends in Information and Communications Technology Applications, NTICT 2017.* /10.1109/ NTICT.2017.7976147

Aggarwal, N., Ahmed, M., Basu, S., Curtin, J.J., Evans, B.J., Matheny, M.E., Nundy, S., Sendak, M.P., Shachar, C., Shah, R.U., & Thadaney-Israni, S. (2020). Advancing artificial intelligence in health settings outside the hospital and clinic. *NAM Perspectives.* 10.31478/202011f

Alshammari, N., Sarker, M.N.I., Kamruzzaman, M.M., Alruwaili, M., Alanazi, S.A., Raihan, M.L., & AlQahtani, S.A. (2022). Technology-driven 5G enabled e-healthcare system during COVID-19 pandemic. *IET Communications*, *16*(5). 10.1049/cmu2.12240

Ara, J., Karim, F.B., Alsubaie, M.S.A., Bhuiyan, Y.A., Bhuiyan, M.I., Bhyan, S.B., & Bhuiyan, H. (2021). Comprehensive analysis of augmented reality technology in modern healthcare system. *International Journal of Advanced Computer Science and Applications*, *12*(6). 10.14569/IJACSA.2021.0120698

Asan, O., Bayrak, A.E., & Choudhury, A. (2020). Artificial intelligence and human trust in healthcare: Focus on clinicians. *Journal of Medical Internet Research*, *22*(6). 10.2196/15154

Attaran, M. (2022). Blockchain technology in healthcare: Challenges and opportunities. *International Journal of Healthcare Management*, *15*(1). 10.1080/20479700.2020.1843887

Bathaee, Y. (2020). Artificial intelligence opinion liability. *Berkeley Technology Law Journal*, *20*(1).

Ben-Ari, D., Frish, Y., Lazovski, A., Eldan, U., & Greenbaum, D. (2017). "Danger, Will Robinson"? Artificial intelligence in the practice of law: An analysis and proof of concept experiment. *Journal of Law Technology*, *23*(2).

Burton, S.L. (2021). Artificial intelligence (AI) and augmented reality (AR): Disambiguated in the telemedicine/telehealth sphere. *Scientific Bulletin*, *26*(1). 10.2478/bsaft-2021-0001

Busch, C., Franceschi, A. De, Luzak, J., Mak, V., Carvalho, J.M., Nemeth, K., Wallis, I.D., Busch, I.C., & Schulte-nölke, H. (2016). The rise of the platform economy: A new challenge for EU consumer law? *Journal of European Consumer and Market Law*, *5*(February).

Cabral, T.S. (2020). Liability and artificial intelligence in the EU: Assessing the adequacy of the current Product Liability Directive. *Maastricht Journal of European and Comparative Law*. 10.1177/1023263X20948689

Casariego-Vales, E., Blanco-López, R., Rosón-Calvo, B., Suárez-Gil, R., Santos-Guerra, F., Dobao-Feijoo, M.J., Ares-Rico, R., & Bal-Alvaredo, M. (2021). Efficacy of telemedicine and telemonitoring in at-home monitoring of patients with COVID-19. *Journal of Clinical Medicine*, *10*(13). 10.3390/jcm10132893

Casey, B., Farhangi, A., & Vogl, R. (2019). Rethinking explainable machines: The Gdpr's "Right to Explanation" Debate and the rise of algorithmic audits in enterprise. *Berkeley Technology Law Journal*, *34*(1).

Char, D.S., Abràmoff, M.D., & Feudtner, C. (2020). Identifying ethical considerations for machine learning healthcare applications. *American Journal of Bioethics*, *20*(11). 10.1080/15265161.2020.1819469

Čerka, P., Grigiene, J., & Sirbikyte, G. (2015). Liability for damages caused by artificial intelligence. *Computer Law and Security Review*. 10.1016/j.clsr.2015.03.008

Coventry, L., & Branley, D. (2018). Cybersecurity in healthcare: A narrative review of trends, threats and ways forward. *Maturitas* (Vol. 113). 10.1016/j.maturitas.2018.04.008

Davenport, T., & Kalakota, R. (2019). The potential for artificial intelligence in healthcare. *Future Healthcare Journal*, *6*(2). 10.7861/futurehosp.6-2-94

Dawson, M., & Szakonyi, A. (2020). Cybersecurity education to create awareness in artificial intelligence applications for developers and end users. *Scientific Bulletin*, *25*(2). 10.2478/bsaft-2020-0012

El-Rashidy, N., El-Sappagh, S., Riazul Islam, S.M., El-Bakry, H.M., & Abdelrazek, S. (2021). Mobile health in remote patient monitoring for chronic diseases: Principles, trends, and challenges. *Diagnostics*, *11*(4). 10.3390/diagnostics11040607

El-Sherif, D.M., Abouzid, M., Elzarif, M.T., Ahmed, A.A., Albakri, A., & Alshehri, M.M. (2022). Telehealth and artificial intelligence insights into healthcare during the COVID-19 pandemic. *Healthcare (Switzerland), 10*(2). 10.3390/healthcare10020385

Erickson, S.M., Outland, B., Joy, S., Rockwern, B., Mire, R.D., & Goldman, J.M. (2020). Envisioning a better U.S. health care system for all: Health care delivery and payment system reforms. *Annals of Internal Medicine, 172*(2), S33–S49. 10.7326/M19-2407

Gerke, S., Minssen, T., & Cohen, G. (2020). Ethical and legal challenges of artificial intelligence-driven healthcare. *Artificial Intelligence in Healthcare.* 10.1016/B978-0-12-818438-7.00012-5

Giuffrida, I. (2019). Liability for AI decision-making: Some legal and ethical considerations. *Fordham Law Review, 88*(2), 439.

Goldstein, M.M., & Bowers, D.G. (2015). The patient as consumer: Empowerment or commodification? Currents in contemporary bioethics. *Journal of Law, Medicine and Ethics, 43*(1). 10.1111/jlme.12203

Guan, J. (2019). Artificial intelligence in healthcare and medicine: Promises, ethical challenges and governance. *Chinese Medical Sciences Journal, 34*(2). 10.24920/003611

Gupta, S., Sharma, H.K., Kapoor, M. (2023). Digital medical records (DMR) security and privacy challenges in smart healthcare system. In *Blockchain for secure healthcare using internet of medical things (IoMT).* Cham: Springer. 10.1007/978-3-031-18896-1_6

Hsiao, J. I.-H. (2019). Patent eligibility of predictive algorithm in second generation personalized medicine. *SMU Science and Technology Law Review, 22*, 23.

Jabłonowska, A., & Pałka, P. (2020). EU consumer law and artificial intelligence. *The Transformation of Economic Law.* 10.5040/9781509932610.ch-006

Kamal, S.A., Shafiq, M., & Kakria, P. (2020). Investigating acceptance of telemedicine services through an extended technology acceptance model (TAM). *Technology in Society, 60.* 10.1016/j.techsoc.2019.101212

Karim, R., Newaz, M.S., & Chowdhury, R.M. (2018). Human rights-based approach to science, technology and development: A legal analysis. *Journal of East Asia and International Law, 11*(1). 10.14330/jeail.2018.11.1.08

Koceski, S., & Koceska, N. (2016). Evaluation of an assistive telepresence robot for elderly healthcare. *Journal of Medical Systems, 40*(5). 10.1007/s10916-016-0481-x

Li, J.P.O., Liu, H., Ting, D.S.J., Jeon, S., Chan, R.V.P., Kim, J.E., Sim, D.A., Thomas, P.B.M., Lin, H., Chen, Y., Sakomoto, T., Loewenstein, A., Lam, D.S.C., Pasquale, L.R., Wong, T.Y., Lam, L.A., & Ting, D.S.W. (2021). Digital technology, telemedicine and artificial intelligence in ophthalmology: A global perspective. *Progress in Retinal and Eye Research, 82.* 10.1016/j.preteyeres.2020.100900

Majeed, A., & Oun Hwang, S. (2022). Data-driven analytics leveraging artificial intelligence in the era of COVID-19: An insightful review of recent developments. *Symmetry, 14*(1). 10.3390/sym14010016

Mandel, J.C., Kreda, D.A., Mandl, K.D., Kohane, I.S., & Ramoni, R.B. (2016). SMART on FHIR: A standards-based, interoperable apps platform for electronic health records. *Journal of the American Medical Informatics Association, 23*(5). 10.1093/jamia/ocv189

Matlani, P., & Londhe, N.D. (2013). A Cloud Computing Based Telemedicine Service. *IEEE EMBS Special Topic Conference on Point-of-Care (POC) Healthcare Technologies: Synergy Towards Better Global Healthcare, PHT 2013.* 10.1109/PHT.2013.6461351

Mehta, N., Pandit, A., & Shukla, S. (2019). Transforming healthcare with big data analytics and artificial intelligence: A systematic mapping study. *Journal of Biomedical Informatics, 100.* 10.1016/j.jbi.2019.103311

Mohamed, S., Png, M.T., & Isaac, W. (2020). Decolonial AI: Decolonial theory as sociotechnical foresight in artificial intelligence. *Philosophy and Technology, 33*(4). 10.1007/s13347-020-00405-8

Murdoch, B. (2021). Privacy and artificial intelligence: Challenges for protecting health information in a new era. *BMC Medical Ethics*, 22(1). 10.1186/s12910-021-00687-3

Naik, N., Hameed, B.M.Z., Shetty, D.K., Swain, D., Shah, M., Paul, R., Aggarwal, K., Brahim, S., Patil, V., Smriti, K., Shetty, S., Rai, B.P., Chlosta, P., & Somani, B.K. (2022). Legal and ethical consideration in artificial intelligence in healthcare: Who takes responsibility? *Frontiers in Surgery*, 9. 10.3389/fsurg.2022.862322

Napi, N.M., Zaidan, A.A., Zaidan, B.B., Albahri, O.S., Alsalem, M.A., & Albahri, A.S. (2019). Medical emergency triage and patient prioritisation in a telemedicine environment: A systematic review. *Health and Technology*, 9(5). 10.1007/s12553-019-00357-w

Pacis, D.M.M., Subido, E.D.C., & Bugtai, N.T. (2018). Trends in telemedicine utilizing artificial intelligence. *AIP Conference Proceedings*, 1933. 10.1063/1.5023979

Painuly, S., Sharma, S., & Matta, P. (2021). Future Trends and Challenges in Next Generation Smart Application of 5G-IoT. In *Proceedings - 5th International Conference on Computing Methodologies and Communication, ICCMC 2021*. 10.1109/ICCMC51019.2021.9418471

Pereira, T., Morgado, J., Silva, F., Pelter, M.M., Dias, V.R., Barros, R., Freitas, C., Negrão, E., de Lima, B.F., da Silva, M.C., Madureira, A.J., Ramos, I., Hespanhol, V., Costa, J.L., Cunha, A., & Oliveira, H.P. (2021). Sharing biomedical data: Strengthening AI development in healthcare. *Healthcare (Switzerland)*, 9(7). 10.3390/healthcare 9070827

Reed, C. (2018). How should we regulate artificial intelligence? *Philosophical Transactions of the Royal Society A: Mathematical, Physical and Engineering Sciences*, 376(2128). 10.1098/rsta.2017.0360

Roin, B.N. (2014). Intellectual property versus prizes: Reframing the debate. *University of Chicago Law Review*, 81(3).

Rustad, M.L., & Koenig, T.H. (2019). Towards a global data privacy standard. *Florida Law Review*.

Sathyanarayana, S., Satzoda, R.K., Sathyanarayana, S., & Thambipillai, S. (2018). Vision-based patient monitoring: a comprehensive review of algorithms and technologies. *Journal of Ambient Intelligence and Humanized Computing*, 9(2). 10.1007/s12652-015-0328-1

Schönberger, D. (2019). Artificial intelligence in healthcare: A critical analysis of the legal and ethical implications. *International Journal of Law and Information Technology*, 27(2). 10.1093/ijlit/eaz004

Smith, H., & Fotheringham, K. (2020). Artificial intelligence in clinical decision-making: Rethinking liability. *Medical Law International*, 20(2). 10.1177/0968533220945766

Sodhro, A.H., & Zahid, N. (2021). Ai-enabled framework for fog computing driven E-healthcare applications. *Sensors*, 21(23). 10.3390/s21238039

Soni, M., & Kumar, D. (2020). Wavelet Based Digital Watermarking Scheme for Medical Images. In *Proceedings - 2020 12th International Conference on Computational Intelligence and Communication Networks, CICN 2020*. 10.1109/CICN49253.2020.9242626

Sullivan, H.R., & Schweikart, S.J. (2019). Are current tort liability doctrines adequate for addressing injury caused by AI? *AMA Journal of Ethics*. 10.1001/amajethics.2019.160

Talal, A.H., Sofikitou, E.M., Jaanimägi, U., Zeremski, M., Tobin, J.N., & Markatou, M. (2020). A framework for patient-centered telemedicine: Application and lessons learned from vulnerable populations. *Journal of Biomedical Informatics*, 112. 10.1016/j.jbi.2020.103622

Thai, M.T., Phan, P.T., Hoang, T.T., Wong, S., Lovell, N.H., & Do, T.N. (2020). Advanced intelligent systems for surgical robotics. *Advanced Intelligent Systems*, 2(8). 10.1002/aisy.201900138

Trenfield, S.J., Awad, A., McCoubrey, L.E., Elbadawi, M., Goyanes, A., Gaisford, S., & Basit, A.W. (2022). Advancing pharmacy and healthcare with virtual digital technologies. *Advanced Drug Delivery Reviews*, 182. 10.1016/j.addr.2021.114098

Ur Rehman, I., Sobnath, D., Nasralla, M.M., Winnett, M., Anwar, A., Asif, W., & Sherazi, H.H.R. (2021). Features of mobile apps for people with autism in a post covid-19 scenario: Current status and recommendations for apps using AI. *Diagnostics*, *11*(10). 10.3390/diagnostics11101923

Venkatesan, M., Mohan, H., Ryan, J.R., Schürch, C.M., Nolan, G.P., Frakes, D.H., & Coskun, A.F. (2021). Virtual and augmented reality for biomedical applications. *Cell Reports Medicine*, *2*(7). 10.1016/j.xcrm.2021.100348

Wachter, S., Mittelstadt, B., & Floridi, L. (2017). Why a right to explanation of automated decision-making does not exist in the general data protection regulation. *International Data Privacy Law*, *7*(2). 10.1093/idpl/ipx005

Wagner, J.K. (2020). The federal trade commission and consumer protections for mobile health apps. *Journal of Law, Medicine and Ethics*, *48*(1_suppl). 10.1177/107311052 0917035

Williams, I., & Bryan, S. (2007). Understanding the limited impact of economic evaluation in health care resource allocation: A conceptual framework. *Health Policy*, *80*(1). 10.1016/ j.healthpol.2006.03.006

Xiang, D., & Cai, W. (2021). Privacy protection and secondary use of health data: Strategies and methods. *BioMed Research International* (Vol. 2021). 10.1155/2021/6967166

Zhang, S., McClean, S.I., Nugent, C.D., Donnelly, M.P., Galway, L., Scotney, B.W., & Cleland, I. (2014). A predictive model for assistive technology adoption for people with dementia. *IEEE Journal of Biomedical and Health Informatics*, *18*(1). 10.1109/ JBHI.2013.2267549

15 Telemedicine

Patient Monitoring and Electronic Healthcare Record Storage

Charul Dewan, T.Ganesh Kumar, and
Sunil Gupta

15.1 INTRODUCTION

Medical care organizations are worried about the ageing global population and the surge in long-term illness cases. As the cases are surging, so is the need for data storage. We need optimized and secure technological platforms to store medical records of patients. Fog computing is the field that evolves in order to provide optimized solutions to the problems of the period. The Industrial Internet of Things (IoT) revolutionizes the methodologies of traditional communication. The path forward is IoT, with the introduction of intelligent devices like mobile, computers, sensors, and many more combined with a wide variety of requirements for applications. By 2025, it is predicted that 41 billion heterogeneous IoT devices will be in use, performing tasks with a range of Quality of Service (QoS) needs. Due to decentralization in fog computing, a large volume of data is divided and filtered at respective fog nodes. Data filtration is being done at fog nodes. Local servers or the fog nodes are being deployed which eases computation problems and can be deployed in any network environment.

Handling large volume of data is a big issue but fog computing can be designed in a way to facilitate data storage. Data is transferred over the public internet; thus, privacy is another issue as sensitive information is being collected at the fog nodes. Dynamic observation of data is not possible with the existing algorithms. Placement of the server is another issue, as it will increase the maintenance cost. Device authentication has to be managed as hackers can use fake IP addresses and can steal sensitive information. Interoperability arises as the algorithms used to handle the private data at fog nodes still suffer from interfacing issues. Data heterogeneity and multimodal data needs to be addressed too.

Cisco's proposal for a new concept, fog computing (FC), runs across the cloud on geo-distributed software to fix the above-mentioned issues. Similarly to CC, at the edge of a network, FC handles latency-sensitive applications. It also performs

DOI: 10.1201/9781003346289-15

latency-tolerant tasks with the aid of powerful fog nodes (FNs) at the cloud infrastructure proxy, quite effectively. In network research, a cloud data center is still being used for deep data analysis. Fog computing enhances cloud computing by extending its capabilities to the edge of the network, as opposed to replacing it. In order to assist end users, cloud and fog essentially offer data, compute, storage, and application services. However, the proximity of the fog to the source and sink, its ability to spread regardless of geography, and, last but not least, its adaptability set it apart from the cloud. Each single bit of data should be delivered to the data center in the case of cloud-based IoT data processing. When the size of the data to be processed increases tremendously (and that is the exact IoT case). The future of automation depends on connected device intelligence in which all devices will interact with each other in order to generate the independent conclusions or consensus, called M2M communication. These clever solutions are essential to Industry 4.0's digital transition and are known as decentralized solution. Compared to centralized solutions, decentralized alternatives offer flexibility and swift decision support. The 802.11ah technology has advanced recently for M2M communication. Real-time communication is required for exchanging machine data, assuring latency, reliability, security, bandwidth, and data privacy across all IoT domains (Bonomi et al., 2012; Chiang, 2015; Chiang et al., 2016a; Tang et al., 2015).

A few challenges in cloud computing are described below, taking into account a variety of IoT scenarios:

- The broad distance here between cloud and edge computers causes delays in dissemination and transmission (Chiang et al., 2016b).
- High computing charge on a single cloud server creates issues in transmission and queuing.
- Increased smart devices have prevented bandwidth requirements from being met.
- Scalability, performance, and technological issues may arise with a large number of smart devices.
- Wireless media between cloud devices and smart devices raises resource management problems.
- Smart devices' heterogeneity property in terms of access technology will cause cloud handling difficulties (Atlam et al., 2018).
- Privacy is a very important thread, as the cloud is open to the public Internet worldwide.
- Each time, computing offloads in the cloud causes a loss of energy and battery life.
- While cloud data storage helps application developers, the confidentiality and authentication requirements of IoT applications should be taken into account (Kumari, 2015).
- For real-time IoT applications, cloud computing is a centralized and dynamic infrastructure.

Cloud computing and cutting-edge technology are enticing for real-time IoT systems; this extended form of cloud computing offers solutions to the problems

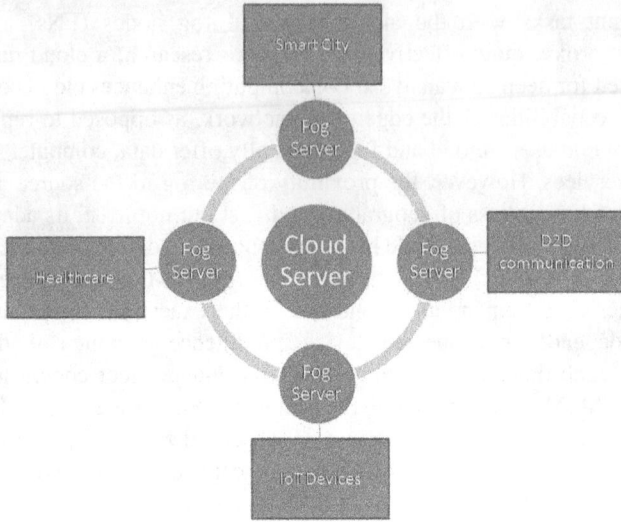

FIGURE 15.1   Generalized view: Cloud, fog, and edge computing IoT technology domains.

that cloud computing faces. Terms like *fog* and *edge computing* are used by the industry parallelly and this has been shown in Figure 15.1. The figure shown below gives a connected view of cloud, edge, and fog computing. Both computing systems introduce resources for storage and processing close to where data arises. Edge computing adds processing to one of a network's devices, complementing fog computing. This item, known as an E-node, is near the data. E-node has higher computational power, resources, and intelligent controllers, such as programmable automation controllers (PACs) (Amin et al., 2017).

In terms of effective and scalable network architecture, fog is a comparatively new concept that develops new challenges. The realization of Industry 4.0 is supposed to develop slowly over the next few years. Challenges such as energy conservation, real-time connectivity, effective use of bandwidth, cache capacity on edge devices, and automated resource allocation are open issues that need to be solved for automation in the future (Bonomi et al., 2014). Without such factors, IoT systems can't meet promised QoS specifications. Scientists must in the future have solutions to these problems for the advancement of industrialization.

15.2   IMPACT OF FOG COMPUTING IN DATA ANALYTICS

Fog computing refers to a decentralized system that is used to handle enormous amounts of data that is being produced either by gadgets or smartphones or driverless cars. As it is a known fact that cloud computing can't handle such massive amounts of data, fog nodes may be the answer. Between the data sources and the cloud are fog nodes. Fog computing definitely yields faster data analytics as fog nodes will take the data from IoT devices and forward it to the sites for analysis and this will reduce the burden on cloud nodes. The impact of fog nodes in data analytics can be observed easily as the issues which were there with cloud computing were addressed and the data analysis was quite frequent.

Data breaching and security issues can be addressed more clearly at the edge nodes compared to cloud servers.

Millions of fog nodes have been introduced between thousands of cloud servers and billions of end devices that are analyzing the data at a faster, secure rate.

15.2.1 REAL-TIME DATA HANDLED WITH IoT

There is an avalanche of real-time data as estimated 41.6 billion devices will be connected by the year 2025. Processing of real-time data needs to be managed as the organizations who are extracting the info from that data have to react to changes occurring in real time. The data that is collected in real time will be converted into useful information and this task will be accomplished without human-to-human interaction and human-to-computer interaction.

15.2.2 FUTURE INTERNET CoT (CLOUD OF THINGS)

Cloud of things is a new term that is a combination of cloud computing and Internet of Things. Cloud of things will give a new dimension to voluminous data as it will handle the data wisely. Internet of Things is used to connect physical devices such as gadgets, wearable, vehicles, and buildings through embedded electronics and is used to send and receive data. Cloud computing is used to store the data on a rental basis without direct active management by the user. Processing large volumes of data and generating intelligent results cannot be handled through IoT alone. Thus, cloud computing and IoT will work in integration and the term coined is *Cloud of Things*. Still, it suffers from various issues like security, privacy, heterogeneity, big data, performance, monitoring, and legal aspects (Atlam et al., 2017).

15.3 INDUSTRIAL INTERNET OF THINGS TECHNOLOGY AND FACILITATORS

Industry 4.0 refers to a variety of important IoT technologies and applications. To take use of IoT and important enabling technologies that support the Industrial Revolution, the evolution phases of Industry 4.0, IoT connectivity technologies, are briefly outlined in this section. By using advancements in networking, digitalization, and new communication technologies, industrial automation is making progress in satisfying corporate and consumer demands. According to market or customer demand, numerous modules are added or removed during the production process, and machines are deftly retooled utilizing advanced communication technology (Atlam et al., 2017). The industry 4.0 world is made up of the Internet of data, the Internet of Things, the Internet of people, and the Internet of facilities. Integration of Industry 4.0 with already-existing digital infrastructure, including social media, corporate networks, smart grids, smart homes, and smart buildings,

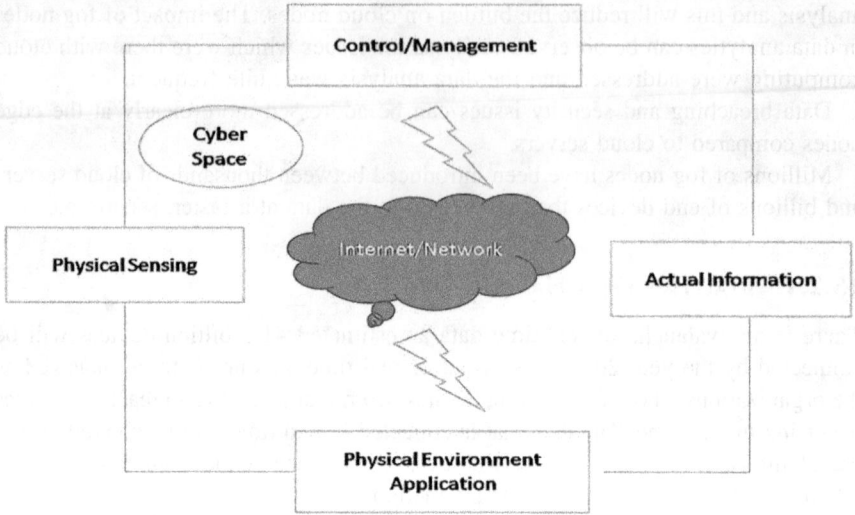

FIGURE 15.2   A cyber physical system architecture.

leading to the development of a CPS system. This revolution, based on CPS, would merge the real and virtual worlds (Cerina et al., 2017). On a computer, a CPS application uses a hierarchical clustering to combine the Internet and its users and this has been depicted in Figure 15.2. It is essentially digitization since these systems combine information technology with electronic and mechanical device modules that communicate through a network. After cataloguing and analyzing huge data, CPS can improve manufacturing of high-quality, low-cost raw materials. As sensor and computing technologies progress, a number of CPSs are starting to appear. To support the development and production cycle of Industry 4.0, CPS evolved into cyber physical production systems (CPPSs) (He et al., 2018; Mohammadi et al., 2018; Mahmud et al., 2018; Yi et al., 2015).

15.4   CHARACTERISTICS OF FOG COMPUTING

1. Since fog devices are located close to IoT devices, that means inside the same local area network, on-campus, thereby IoT software is handled locally (Kumari et al., 2018).
2. End-proximity: Fog systems are near to end users in addition to being close to IoT devices.
3. Quicker response: Making a quick conclusion is possible since the processed data doesn't need to travel very far and the results are sent to nearby businesses.
4. Location Aware: The Fog devices are location-aware since they are distributed locally on a small scale close to the sensing devices (Vora et al., 2018a).
5. Mobility Support: The majority of Internet of Things (IoT) devices are portable. With its mobile devices, fog enables this mobility (Vora et al., 2018b).

6. Massive in size: The number of fog endpoints is in the billions, just like the number of IoT devices dispersed throughout vast geospacer (Vora et al., 2017).

7. Wireless communication predominates: Since wireless connections are used to connect IoT devices to fog nodes, fog architecture is adaptable and ubiquitous.

15.5 TAXONOMY OF FOG DATA ANALYTICS COMMUNICATION

Outlines a thorough taxonomy of FDA interaction, including current methods for data collecting and storage, data processing and analytics, FN accessibility, data distribution, QoE/QoS, security, and privacy. It stands for a novel process framework for FDA communication where heterogeneous multi-modal data from multiple IoT devices is gathered. Preprocessed data are sent for real-time scrutiny, after which they are processed or stored in accordance with their needs. The data intelligence on the processed data then happens (Cao et al. (2015).

15.5.1 COLLECTION AND STORAGE OF DATA

Massive amounts of real-world data are now being collected far more frequently because of the development of the Internet, the Internet of Things, and social media. In the past, there was a social sensing system where respondents submitted information in real time and acted as sources. The additional information that these social sensors generated needed careful pre-processing.

Either a distributed or centralized architecture can be used with fog or a combination of the two. Every fog node in a centralized design works under a central node. The management of a centralized fog becomes increasingly challenging as more linked devices are added. Fog technology communicates using P2P. A number of fog devices share the processing duties. Every fog node connects with one another for a variety of reasons (example: job distribution, self-organization, peer discovery, etc.).

The fog mediates three types of communication, which is listed as under:

- Machine-to-Machine: One "thing" generates data, which another "thing" consumes.
- Machine-to-People: Humans absorb the data created by "things" and vice versa.
- People-to-People: A human consumes data that a human has produced (Chen et al., 2016).

These encounters can last anywhere from a few seconds to several days. Real-time apps, for instance, have interactions that range from a few seconds to a few minutes. The transactional analytics, however, could take many days. As a result, the fog design may handle many storage types. It should have temporary memories for real-time interactions, and semi-permanent storage is needed for interactions that last longer (Gia et al., 2015).

15.6   DATA PROCESSING ARCHITECTURE IN THE FOG

IoT produces a ton of data. The cloud might not be the ideal IoT solution for maintaining the standard of real-time data processing. It is obvious that the IoT requires a different type of architecture in order to meet its latency and mobility problems. In this way, IoT data processing often fits the fog computing architecture.

- Segregation: Multiple IoT applications share the same resources because fog supports a multi-tenancy approach. Segregating entails clearly identifying and separating the data that are exclusive to each application (Dubey et al., 2015).
- Aggregation: Aggregation is the process of accumulating the same application data over a period of time in order to gain insight from the data.
- Data Encryption: Plain raw data acquired from the sensor/"things" is encrypted to ensure privacy and security requirements.

15.6.1   CRITERIA USED IN OFFLOADING

Offloading can be defined as the transfer of the tasks that are using up the device resources onto an external platform such as a grid or cloud (Akherfi et al., 2018a). Criteria which is used in taking decision regarding offloading is dependent on:

1. **Server specifications:** It includes speed of the processor, memory availability, and storage availability (Akherfi et al., 2018b).
2. **Network Specifications:** It includes which network one is using, like 3G, Wi-Fi, Wimax, and how much bandwidth is available.
3. **Application Specifications:** It includes execution time. The offloading decision is also dependent on user preferences and data confidentiality.

15.6.2   RESOURCE CONSTRAINT

Resource constraint refers to limited availability of input so there is difficulty in completing a particular task. When we discuss offloading, then the environment is resource-constrained. Resource-constraint in offloading environments includes data encryption, privacy preservation, vulnerabilities, threats, and attacks.

Data encryption means data will be encrypted or coded in such a form that people who have access to a secret key can only access it. But if the data is transmitted in the form of plaintext on Internet, then there can be security issues and will lead to eavesdropping. Privacy preservation means data transmitted between two parties should be secure or data can be leaked, which will interfere with offloading. Vulnerabilities mean weaknesses in the system that will breach information security and related information. Threats can damage the computer system, resources, etc. Attacks are also a kind of information security threat that can cause potential damage to files, data resources, and can obtain or reveal information by unauthorized access.

15.6.3 Latency Requirement

Latency can be defined as the delay between a client request and the response given by a cloud service provider. Multiple factors like architecture, topology, network traffic, hop count, etc. of the network determines the varying values of latency (Moore et al., 2017). As one moves from cloud computing technology to edge/fog computing, latency values drops substantially. The delay, which was in milliseconds, reduced to sub-seconds when the technology was shifted from cloud to edge. It been depicted in Figure 2.2, where fog nodes are analyzing the data and cloud servers are handling business data.

15.6.4 Load Balancing

Load balancing refers to the traffic distribution to different nodes so as to speed up the processing and throughput will be enhanced. Earlier, data from all the IoT devices was handled by the cloud server. There was a delay in response time when a huge amount of data was transferred from devices to the cloud server, processing was taking place at servers, and the result was transferred back to the devices. So, there was a need to balance the load at servers because a delay in response time is not applicable in the healthcare industry. Load balancing algorithms are being used at fog nodes, which will help in managing the network traffic. Results will be delivered faster as the traffic will be diverted to multiple fog nodes that are placed between devices and cloud servers. The data will be processed at these fog nodes and compiled data will be transferred to cloud servers, which will now process the data faster. But, choosing the right load balancing algorithm will be an issue. Thus, a load balancing algorithm should be chosen appropriately.

15.7 IOT MIDDLEWARE TECHNOLOGY

15.7.1 Cloudlet

Cloudlet refers to a group of computers that provides cloud services to users over a network who are located in close proximity. Cloudlet uses local area network and differs from cloud computing in many ways. Cloud computing services are provided by public cloud, whereas cloudlet services are self-managed or employed by localized users or businesses. Cloudlet has been designed basically to support mobile devices, gadgets like wearables, and smartphones. Thus, a cloudlet acts as a middle layer between cloud platforms and IoT applications (Bangui et al., 2018).

15.7.2 Micro Data Center

A micro data center is defined as a secure technology that provides infrastructure for UPS, servers, etc., and supports various software devices. Micro data centers will compute the data locally, process it at the local data center, and results will be displayed faster. This is done to achieve optimized results.

15.7.3 NANO DATA CENTER

NaDa (nano data center) is a concept that is based on distributed computing. These data centers are basically designed to reduce carbon footprints and they consist of small-sized interconnected data centers that are placed directly in offices or buildings. They also compute the data locally and it is a decentralized approach that provides low-latency cloud computing to its users.

15.7.4 DELAY TOLERANT NETWORK

Delay or disruption tolerant network (DTN) is a mechanism related to computer networks where transmission of data from source to destination is done without interruption. When we talk of mobile data communication, terrestrial environments where data communication is difficult because of different operating systems and protocols, then we use DTN, as it is used to address issues where Internet connectivity is discontinuous. Thus, it addresses the issues of heterogeneous networks and works on store and forward policy.

In cloud computing or IoT applications, where different devices connect and have compatibility issues, then to obtain efficient transmission in DTN a new protocol has been devised. This protocol is named PROPHET (Scheduling-Probabilistic Routing Protocol using History of Encounters and Transitivity) (Mao et al., 2019).

15.8 CHALLENGES IN FOG COMPUTING

The various challenges being faced in fog computing are as follows.

15.8.1 NETWORK MANAGEMENT

There are too many devices and billions of active networks, making it difficult for network management to control the network using hardware and necessitating "Softwarization" (Vaquero & Rodero-Merino, 2014). Fog on the heterogeneous systems is homogeneously simple to implement thanks to automated scripts and virtual machines.

- Resource Management: Managing the fog architecture's resources, including the extent of virtualization as well as preventing device deadlocks, is a major challenge as demand for the system's resources will be strong, so a good resource management strategy must be maintained (Vyas & Gupta, 2022).
- Resource Allocation: Resource allocation in the fog architecture is a challenging problem due to the fact that the devices involved are heterogeneous and managed by virtualization rates, making it difficult to assign specific resources. Therefore, resource distribution must be analyzed as it can cause the entire system to cease.

- Task Timing: An integral component of every distributed system is task scheduling. As a result, the device may operate a range of virtual machines across a distributed fog architecture with the least amount of conflict possible.
- Job Offloading: The issues of task offloading, i.e., the architecture's understanding of the workflow and support for reassigning the job to another fog node, in the case of failure, arise because the fog architecture facilitates the use of the distributed computing architecture. For fault tolerance, it should also ensure effective process replication (Ahmad et al., 2016; Stojmenovic, 2014).
- Inventory Issues: The supply of fog capital is not very large. Proper data backup (to the cloud or local data centers) policy is required to tackle the continuous flow of data streaming (Vyas, 2023).

15.9 CONCLUSION

In this book chapter, we have discussed a software strategy and the challenges faced during the management of medical records. Telemedicine generates a lot of data and cloud computing, and, along with fog computing, offers the features of availability and performance by storing the data close to the applications and devices. The data analytics interface has a wide range of applications in smart cities, infrastructure, satellite imaging, intelligent transport network, and smart grid applications. Fog data analytics is the best way to develop the full benefits of these devices in modern society. As the survey results indicated, efficient and effective data storage and data processing algorithms are needed for adequate analytics as we move forward with this groundbreaking technology.

REFERENCES

Ahmad, M., Amin, M.B., Hussain, S., Kang, B.H., Cheong, T., & Lee, S. (2016). Health fog: A novel framework for health and wellness applications. *The Journal of Supercomputing*, 72(10), 3677–3695.

Akherfi, K., Gerndt, M., & Harroud, H. (January 2018a). Mobile cloud computing for computation offloading: Issues and challenges. *Applied Computing and Informatics*, 14(1), 1–16

Akherfi, K., Gerndt, M., & Harroud, H. (January 2018b). Mobile cloud computing for computational offloading: Issues and challenges. *Applied Computing and Informatics*, 14(1), 1–16.

Amin, A., Riyaz, S., Ali, A., & Paul, Z. (2017). Review of IoT data analytics using big data, fog computing and data mining. *International Journal of Computer Science and Mobile Computing*, 6, 33–39.

Atlam, H.F., Alenzei, A., Alharthi, A., & Walters, R.J. (2017). Internet of Things: Integration of Cloud Computing with Challenges and Open Issues. *IEEE Conference on iThings*.

Atlam, H.F., Walters, R.J., & Wills, G.B. (2018). Fog computing and the Internet of Things: A review. *Big Data and Cognitive Computing*, 2(2), 10.

Bangui, H., Rakrak, S., Raghay, S., & Buhnova, B. (2018). *Moving to the Edge-Cloud-of-Things: Recent Advances and Future Research Directions*. MDPI.

Bonomi, F., Milito, R. et al. (2012). Fog Computing and Its Role in the Internet of Things. In *Proceedings of MCC'12*, Helsinki, Finland.

Bonomi, F., Milito, R., Natarajan, P., & Zhu, J. (2014). Fog computing: A platform for Internet of Things and analytics. In *Big data and internet of things: A roadmap for smart environments*, (pp. 169–186). Springer.

Cao, Y., Hou, P., Brown, D., Wang, J., & Chen, S. (2015). Distributed Analytics and Edge Intelligence: Pervasive Health Monitoring at the Era of Fog Computing. In *ACM Proceedings of the Workshop on Mobile Big Data* (pp. 43–48).

Cerina, L., Notargiacomo, S., Paccanit, M.G., & Santambrogio, M.D. (2017). A Fog-Computing Architecture for Preventive Healthcare and Assisted Living in Smart Ambients. In *Research and Technologies for Society and Industry (RTSI), 2017 IEEE 3rd International Forum on* (pp. 1–6).

Chen, N., Chen, Y., You, Y., Ling, H., Liang, P., & Zimmermann, R. (2016). Dynamic Urban Surveillance Video Stream Processing Using Fog Computing. In *Multimedia Big Data (BigMM), IEEE Second International Conference on* (pp. 105–112).

Chiang, M. (2015). Fog Networking: An Overview on Research Opportunities. http://www.princeton.edu/~chiangm/FogResearchOverview.pdf

Chiang, M., et al. (2016a). Fog and IoT: An overview of research opportunities. *IEEE Internet of Things Journal*, *3*(6), 854–864. 10.1109/JIOT.2016.2584538

Chiang, M., et al. (2016b). Fog and IoT: An Overview of Research Opportunities. *IEEE Internet of Things Journal*. 10.1109/JIOT.2016.2584538

Dubey, H., Yang, J., Constant, N., Amiri, A.M., Yang, Q., & Makodiya, K. (2015). Fog Data: Enhancing Telehealth Big Data Through Fog Computing. In *ACM Proceedings of the ASE BigData & Social Informatics 2015* (p. 14).

Gia, T. N., Jiang, M., Rahmani, A.-M., Westerlund, T., Mankodiya, K., Liljeberg, P., & Tenhunen, H. (2015). Fog computing in body sensor networks: An energy efficient approach. In *Proc. IEEE Int. Body Sensor Netw. Conf.(BSN)* (pp. 1–7).

He, J., Wei, J., Chen, K., Tang, Z., Zhou, Y., & Zhang, Y. (2018). Multitier fog computing with large-scale IoT data analytics for smart cities. *IEEE Internet of Things Journal*, *5*(2), 677–686.

Kumari, S. (2015, December 28). Agility on Cloud – A Vital Part of Cloud Computing. (Sysfore Blog) Retrieved July 2017, from http://blog.sysfore.com/agility-on-cloud-a-vital-part-of-cloud-computing/

Kumari, A., Tanwar, S., Tyagi, S., & Kumar, N. (2018). Fog computing for healthcare 4.0 environment: Opportunities and challenges. *Computers & Electrical Engineering*, *72*, 1–13.

Mahmud, R., Kotagiri, R., & Buyya, R. (2018). Fog Computing: A Taxonomy, Survey and Future Directions. *Internet of Everything: Algorithms, Methodologies, Technologies and Perspectives*, 103–130.

Mao, Y., Zhou. C., Ling, Y., & Lloret, J. (2019). *An optimized probabilistic delay tolerant network (DTN) routing protocol based on scheduling mechanism for Internet of Things (IoT)*. MDPI.

Mohammadi, M., Al-Fuqaha, A., Sorour, S., & Guizani, M. (2018). Deep Learn- ing For IoT Big Data and Streaming Analytics: A Survey. *IEEE Communications Surveys Tutorials*, 1–1.

Moore, A.W., Andreea Popescu, D., & Zilberman, N. (Nov. 2017). Characterizing the impact of network latency on cloud-based applications' performance. ISSN 1476-2986.

Stojmenovic, I. (2014). Fog Computing: A Cloud to the Ground Support for Smart Things and Machine-to-Machine Networks. In *Telecommunication Networks and Applications Conference (ATNAC), 2014*, Australasian (pp. 117–122).

Tang, B., Chen, Z., et al. (2015). A Hierarchical Distributed Fog Computing Architecture for Big Data Analysis in Smart Cities. In *ACM Proceedings of the ASE Big Data; Social Informatics*, New York, NY, USA (pp. 28:1–28:6).

Vora, J., Tanwar, S., Tyagi, S., Kumar, N., & Rodrigues, J.J.P.C. (2017). FAAL: Fog Computing-Based Patient Monitoring System for Ambient Assisted Living. In *19th IEEE International Conference on e-Health Networking, Applications and Services, Healthcom*, Dalian, China, (pp. 1–6).

Vora, J., Italiya, P., Tanwar, S., Tyagi, S., Kumar, N., Obaidat, M.S., & Hsiao, K. (2018a). Ensuring Privacy and Security in E-health Records. In *2018 International Conference on Computer, Information and Telecommunication Systems, CITS, Alsace*, Colmar, France (pp. 1–5).

Vora, J., DevMurari, P., Tanwar, S., Tyagi, S., Kumar, N., & Obai-dat, M.S. (2018b). Blind Signatures Based Secured E-Healthcare System. In *2018 International Conference on Computer, Information and Telecommunication Systems, CITS, Alsace*, Colmar, France (pp. 1–5).

Vyas, S. (2023). Extended reality and edge AI for healthcare 4.0: Systematic study. In *Extended Reality for Healthcare Systems* (pp. 229–240). Academic Press.

Vyas, S., & Gupta, S. (2022). Case study on state-of-the-art wellness and health tracker devices. In *Handbook of Research on Lifestyle Sustainability and Management Solutions Using AI, Big Data Analytics, and Visualization* (pp. 325–337). IGI Global.

Yi, S., Li, C., & Li, Q. (2015). A Survey of Fog Computing: Concepts, Applications and Issues. In *ACM Proceedings of the 2015 Workshop on Mobile Big Data* (pp. 37–42).

16 Gastric Cancer Diagnosis Using Machine Learning Techniques
A Survey

Danish Jamil, Sellappan Palaniappan,
Susama Bagchi, Sanjoy Kumar Debnath,
Chetna Monga, Shikha Mittal, Suvarna Sharma,
and Tejinder Kaur

16.1 INTRODUCTION

Gastric cancer is a significant global health concern, ranking as the fifth most prevalent cancer and the third primary cause of cancer-related mortality. It was predicted that there would be 1.03 million new cases and 782,685 deaths worldwide in 2020. There is a significant geographical disparity in the prevalence of stomach cancer, with the highest rates being seen in Eastern Asia and the lowest being shown in Northern America and Oceania (Jamil et al., 2022; Xia & Aadam, 2022). In recent years, the use of artificial intelligence (AI) and machine learning (ML) methods has shown improvement in both the diagnostic process and the prognosis for patients suffering from gastric cancer (Bagchi et al., 2020; Mohammad Reza Afrash & Kazemi-Arpanahi, 2022). Recent advances in AI and ML have demonstrated promising results and have the potential to revolutionize the healthcare system. These tools enable the rapid, precise analysis of massive data sets, allowing medical personnel to get actionable insights for better patient care. To guarantee the safe and ethical use of new technologies in healthcare, we will also stress the need to resolve these concerns (Jamil et al., 2022). Several studies have shown potential of ML in improving the accuracy of gastric cancer (GC) diagnosis and prognosis. One study used a deep learning algorithm to accurately predict the survival outcomes of GC patients (Conti et al., 2023). (Shao et al., 2021) developed a radionics-based ML model that achieved an accuracy of 85.7% in distinguishing between benign and malignant gastric tumors, outperforming traditional diagnostic methods. In addition, (Akcay et al., 2020) used a deep learning algorithm to predict the survival outcomes of GC patients, achieving an accuracy of 768.8% compared to conventional prognostic models. For instance, (Baradaran Rezaei et al., 2022) developed an ML model to

DOI: 10.1201/9781003346289-16

predict the response of gastric cancer patients to chemotherapy, which achieved an accuracy of 769.9%. In another study, (Bang et al., 2021) developed an ML model to predict the optimal surgical approach for gastric cancer patients, with an accuracy of 78.6%. ML is an innovative branch of AI that analyzes data to improve computer performance without explicit programming. In the medical field, ML has the potential to revolutionize diagnosis and treatment methods, leading to diagnostics that are more accurate, personalized treatments, and provide better patient outcomes (Sethi et al., 2021). Support vector machines (SVMs) and random forests have been used to analyze radiographic images for GC detection. ML algorithms can also accurately predict treatment response and patient outcomes based on clinical and genetic factors (Tabari et al., 2023). One major challenge is the availability and quality of data (UmaMaheswaran et al., 2022), especially for rare diseases like GC. Another challenge is the interpretability of ML models, which are often complex and difficult to understand (Sachdeva et al., 2022).

To address these challenges, efforts are being made to develop new approaches, such as liquid biopsy, which is a minimally invasive method of detecting cancer by analyzing tumor-specific DNA, RNA, and proteins in the blood. Further, ML models that are easier to understand should be created, such as decision trees and rule-based systems. Clear ethical and legal frameworks for the use of ML in healthcare are also required, including norms for data protection, bias, and responsibility (Nayak et al., 2021; Verma & Verma, 2021). Despite encouraging results, additional research is required to surmount the limitations and obstacles associated with the use of machine learning for diagnosing gastric cancer. To address these issues, potentially improve gastric cancer diagnosis and patient outcomes, and ultimately result in significant healthcare cost savings, it is crucial to employ robust data collection techniques and rigorous evaluation and validation of machine learning models before their clinical application. Nonetheless, still issues are associated with this method that must be addressed (Kavas et al., 2023; Shehab et al., 2022). In light of the promising performance of machine learning in enhancing GC diagnosis, it is essential to address the challenges associated with this method. This research aims to investigate how AI and ML techniques can be optimized for the early detection of GC. These are the objectives of this study:

- To examine the fundamental concepts underlying AI, ML, big data, and deep learning (DL) and their implications for healthcare.
- To investigate the relevant issues and problems for healthcare professionals and physicians when using AI for the diagnosis of gastric cancer.
- Evaluate the use of AI in the diagnosis and prognosis of gastric cancer.
- To emphasize the significance of early gastric cancer detection for effective treatment and improved patient outcomes.
- Using AI and ML techniques, investigate and develop more optimized approaches for the early diagnosis of gastric cancer (Vyas, S., 2023).

TABLE 16.1

Two Types of the Most Popular ML Approaches, and Their Benefits and Drawbacks

Method	Benefit	Drawback
Supervised Learning	There are notions of output that occur in the learning process. It performs classification and regression functions and improve the results.	A labeled data set is usually required in the initial phase which involves a training phase.
Unsupervised Learning	There are some methods where no training data is needed, and automatic labeling of the training data set saves a lot of time, that would have been wasted on manual classification.	In the learning process without notions of output, it does not provide for the calculation or analysis of new sample data. Also, outliers dramatically affect the results obtained.

16.1.1 OBSTACLES RELATING TO MACHINE LEARNING FOR GASTRIC CANCER DIAGNOSIS

Large and diverse data sets are required to accurately train and validate machine learning models for gastric cancer diagnosis (GS Pradeep et al., 2021). Additionally, there is a risk of bias in training data, which can lead to inaccurate or unreliable models. More rigorous validation methods, such as cross-validation or bootstrapping, can also be used to ensure that models are reliable and unbiased (Adlung et al., 2021). However, current diagnostic techniques for gastric cancer require years of medical practice and rigorous testing, making it difficult for doctors to assess its threat to patients (Gillen, 2021; Mortezagholi et al., 2019), as seen in Table 16.1. DM techniques can help pathologists to reduce their workload and decision-making process.

The paper highlights the future of AI and ML in clinical cancer research, particularly GC diagnosis and treatment. AI and ML have potential for gastric cancer diagnosis, but further research is needed to overcome limitations (Conti et al., 2023; Jamil, 2022).

16.2 TRADITIONAL METHODS VERSUS AI-BASED METHODS FOR GASTRIC CANCER DIAGNOSIS

AI has shown great promise in the early detection, diagnosis, and treatment of gastric cancer. In this paper, we will discuss the proper use of AI in GC and how it can be integrated into existing medical processes to improve early detection, accurate diagnosis, and personalized treatment. The knowledge discovery in database process (KDD) is a widely used process for discovering valuable insights from data. In the case of gastric cancer, KDD is used to process a large amount of medical data, including patient records, diagnostic reports, and medical images.

Improved early detection and diagnosis are possible with the help of AI-powered tools (Chandrakar & R. Saini, 2015; Gupta, S., et.al. 2023). AI-powered diagnostic systems analyze medical pictures and diagnostic data to improve the diagnostic precision and turnaround time. Differentiating between benign and malignant tumors and detecting tumors in their earliest stages falls under this category. AI-powered technologies tailor treatments to each patient's unique genetic mutations, age, and health status (Lysaght et al., 2019). This involves making decisions on which medical interventions, including surgery, chemotherapy, or radiation treatment, will provide the best chance of recovery. Knowledge-based systems, a kind of CDSS, have the potential to enhance the early identification and diagnosis of illnesses like GC, resulting in improved patient outcomes (Bhatia et al., 2021). With the ability to analyze and extract useful information from large amounts of data and integrate it with the medical knowledge of healthcare professionals, decision support systems (DSS) and CDSS are gaining popularity in the healthcare industry (Cohen, 2020). Endoscopy, biopsies, and histopathology reports are the tried-and-true procedures for diagnosing stomach cancer. There are limits to these approaches despite their widespread use and positive impact on stomach cancer detection and therapy. In contrast, AI-based approaches to diagnosing stomach cancer may be able to get around some of these restrictions. Another study developed a deep learning model to predict the risk of lymph node metastasis in GC and achieved an accuracy of 87.5% (Bhatia et al., 2021). Overall, AI-based methods for GC diagnosis have shown promising results and have the potential to improve the accuracy and efficiency of diagnosis, leading to better patient outcomes.

The process described in Figure 16.1 shows how to use a clinical decision support system (CDSS) that analyzes patient data using AI to help healthcare practitioners make better patient care choices. To analyze user data, the CDSS uses personal data, medical testing, family history, environmental variables, and domain expertise. The user – a healthcare practitioner or patient – enters clinical data into the CDSS. Clinical data includes test findings, medical pictures, and patient history. The CDSS analyzes data using AI and domain knowledge after input. The CDSS logic may incorporate illness information, treatment alternatives, and patient care best practices. Based on the patient's medical circumstances, the CDSS provides more accurate and relevant information. Finally, the CDSS analyzes clinical data and incorporates domain knowledge to provide risk, treatment, and follow-up information. AI and CDSS may enhance patient outcomes by providing healthcare practitioners with more accurate and relevant information to make treatment choices (Iqbal et al., 2021; Vyas, S., et al., 2023). As shown in Figure 16.2, machine learning and AI determine stomach cancer. Patient records are collected and cleaned for correctness and dependability. GC is determined by selecting data attributes. A machine learning system classifies stomach cancer based on these criteria. An AI-based decision support system analyzes the machine learning algorithm output to help clinicians make correct diagnoses. Finally, deep learning analyzes medical photos to diagnose stomach cancer. This technique accurately diagnoses stomach cancer, allowing physicians to deliver the best treatment (Kelly et al., 2019). As seen in Table 16.2, it shows the benefits and drawbacks of machine learning algorithms.

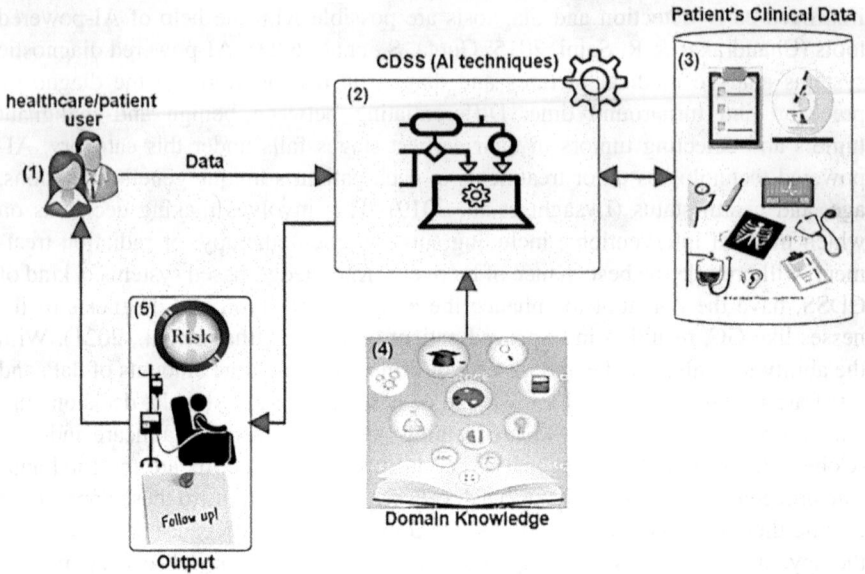

FIGURE 16.1   CDSS process in the development of gastric cancer (Iqbal et al., 2021).

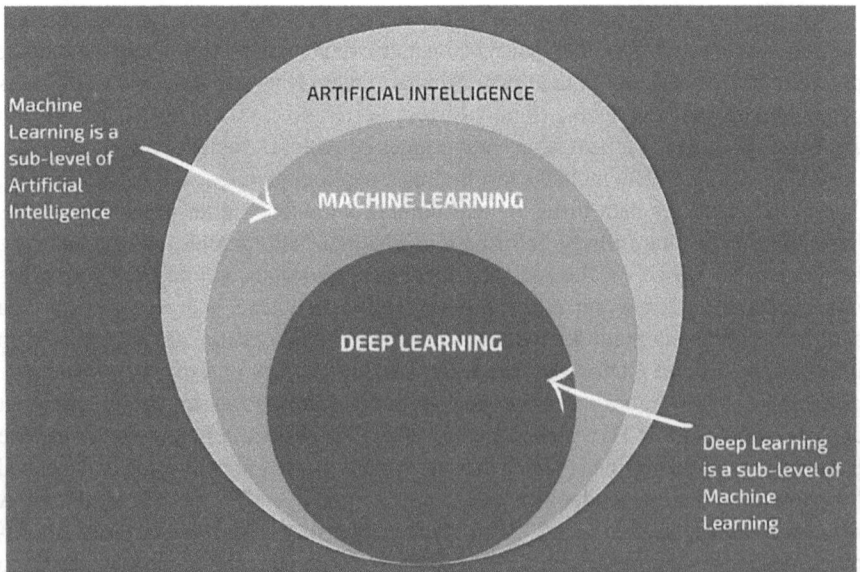

FIGURE 16.2   Integration of artificial intelligence, machine learning, and deep learning in GC diagnosis (Kelly et al., 2019).

TABLE 16.2
Benefits and Drawbacks of Described Algorithms

Method	Benefit	Drawback
KNN	This is an incredibly simple method to implement. It has a low computing cost because of the training phase.	Storage space is needed for complex projects. Sensitivity to noisy databases.
Naïve Bayes Bayesian Networks	This method is straightforward to execute. The method is significantly more accurate in high-dimensional data sets.	Whenever factors are interdependent, low accuracy is obtained.
Logistic Regression	Better accuracy vs other classifiers. It is easy to determine the underlying relationship between dependent and independent variables.	When outliers are present, the results vary significantly. Effectiveness of the classifier is dependent on the kind of data set, which renders it unpredictable.
Neural Network	The identifying of meaningful interconnections between dependent and independent variables is very straightforward.	There is a high likelihood of local minima. There is a significant likelihood of overfitting problems.

16.3 ROLE OF GASTRIC CANCER IN ML-BASED, KNOWLEDGE-BASED, AND MEDICAL DECISION SUPPORT SYSTEMS

ML-based and knowledge-based decision support systems have the potential to greatly improve the accuracy and speed of GC diagnosis. ML-based DSS can analyze large amounts of data and identify patterns in the data that may not be apparent to human diagnosticians, improving diagnostic accuracy and reducing the workload of pathologists. On the other hand, knowledge-based DSS can incorporate expert knowledge and guidelines to assist medical practitioners in their decision-making process, ensuring consistency in diagnosis and treatment (Shanbehzadeh et al., 2022). Both approaches have their advantages and limitations. While ML-based DSS can learn and adapt to new data, it may not be able to explain its decision-making process, which could raise concerns among medical practitioners. On the other hand, knowledge-based DSS may require constant updates and manual input to keep up with new developments in the field (Trowbridge & Weingarten, 2001). A hybrid decision support system that combines both approaches could potentially overcome the limitations of each and provide a more accurate and efficient diagnosis of GC. One potential research gap in the diagnosis of GC could be the development of such a hybrid system. Additionally, further research could be conducted to evaluate the reliability and effectiveness of these systems in clinical settings and to determine the optimal combination of expert knowledge and machine learning algorithms. Table 16.3 lists some pros and cons of ML-based and knowledge-based DSS. To overcome these challenges, researchers can explore new data acquisition and standardization methods, collaborate with medical

TABLE 16.3

Comparative Analysis of ML and Knowledge-Based DSS in the Diagnosis of GC

Decision Support System	Strengths	Limitations	Diagnosis of Gastric Cancer
ML-based DSS	High accuracy, speed in data analysis, pattern recognition, ability to handle large and complex data sets.	Requires high-quality larger data set for training and may be prone to overfitting.	Can assist pathologists in reducing workload while increasing diagnostic accuracy and aid medical practitioners in their decision-making process.
Knowledge-based DSS	Incorporates expert knowledge and domain-specific rules into decision-making processes.	Limited by quality and availability of expert knowledge and rules- limited ability for new adaptation.	Can assist in providing personalized treatment recommendations based on patient data and expert knowledge practitioners in their decision making.
Medical DSS	Can provide a wide range of clinical decision-making support.	Limited by quality and availability of the medical data.	Can provide clinical decision-making support for the GC diagnosis.

professionals to ensure data accuracy and relevance, and develop transparent and explainable models to increase practitioner acceptance. Further research is needed to optimize these systems and ensure clinical efficacy.

16.4 RELATED WORK

This research aims to analyze various papers focusing on early GC detection and proposed solutions. Although the studies differ in their objectives, they all emphasize the importance of accurately diagnosing this disease at an early stage. AI is transforming the diagnosis and treatment of diseases, including cancer detection. The use of AI in medical applications offers many benefits, including improved interpretation of medical images, targeted interventions, optimized drug usage, and personalized treatment. This research seeks to develop a framework for predicting GC (Unhelker et al., 2022). Recent advances in computational approaches have demonstrated promising potential in multiple aspects of clinical cancer research, including improving the accuracy of predictive forecasts and disease staging. Machine learning and deep learning are emerging subfields of AI that hold great potential in histological image analysis, specifically in endoscopic image analysis-based computer-aided cancer detection (Unhelker et al., 2022). The literature on the GC diagnosis suggests that several gaps need to be addressed. One gap is the limited accuracy of traditional diagnostic methods, such as endoscopy and biopsy, which can lead to missed diagnoses or delayed treatment. There is also a lack of uniformity in diagnostic criteria and reporting, which make it difficult to compare findings from different research and institutions. In addition, more effective risk stratification and screening methods are required to identify those at high risk so that they may get therapy as soon as possible. Improving the precision of diagnosis and informing individualized treatment strategies requires the creation of bio-markers and genetic testing. Another gap is the limited access to healthcare resources, particularly in low- and middle-income countries, which can result in disparities in cancer diagnosis and treatment outcomes. There is a need to develop cost-effective and scalable diagnostic tools and strategies that can be used in resource-limited settings. Overall, the literature suggests that several gaps in the GC diagnosis need to be addressed through further research and innovation. By improving the accuracy, standardization, and accessibility of diagnostic tools and strategies, we can improve outcomes for GC patients and reduce the global burden of this disease.

16.5   LITERATURE REVIEW

In the paper by (Tama et al., 2020), researchers built an ensemble model to help in the diagnosis of cardiovascular problems in their study. The model was developed using diverse classification techniques, including random forest, gradient boosting, and extreme gradient boosting. In another study, (Paragliola & Coronato, 2021) devised a hybrid model that uses ECG data to predict the likelihood of cardio-vascular problems in hypertensive patients. Combination of short-term memory and convolutional neural network to predict hypertension. In another study, authors (Terrada et al., 2020) developed a method that combines ANN, AdaBoost, and decision tree algorithms to diagnose cardiovascular illness. (Yasar et al., 2019) developed a method that combines two SVM models for predicting cardiac arrest. Separate models handled both feature selection and prediction. SVM hyperpara-meter tweaking was used to obtain the best results for both models. (Mohan et al., 2019) developed a system that predicts when a patient's cardiac disease will manifest using random forest and linear approaches. In the study (Casal-Guisande et al., 2022), an intelligent clinical decision support system was developed to prevent breast cancer detection. ML algorithms were applied to public data sets to compare metrics. This study (Casal-Guisande et al., 2022) developed a soft computing-based medical decision support system using fuzzy cognitive mapping (FCM) to help doctors select the most effective course of treatment for each patient. In a study by (Mohammed et al., 2022), data mining methods and disease-related features were used to accurately predict and diagnose stomach cancer, with the SVM algorithm producing the highest quality classifications. (Mohammed et al., 2022) proposed a deep-learning automated approach for classifying brain tumors into four subtypes: T1W, T1CE, T2W, and Flair. The experimental results showed that the proposed technique significantly outperformed the baseline on the selected data sets, but it simplified several crucial characteristics that affect the system's reliability. Future experiments will be undertaken using more up-to-date deep learning techniques and the BRATS 2019 data sets, as in Table 16.4, which

TABLE 16.4

ML-Based Decision Support Systems for Cancer Diagnosis and Prognosis

Title	Methodology	Main Findings
Ensemble model for detecting cardiac conditions	Random forest, gradient boosting, extreme gradient boosting	Successful in detecting cardiac conditions
Hybrid model for predicting cardiovascular problems in hypertensive individuals	Extended short-term memory network, convolutional neural network	Effective in predicting likelihood of cardiovascular problems
Method for diagnosing cardiovascular illness	ANN, AdaBoost, decision tree	Successful in diagnosing cardiovascular illness
Method for predicting cardiac arrest	SVM	Effective in predicting cardiac arrest
System for forecasting manifestation of cardiac disease	Random forest, linear approaches	Successful in forecasting manifestation of cardiac disease
Clinical decision support system for preventative detection of breast cancer	Machine-learning techniques for binary classification	Successful in preventative detection of breast cancer
Fuzzy cognitive mapping-based medical decision support system	Fuzzy cognitive mapping	Effective in aiding doctors in selecting the most effective course of treatment
Data mining for accurate prediction and diagnosis of stomach cancer	SVM	SVM produced the highest quality classifications in test samples
Deep learning automated approach for classifying	Deep learning	Effective in classifying brain tumors into subtypes

summarizes ML-based decision support systems for cancer diagnosis and prognosis. Even with all the progress that has been made, there are still some research gaps that need to be filled if we want to see the DSS's performance and impact on cancer care. Some of the identified gaps include the need for more comprehensive and diverse data sets, the development of more accurate and robust models, and the consideration of ethical and legal issues associated.

Future research should focus on improving the performance and applicability of decision support systems (DSS) in the cancer domain. Advanced machine learning techniques, such as deep learning, reinforcement learning, and transfer learning, can be used to enhance the accuracy and robustness of DSS models. Additionally, incorporating additional types of data such as genomic and proteomic data can improve the predictive power of DSS. Finally, yet importantly, user-centered design concepts may be used to improve DSS interfaces for healthcare practitioners. Decision support systems (DSSs) based on machine learning have seen much research and application in the fields of cancer diagnosis, prognosis, treatment planning, and follow-up. These studies demonstrate the use of various machine learning techniques such as random forest, gradient boosting, extreme gradient boosting, extended short-term memory network, convolutional neural network, AdaBoost, decision tree, and fuzzy cognitive mapping. These techniques have been applied to data sets of various sizes and formats to address different cancer-related problems.

16.6 RESULTS AND DISCUSSION

The use of AI in the diagnosis and treatment of cancer has enormous potential to revolutionize cancer care. Deep learning models can identify patterns and make predictions based on large data sets, enabling them to assist clinicians in making more informed decisions regarding cancer treatment (Casamayor et al., 2018; Paragliola & Coronato, 2021). Several studies have evaluated the efficacy of deep learning methods in characterizing stomach cancer H- and E-stained histopathology slides (Das & Dash, 2022). The findings of these studies have demonstrated that convolutional neural network designs outperform traditional image examination techniques that involve the manual calculation of handcrafted features (Bajaj & Arora, 2013). While previous research has shown promising results in using machine learning algorithms to predict and diagnose GC, further research is needed to fully realize the potential of ML techniques in early detection and to ensure that ML algorithms are unbiased, transparent, and accessible to healthcare providers. Addressing these areas of research can lead to significant improvements in patient outcomes and revolutionize cancer care. As seen in Table 16.5, which summarizes ML and DL techniques used in GC that need more research, current AI research has made significant achievements, highlighting promising outcomes. Deep learning models have demonstrated superior outcomes in terms of accuracy, sensitivity, and specificity, with various models achieving AUROC values ranging from 0.741 to 0.9877. AI in healthcare has the potential to improve patient outcomes, but ongoing research is needed to refine techniques and optimize performance.

TABLE 16.5
Various ML and DL techniques used in the gastric cancer diagnosis

Challenges	Techniques	Accuracy
We need to fix the overfitting issue and enhance the current data collection	A neural network with a high degree of connectivity	Accuracy = 0.9679, Recall = 0.9492 F1score = 0.9470
This may be intermediate between intraepithelial neoplasia and normal gastric mucosa depending on glandular context and the degree of SHG.	Nonlinear optical microscopy with several probe modes	Neoplasia inside the epithelium is 595.64 m Metaplasia of the intestinal tissue is 308.30
They have merely given their findings based on a review of the literature.	Decision tree, artificial neural network	By comparing ANN to a decision tree, we get a value of 0.912.

(Continued)

TABLE 16.5 (Continued)
Various ML and DL techniques used in the gastric cancer diagnosis

Challenges	Techniques	Accuracy
The suggested method requires extensive, high-volume picture data sets for training.	In-depth convolutional AlexNet	Accuracy for cancer classification = 0.6990 Accuracy for necrosis detection = 0.8144
Regardless of the presence or absence of lymph node metastases, the EGC-defined agreement is that at the time of diagnosis, the cancer is confined to the submucosa and thermucosa.	Systematic approach based on the Delphi technique	Accuracy = 0.9718 Specificity = 0.9744 Sensitivity = 0.9688 F1 Score = 0.91
Fourteen cases couldn't be predicted using the sigmoid kernel-based SVM.	High-Performance Liquid Chromatography-Mass Spectrometry, and a Linear-SVM	Accuracy = 0.9718 Specificity = 0.9744 Sensitivity = 0.9688 F1
The TCGA data set had not been approved at that point.	Naive Bayesian, Correlation Coefficient of Pearson	AUC = 0.741 F1 Accuracy = 0.82 Specificity = 0.749 Sensitivity = 0.732
A noninvasive device capable of predicting LN metastasis in GC development is required for this technique.	Random forest algorithm in a Radiomics Model	AUC = 0.837 Accuracy = 0.931 Specificity = 0.881 Sensitivity = 0.921
Convolutional Neural Network, Computer Aided Design System, Deep Learning	High success rates are likely attributable to the fact that almost none of the EGC photos.	Accuracy = 0.9877 Sensitivity = 0.98 Specificity = 1.0 AUC = 0.875 F1 PPV = 1.0

16.7 MAJOR OBSTACLES TO ARTIFICIAL INTELLIGENCE IN GASTRIC CANCER DIAGNOSIS

Machine learning (ML) has gained widespread adoption in various industries, including healthcare, leading to a growing need to understand its complexities, such as dealing with irrelevant features, uncertainty, computational complexity, dynamic nature, and computational time (Bagchi et al., 2017). Potential areas of research in healthcare include personalized treatment, addressing data loss during pre-processing, collecting clinical data for scientific purposes, automation for junior expert users (Fawad Salam Khan et al., 2020). In the case of GC diagnosis, some of the challenges using ML techniques include:

  i. Identifying relevant features that accurately predict the presence of GC
  ii. Dealing with uncertainty and missing data in the data set

iii. Balancing the trade-off between computational complexity and accuracy

iv. Addressing the computational time required for the prediction.

This paper elaborates on the use of ML techniques for GC diagnosis and treatment highlighting the potential impact of knowledge discovery through ML on medical practitioners' workload and time, as well as the importance of nutritional status for GC patients. At the same time, it acknowledges the need for further research to fully understand its impact on medical practitioners and patients.

16.8 FUTURE RESEARCH AND DEVELOPMENTAL CHALLENGES

Hybrid systems that integrate knowledge-based and machine learning techniques are the focus of future research in the field of decision-support systems for stomach cancer detection. In addition, imaging and genetic data, among others, might be investigated to enhance diagnostic precision and facilitate early diagnosis (Shim et al., 2002). Ethical issues, such as patient privacy and informed consent, should be at the forefront of future research into the development and use of decision support systems in clinical settings for the detection of stomach cancer (Loeb, 2021). Future research in these areas may ultimately lead to more reliable GC diagnosis, earlier detection of the disease, and improved patient outcomes.

16.9 CONCLUSION

Successful treatment and better patient outcomes are dependent on the early identification and correct diagnosis of stomach cancer. Clinical investigation of AI-based approaches such as machine learning, deep learning, and data mining are warranted because of their significance to aid medical professionals and lighten the burden of pathologists. Establishing clear norms and best practices are crucial for ensuring the ethical and responsible usage of AI and big data analytics. Predictive models for early gastric cancer detection and subtype classification have been developed using AI-based approaches, with promising results including increased diagnosis accuracy and decreased pathologist's effort. Application of ethical best practices, overcoming difficulties, and prioritization of continuous research and developments are required in transforming the way stomach cancer and other fatal illnesses are diagnosed and treated. There is a great potential for AI and big data analytics to revolutionize the healthcare sector by creating better diagnostic tools to dramatically enhance patient outcomes and completely transform the medical industry.

REFERENCES

Adlung, L., Cohen, Y., Mor, U., & Elinav, E. (2021). Machine learning in clinical decision making. *Med*, 2(6), 642–665.

Afrash, M.R., Shanbehzadeh, M., & Kazemi-Arpanahi, H. (2022). Design and development of an intelligent system for predicting 5-year survival in gastric cancer. *Clinical Medicine Insights: Oncology*, 16(1), 1–13. 10.1177/11795549221116833

Akcay, M., Etiz, D., & Celik, O. (2020). Prediction of survival and recurrence patterns by machine learning in gastric cancer cases undergoing radiation therapy and chemotherapy. *Advances in Radiation Oncology*, *5*(6), 1179–1187.

Bagchi, S., Huong, A., & Tay, K. G. (2017). Investigation of different spatial filters performance toward mammogram de-noising. *Int. J. Integr. Eng*, *9*(3), 49–53.

Bagchi, S., Mohd, M. N. H., Debnath, S. K., Nafea, M., Suriani, N. S., & Nizam, Y. (2020, September). Performance Comparison of Pre-trained Residual Networks for Classification of the Whole Mammograms with Smaller Dataset. In *2020 IEEE Student Conference on Research and Development (SCOReD)* (pp. 368–373). IEEE.

Bajaj, K., & Arora, A. (2013). Dimension reduction in intrusion detection features using discriminative machine learning approach. *International Journal of Computer Science Issues (IJCSI)*, *10*(4), 324.

Bang, C. S., Ahn, J. Y., Kim, J.-H., Kim, Y.-I., Choi, I. J., & Shin, W. G. (2021). Establishing machine learning models to predict curative resection in early gastric cancer with undifferentiated histology: development and usability study. *Journal of Medical Internet Research*, *23*(4), e25053.

Baradaran Rezaei, H., Amjadian, A., Sebt, M. V., Askari, R., & Gharaei, A. (2022). An ensemble method of the machine learning to prognosticate the gastric cancer. *Annals of Operations Research*, 1–42.

Bhatia, S., Dubey, A. K., Chhikara, R., Chaudhary, P., & Kumar, A. (2021). *Intelligent healthcare: Applications of AI in ehealth*. Springer International Publishing. https://books.google.com.pk/books?id=M7I2EAAAQBAJ

Casal-Guisande, M., Comesaña-Campos, A., Dutra, I., Cerqueiro-Pequeño, J., & Bouza-Rodr\'\iguez, J.-B. (2022). Design and development of an intelligent clinical decision support system applied to the evaluation of breast cancer risk. *Journal of Personalized Medicine*, *12*(2), 169.

Casamayor, M., Morlock, R., Maeda, H., & Ajani, J. (2018). Targeted literature review of the global burden of gastric cancer. *Ecancermedicalscience*, *12*.

Chandrakar, O., & Saini, J.R. (2015). Predicting Examination Results using Association Rule Mining. *International Journal of Computer Applications*. 10.5120/20298-2330

Cohen, S. (2020). *Artificial intelligence and deep learning in pathology*. Elsevier Health Sciences. https://books.google.com.pk/books?id=CdPoDwAAQBAJ

Conti, C. B., Agnesi, S., Scaravaglio, M., Masseria, P., Dinelli, M. E., Oldani, M., & Uggeri, F. (2023). Early gastric cancer: Update on prevention, diagnosis and treatment. *International Journal of Environmental Research and Public Health*, *20*(3), 2149.

Das, M., & Dash, R. (2022). A comparative study on performance of classification algorithms for breast cancer data set using WEKA tool. In *Intelligent Systems* (pp. 289–297). Springer.

Ghantasala, G. P., Kumari, N. V., & Patan, R. (2021). Cancer prediction and diagnosis hinged on HCML in IOMT environment. In *Machine learning and the internet of medical things in healthcare* (pp. 179–207). Academic Press.

Gillen, S. (2021). Advancing early gastric cancer detection. *FEBS Open Bio*, *11*(7), 1812–1813.

Gupta, S., Sharma, H.K., Kapoor, M. (2023). Digital medical records (DMR) security and privacy challenges in smart healthcare system. In *Blockchain for Secure Healthcare Using Internet of Medical Things (IoMT)*. Cham: Springer. 10.1007/978-3-031-18896-1_6

Iqbal, M. J., Javed, Z., Sadia, H., Qureshi, I. A., Irshad, A., Ahmed, R., Malik, K., Raza, S., Abbas, A., Pezzani, R., & others. (2021). Clinical applications of artificial intelligence and machine learning in cancer diagnosis: Looking into the future. *Cancer Cell International*, *21*(1), 1–11.

Jamil, D. (2022). Diagnosis of gastric cancer using machine learning techniques in healthcare sector: A survey. *Informatica*, *45*. 10.31449/inf.v45i7.3633

Jamil, D., Palaniappan, S., Lokman, A., Naseem, M., & Zia, S.S. (2022). Diagnosis of Gastric Cancer Using Machine Learning Techniques in Healthcare Sector: A Survey. *Informatica*.

Jamil, D., Palaniappan, S., Zia, S. S., Lokman, A., & Naseem, M. (2022). Reducing the risk of gastric cancer through proper nutrition: A meta-analysis. *International Journal of Online & Biomedical Engineering*, *18*(7).

Jamil, D., Palaniappan, S., Debnath, S. K., Naseem, M., Bagchi, S., & Lokman, A. (2023). Prediction model for gastric cancer via class balancing techniques. *International Journal of Computer Science and Network Security*, *23*(01), 53–63. https://doi.org/ http://paper.ijcsns.org/07_book/202301/20230108.pdf

Kavas, P. Ö., Bozkurt, M. R., Kocayiğit, İ, & Bilgin, C. (2023). Machine learning-based medical decision support system for diagnosing HFpEF and HFrEF using PPG. *Biomedical Signal Processing and Control*, *79*, 104164.

Kelly, C. J., Karthikesalingam, A., Suleyman, M., Corrado, G., & King, D. (2019). Key challenges for delivering clinical impact with artificial intelligence. *BMC Medicine*, *17*(1), 1–9.

Khan, F. S., Mohd, M. N. H., Khan, M. D., & Bagchi, S. (2020, September). Breast cancer histological images nuclei segmentation using mask regional convolutional neural network. In *2020 IEEE Student Conference on Research and Development (SCOReD)* (pp. 1–6). IEEE.

Loeb, G. E. (2021). A new approach to medical diagnostic decision support. *Journal of Biomedical Informatics*, *116*, 103723.

Lysaght, T., Lim, H.Y., Xafis, V. et al. (2019). AI-assisted decision-making in healthcare. *Asian Bioethics Review*, *11*, 299–314. 10.1007/s41649-019-00096-0

Mohammed, M. A., Ghani, M. K. A., Arunkumar, N., Hamed, R. I., Mostafa, S. A., Abdullah, M. K., & Burhanuddin, M. A. (2022). Retraction Note: Decision support system for nasopharyngeal carcinoma discrimination from endoscopic images using artificial neural network. *The Journal of Supercomputing*, 1–2.

Mohan, S., Thirumalai, C., & Srivastava, G. (2019). Effective heart disease prediction using hybrid machine learning techniques. *IEEE Access*, *7*, 81542–81554.

Montserrat, C., Morlock, R., Maeda, H., & Jaffer, A. (2018). Targeted literature review of the global burden of gastric cancer. *Ecancermedicalscience*, *12*.

Mortezagholi, A., Khosravizadehorcid, O., Menhaj, M. B., Shafigh, Y., & Kalhor, R. (2019). Make intelligent of gastric cancer diagnosis error in Qazvin's medical centers: Using data mining method. *Asian Pacific Journal of Cancer Prevention*, *20*(9), 2607–2610. 10.31557/APJCP.2019.20.9.2607

Nayak, J., Favorskaya, M. N., Jain, S., Naik, B., & Mishra, M. (2021). *Advanced machine learning approaches in cancer prognosis: Challenges and applications*. Springer International Publishing. https://books.google.com.pk/books?id=jaAw EAAAQBAJ

Paragliola, G., & Coronato, A. (2021). An hybrid ECG-based deep network for the early identification of high-risk to major cardiovascular events for hypertension patients. *Journal of Biomedical Informatics*, *113*, 103648.

Sachdeva, M., Sharma, H. K., Kumar, A., Bansal, A., & Saluja, K. (2022, November). Applying Neural Imaging and ML to OCD Severity Prediction. In *2022 Seventh International Conference on Parallel, Distributed and Grid Computing (PDGC)* (pp. 402–406). IEEE.

Sethi, M., Ahuja, S., & Kukreja, V. (2021, August). An empirical study for the deep learning models. In *Journal of Physics: Conference Series* (Vol. 1950, No. 1, p. 012071). IOP Publishing.

Shanbehzadeh, M., Kazemi-Arpanahi, H., & Nopour, R. (2022). Design of clinical decision support system to diagnose breast cancer: An approach using data mining. *Payavard Salamat, 16*(2), 145–158.

Shao, M., Niu, Z., He, L., Fang, Z., He, J., Xie, Z., Cheng, G., & Wang, J. (2021). Building radiomics models based on triple-phase CT images combining clinical features for discriminating the risk rating in gastrointestinal stromal tumors. *Frontiers in Oncology, 11*, 737302.

Shehab, M., Abualigah, L., Shambour, Q., Abu-Hashem, M. A., Shambour, M. K. Y., Alsalibi, A. I., & Gandomi, A. H. (2022). Machine learning in medical applications: A review of state-of-the-art methods. *Computers in Biology and Medicine, 145*, 105458.

Shim, J. P., Warkentin, M., Courtney, J. F., Power, D. J., Sharda, R., & Carlsson, C. (2002). Past, present, and future of decision support technology. *Decision Support Systems, 33*(2), 111–126.

Tabari, A., Chan, S. M., Omar, O. M. F., Iqbal, S. I., Gee, M. S., & Daye, D. (2023). Role of machine learning in precision oncology: Applications in gastrointestinal cancers. *Cancers, 15*(1), 63.

Tama, B. A., Im, S., & Lee, S. (2020). Improving an intelligent detection system for coronary heart disease using a two-tier classifier ensemble. *BioMed Research International, 2020*.

Terrada, O., Hamida, S., Cherradi, B., Raihani, A., & Bouattane, O. (2020). Supervised machine learning based medical diagnosis support system for prediction of patients with heart disease. *Advances in Science, Technology and Engineering Systems Journal, 5*(5), 269–277.

Trowbridge, R., & Weingarten, S. (2001). Chapter 53. Clinical Decision Support Systems. *Internet WWW], Available: United States Department of Health & Human Services Agency for Healthcare Research and Quality Website.*

UmaMaheswaran, S. K., Kaur, G., Pankajam, A., Firos, A., Vashistha, P., Tripathi, V., & Mohammed, H. S. (2022). Empirical analysis for improving food quality using artificial intelligence technology for enhancing healthcare sector. *Journal of Food Quality, 2022*.

Unhelker, B., Pandey, H. M., & Raj, G. (2022). *Applications of artificial intelligence and machine learning: Select proceedings of ICAAAIML 2021.* Springer Nature Singapore. https://books.google.com.pk/books?id=60qJEAAAQBAJ

Verma, V. K., & Verma, S. (2021). Machine learning applications in healthcare sector: An overview. *Materials Today: Proceedings.*

Vyas, S. (2023). Extended reality and edge AI for healthcare 4.0: Systematic study. In *Extended Reality for Healthcare Systems* (pp. 229–240). Academic Press.

Vyas, S., Gupta, S., & Shukla, V. K. (2023, March). Towards Edge AI and Varied Approaches of Digital Wellness in Healthcare Administration: A Study. In *2023 International Conference on Computational Intelligence and Knowledge Economy (ICCIKE)* (pp. 186–190). IEEE.

Xia, J. Y., & Aadam, A. A. (2022). Advances in screening and detection of gastric cancer. *Journal of Surgical Oncology, 125*(7), 1104–1109.

Yasar, A., Saritas, I., & Korkmaz, H. (2019). Computer-aided diagnosis system for detection of stomach cancer with image processing techniques. *Journal of Medical Systems, 43*(4), 1–11.

17 Blockchain and Artificial Intelligence in Telemedicine and Remote Patient Monitoring

Harsh Bansal, Divya Gupta, and Darpan Anand

17.1 INTRODUCTION

Today's healthcare industry is characterized by a constantly changing environment brought on by advances in technology, changing patient needs, evolving laws and regulations, and rising health issues around the globe. The use of modern approaches that make use of technologies such as artificial intelligence and blockchain can overcome the shortcomings of traditional healthcare, such as limited data security, limited interoperability, lack of personalization, limited remote monitoring capabilities, challenges in telemedicine consultations, and manual administrative processes, thus enhancing the delivery of healthcare services. The present-day healthcare system depends significantly on telemedicine and remote patient monitoring, which are widely acknowledged for their value in boosting patient care, increasing access to healthcare services, streamlining healthcare delivery, and enabling remote monitoring of patients' health conditions, especially for those with chronic diseases. Blockchain and artificial intelligence (AI) coupled with one another could revolutionize telemedicine and remote patient monitoring by improving data security, interoperability, and personalized healthcare. This chapter outlines how blockchain, which offers a decentralized and immutable ledger for storing health data, might improve data security and privacy, enable consent-based data exchange, and foster interoperability in telemedicine and remote patient monitoring. This chapter also examines AI's use in telemedicine and remote patient monitoring in order to create insights and enhance clinical decision making, AI algorithms can analyze massive volumes of health data, including data from wearable devices, electronic health records, and patient-generated data. With the potential to enhance patient care and results, the convergence of blockchain with AI can create novel approaches to health data management, diagnosis, treatment, remote monitoring, and telemedicine.

DOI: 10.1201/9781003346289-17   **279**

In recent years, the healthcare sector has made significant progress using advanced technologies like blockchain and artificial intelligence (AI) (Tagde et al., 2021). These advancements might completely transform how healthcare is offered, particularly in fields like telemedicine and remote patient monitoring (Haleem et al., 2021).

Virtual consultations, remote diagnosis and monitoring, telepharmacy, and telepsychiatry are just a few of the many services included in telemedicine (Wootton, 1996). Telemedicine has the potential to lower healthcare expenses by reducing the need for in-person visits to healthcare facilities, especially in rural areas (Chen et al., 2021). A crucial part of telemedicine is remote patient monitoring (RPM), which involves using technology to check on patients' health remotely, typically from their place of residence (Malasinghe et al., 2019). To gather and send data on patients' vital signs, symptoms, medication adherence, and other health-related information, RPM may use a variety of devices, including wearables, sensors, and mobile apps (Hilty et al., 2021). Healthcare professionals can utilize this information to track patients' health, identify possible problems before they become serious, and act quickly when they do. In terms of treating chronic illnesses, lowering hospital readmissions, and enhancing patient outcomes, RPM has demonstrated potential (El-Rashidy et al., 2021).

The limitations of traditional methods used in telemedicine and RPM comprise data breaches and tampering, difficulty exchanging data where fragmented and incomplete patient data hinder decision making, inability to properly analyze large amounts of health data for personalized recommendations, limited ability to collect and transmit real-time health data from patients, and limited access to real-time visual assessments and communications (Ahmad et al., 2021). Additionally, time-consuming and error-prone manual administrative tasks including paperwork, transactions, and payments may reduce the effectiveness and cost-effectiveness of telemedicine and remote patient monitoring. To overcome the drawbacks of conventional telemedicine and remote patient monitoring techniques, blockchain, and artificial intelligence technology can be utilized (Jabarulla et al., 2021).

Blockchain technology can provide a decentralized, immutable, and transparent ledger that can securely store and manage healthcare data and help make decisions by using relevant consensus mechanisms (Bansal et al., 2022). This can contain patient permission forms, clinical trial data, electronic health records (EHRs), and other private information (Mc Cord et al., 2019). By utilizing sophisticated cryptographic methods, eliminating the need for middlemen, and providing audit trails for data access and alteration, blockchain can improve data privacy and security, which can help facilitate interoperability, data exchange, and patient consent management (Ghosh et al., 2023). Data breaches, unauthorized access, and manipulation of healthcare data can all be avoided, which is important for the healthcare sector (Shi et al., 2020).

On the other hand, AI can analyze enormous volumes of healthcare data to uncover patterns, trends, and insights that can help guide clinical judgment, create individualized treatment plans, and anticipate disease. To assist in diagnosis, treatment planning, monitoring, and population health management, AI algorithms can analyze EHRs, medical imaging, genomics data, wearable devices, and other healthcare data

(Bohr et al., 2020). AI can also automate regular administrative operations like appointment scheduling, inventory management, and claim processing, which will boost operational efficiency and reduce costs (Edlich et al., 2019).

Combining blockchain and AI in healthcare can enable secure and interoperable data interchange, which can make it easier for patients, payers, researchers, and healthcare professionals to share information. While AI may use shared data to give insights and decision assistance, blockchain can provide a trust layer that protects data integrity, consent management, and data provenance (Tagde et al., 2021). To find patterns and trends in illness prevalence, treatment efficacy, and adverse events, for instance, AI algorithms may analyze data from numerous sources, such as EHRs from various hospitals. This information can then be utilized to improve clinical recommendations and enhance patient outcomes (Jiang et al., 2017).

To address issues such as regulatory and legal considerations, data standardization, interoperability, and scalability, ensuring data privacy, security, and ethical use of AI are critical concerns and to realize the full potential of blockchain and AI in healthcare, collaborations between healthcare organizations, technology providers, policymakers, and other stakeholders are crucial.

17.2 RELATED HEALTHCARE PROJECTS USING BLOCKCHAIN AND AI

This section reviews some significant telemedicine and remote patient monitoring projects that use blockchain and AI (Priya et al., 2020) as follows:

1. doc.ai: doc.ai is a telemedicine platform that enables personalized, decentralized, and secure health data exchange using AI and blockchain. Their blockchain-based network enables users to safely save and share their health data, and they provide a variety of AI-powered health modules that offer insights into different health issues (Doc.Ai, 2018).

2. Medicalchain: Medicalchain is a blockchain-based telemedicine platform that employs AI to analyze data and keep track of patients. Patients can authorize access to their health data to healthcare professionals using their blockchain-based safe and transparent health record management service, and their AI algorithms analyze the data to create individualized care plans (Medicalchain, 2018).

3. DeepMind Health: DeepMind Health, a division of Alphabet Inc. (Google's parent company), develops technology for remote patient monitoring and data analysis in the healthcare industry. While blockchain assures data privacy and security, their AI algorithms examine health data to deliver predictive insights for early disease identification and personalized treatment recommendations (DigiDyl, 2021).

4. Coral Health: Coral Health is a blockchain-based platform that uses AI for remote patient monitoring and the analysis of health data. They offer healthcare practitioners solutions to safely exchange, examine, and capitalize on health data while utilizing AI algorithms to produce insights for individualized patient treatment (Coral Health, 2021).

5. Healthereum: Healthereum is a blockchain-based platform that employs AI to engage patients and conduct remote patient monitoring. Patients can safely maintain their health records on their platform, get rewards for completing health-related tasks, and engage in telemedicine consultations with medical professionals (Tech, 2021).

6. Health Nexus: To improve patient outcomes, Health Nexus is a blockchain-based platform that makes use of AI and remote monitoring. Their platform enables the safe exchange of health data, AI-powered analytics for patient monitoring, and individualized care plans for the treatment of chronic illnesses.

7. BurstIQ: BurstIQ is a platform for exchanging health data driven by blockchain and AI that enables remote patient monitoring and individualized care management. LifeGraph makes it possible to securely exchange and analyze health data, and it makes use of AI algorithms to produce insights for remote patient monitoring and care coordination (Burstiq, 2017).

8. Medicohealth: Medicohealth is a blockchain-based telemedicine platform with predictive analytics and AI for remote patient monitoring. Secure video consultations, remote health parameter monitoring, and AI-powered analytics for early disease detection and individualized treatment regimens are all made possible by their platform (Medicohealth, 2018).

9. Solve.Care: Solve.Care is a blockchain-based healthcare platform that leverages the Care.Protocol consensus algorithm and integrates tele-medicine and remote patient monitoring. A novel consensus method called Care.Protocol enables decentralized decision making and agree-ment across many participants in the healthcare ecosystem, including patients, providers, employers, insurers, and others (Solve.Care, 2018).

10. MedCredits: MedCredits is a blockchain-based telemedicine network that uses the Proof of Medicine (PoM) method of consensus. To ensure the correctness and quality of medical information exchanged on the net-work, PoM enables doctors to validate and certify medical cases on the blockchain (Todaro, 2018).

Table 17.1 shows a parametric comparison of the discussed projects in the field of telemedicine and remote patient monitoring that leverage AI and blockchain.

17.3   APPLICATIONS OF BLOCKCHAIN AND AI IN HEALTHCARE

Blockchain and AI have the potential to improve the efficiency and functionality of current healthcare delivery solutions. Following are some potential blockchain and AI uses in the healthcare industry (Ahmad et al., 2021; Jabarulla et al., 2021; Tagde et al., 2021):

• Patient Identity and Authentication: In healthcare, blockchain can act as a decentralized and unchangeable repository for patient identity and authentication. For safe authentication during telemedicine consultations, remote patient monitoring, and other virtual healthcare interactions,

TABLE 17.1

Comparison of Healthcare Projects That Use Blockchain and AI

Project Name	Key Features	Consensus Mechanism	Smart Contracts	Immutability	Predictive Analytics
doc.ai	Personalized health data exchange, AI powered health modules	Not specified	Yes	Yes	Yes
Medicalchain	Secure health record management, data sharing, AI powered care plans	Proof of Concept	Yes	Yes	Yes
DeepMind Health	Tools for remote patient monitoring, health data analysis	Not specified	Yes	Yes	Yes
Coral Health	Secure health data exchange, AI-powered health data analytics	Not specified	Yes	Yes	Yes
Healthereum	Patient engagement, remote patient monitoring, telemedicine consultations	Proof of Stake	Yes	Yes	No
Health Nexus	Health data exchange, remote monitoring, personalized care plans	Proof of Stake	Yes	Yes	Yes
BurstIQ	Health data exchange, remote patient monitoring, personalized care management	Not specified	Yes	Yes	Yes
Medicohealth	Telemedicine consultations, remote patient monitoring, predictive analytics	Not specified	Yes	Yes	Yes
Solve.Care	Telemedicine consultations, remote patient monitoring, care coordination	Care Protocol	Yes	Yes	Yes
MedCredits	Telemedicine consultations, medical case verification, data exchange	Proof of Medicine (PoM)	Yes	Yes	No

patients may have a distinct digital identity on the blockchain. In order to enhance patient identification verification and authentication, AI can analyze patient data, such as facial recognition or voice recognition.

- Medical Supply Chain Management: Blockchain can improve the visibility and traceability of pharmaceuticals and medical supplies. Authenticity, quality, and safety can be guaranteed by using blockchain to record the movement of medical supplies, drugs, and other healthcare products in an unchangeable and transparent way. Data from the supply chain, including inventory levels, pattern of demand, and logistics information, to optimize supply chain operations and ensure timely delivery of critical medical resources.

- Patient Consent and Privacy Management: Blockchain can offer a decentralized and transparent platform in healthcare, especially in telemedicine and remote patient monitoring for patient consent and privacy management. The blockchain can be used to store patient consent for exchanging health data, giving patients control over their data and enabling safe and auditable data sharing. In order to improve patient privacy and compliance, AI can help manage patient permission preferences and make sure that data is shared in accordance with those preferences and applicable laws.

- Telemedicine and RPM Marketplace: Blockchain can help create a decentralized market place for telemedicine and RPM, where patients may interact with healthcare professionals and use telemedicine and RPM services. In order to ensure patient privacy and data security, blockchain can offer a secure and transparent platform for patients to search for, schedule, and pay for telemedicine appointments or remote monitoring services. In order to enhance patient experience and access to care, AI can offer recommendations and personalized matching based on patient preferences and healthcare provider expertise.

- Fraud Detection and Prevention: AI can analyze big data sets from health management systems to identify and stop fraud, waste, and abuse in the delivery of healthcare services. AI is capable of analyzing patterns, trends, and abnormalities in data to spot potential forgeries like billing fraud or misuse of telemedicine services. For the purpose of healthcare fraud detection and prevention, blockchain can offer a transparent and unchangeable record of all transactions that can be audited and confirmed.

- Medical Records Management: Electronic health records (EHRs) management can be done on a secure, decentralized platform using blockchain technology. Authorized healthcare professionals can securely access and update EHRs stored on the blockchain, preserving data integrity, interoperability, and patient privacy. EHRs and other pertinent patient data can be analyzed with the help of AI to provide insights for diagnosis, treatment planning, and remote monitoring, enhancing professional judgment and patient care.

- Patient Engagement and Education: AI can offer patients in healthcare personalized health information and education, assisting them in better understanding their medical situations and making defensible decisions. Chatbots or virtual health assistants powered by AI can communicate with patients, deliver educational materials, respond to inquiries, and provide advice on self-care, medication adherence, and lifestyle changes. Patients may access and verify health information on a safe and transparent platform thanks to blockchain, which also increases patient empowerment and involvement.

These technologies have the power to completely transform the way healthcare is provided by boosting patient care, data security, and privacy, and offering novel and innovative approaches to delivering virtual healthcare services, as shown in Figure 17.1. We can anticipate future developments and applications of blockchain and AI in healthcare as the industry develops, which will improve patient outcomes and strengthen healthcare delivery.

17.4 PATIENT CENTRIC FRAMEWORK USING BLOCKCHAIN AND AI IN TELEMEDICINE AND REMOTE PATIENT MONITORING

Components of blockchain and artificial intelligence such as decentralization, distributed ledger technology (DLT), consensus mechanisms, smart contracts, immutability, machine learning, deep learning, and predictive data analytics (Tagde et al., 2021) can revolutionize healthcare when combined with telemedia and remote patient monitoring. The following could serve as the structure of a patient-centric framework for telemedicine and remote patient monitoring:

- Decentralized Health Data Ownership: Patients should be able to securely store, manage, and share their data via a decentralized system, such as a blockchain, and should have control over their health data. As a result, data privacy and permission would be guaranteed and patients could grant healthcare practitioners access to their health data as needed.
- Distributed Ledger for Health Data: To guarantee data security, integrity, and transparency, health data, including electronic health records, wearables data, and remote patient monitoring data, can be kept in a distributed ledger. By doing so, the healthcare ecosystem's various stakeholders would be able to exchange data securely while maintaining patient privacy.
- Consensus Mechanisms for Data Validation: The reliability and precision of health data maintained in the distributed ledger can be verified using the consensus method. This would ensure that all parties in the network have the same view of the data, eliminating the need for intermediaries and enhancing trust in the data.
- Automated Processes with Smart Contracts: Smart contracts can be used to automate tasks like appointment scheduling, notifications from remote monitoring, and payment processing. For instance, a smart contract can automatically warn the healthcare professional when a patient's vital signs fall below certain levels so that they can start the necessary interventions.

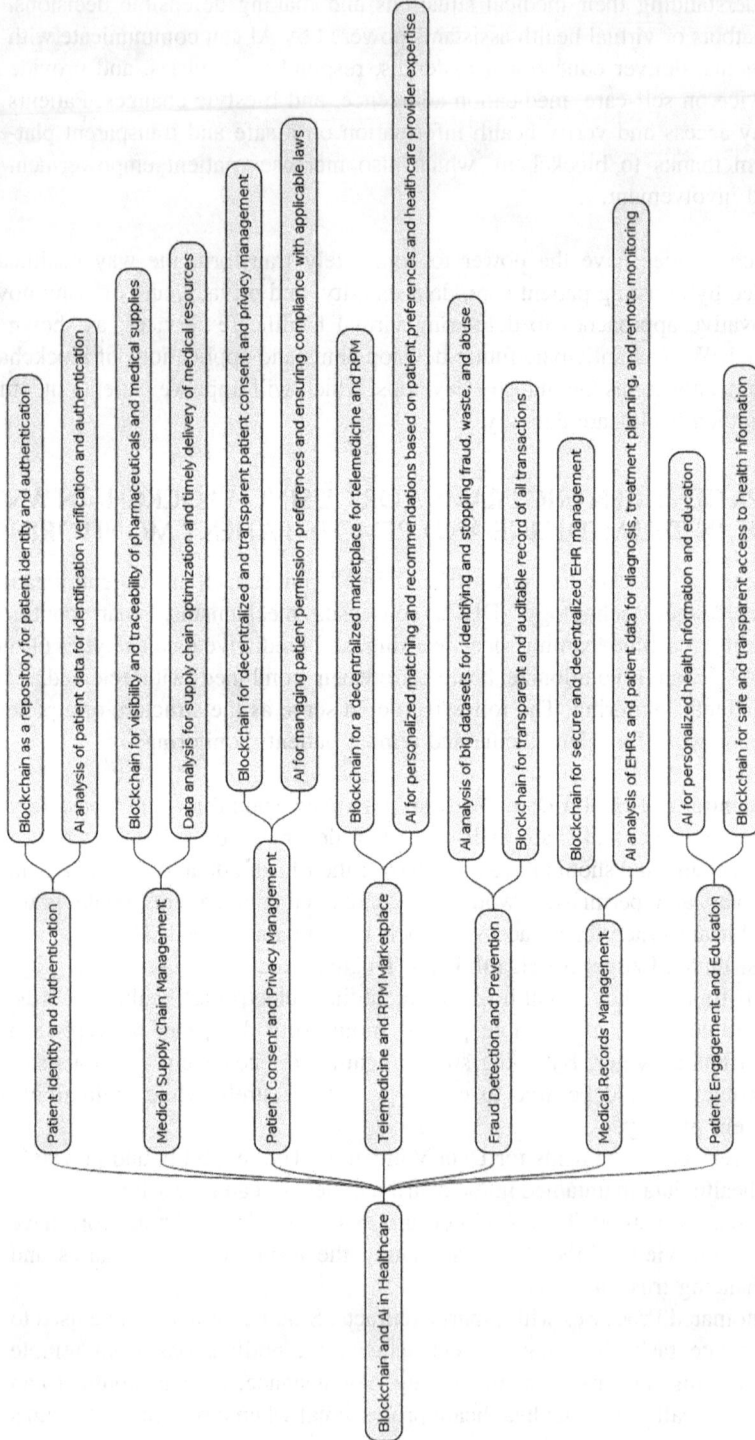

FIGURE 17.1 Potential applications of blockchain and AI in healthcare.

This would improve patient care, speed up the administrative process, and decrease delays.

- Immutable Health Data: To prevent tampering or alteration, health information entered into the distributed ledger should be immutable which is possible with blockchain. This would offer an auditable record of data modifications and verify the reliability and integrity of patient data.
- Machine Learning and Deep Learning for Predictive Analytics: Machine learning and deep learning algorithms are capable of analyzing the enormous amounts of health data gathered from remote patient monitoring devices to find patterns, trends, and insights. These algorithms can predict potential health issues, detect anomalies, and generate personalized recommendations for patients, empowering them to take proactive steps toward managing their health.
- Predictive Data Analytics for Early Intervention: By identifying patients who are at risk of developing chronic diseases or experiencing adverse events, predictive data analytics can enable early intervention and preventative therapy. In order to enhance patient outcomes and save healthcare costs, healthcare practitioners can use this information to intervene early, offer personalized interventions, and optimize treatment activities.
- Patient Empowerment and Engagement: Empowerment and engagement of the patient should be given top priority in patient-centric telemedicine and remote patient monitoring. Patients should actively participate in making decisions about their care, have access to their health information, and have the training and resources they need to effectively manage their health. Patients would be able to take charge of their health, make educated decisions, and actively participate in their care process.

Integrating these elements in Figure 17.2 can revolutionize healthcare by improving patient outcomes, data security, interoperability, and transparency. Telemedicine

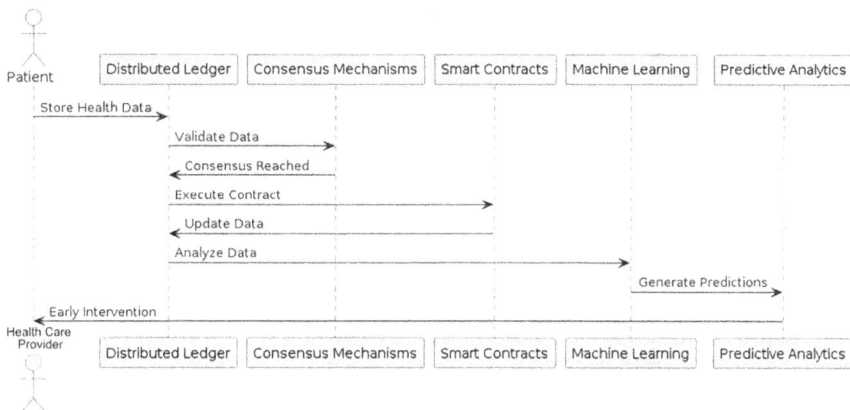

FIGURE 17.2 Patient-centric framework for healthcare using blockchain and artificial intelligence.

and remote patient monitoring can further enhance the advantages of these technologies by enabling remote care, personalized treatment, and early disease identification, resulting in a more effective and patient-centric healthcare system.

17.4.1 CASE STUDY: ACCESS TO RURAL HEALTHCARE THROUGH TELEMEDICINE AND REMOTE PATIENT MONITORING

A patient with limited access to medical care lives in a remote rural area. Assume that the patient has asthma, which needs to be regularly managed and monitored.

17.4.1.1 Framework Implementation

As shown in Figure 17.3, the patient owns and securely stores their health data in a system based on a blockchain. It enables the patient to provide access to their data to their healthcare professional and other important parties. A patient's health information, such as their symptoms of asthma, use of medications, and results of lung function tests, is maintained in a distributed ledger to ensure data security and accessibility. Consensus mechanisms verify the correctness and dependability of patient health data, ensuring that all network stakeholders have a consistent understanding of that patient's data without the need for manual intervention. Telemedicine discussions between a patient and healthcare practitioner are made possible via smart contracts. The smart contract automates the process of booking virtual consultations and contains terms like *consultation cost, scheduling,* and *reimbursement.* The distributed ledger's immutable nature ensures data integrity and secures patient health information from unauthorized modification. In order to forecast probable asthma exacerbations and trigger alarms, machine learning and deep learning algorithms analyze patients' health data in real time, detecting patterns and trends. Predictive data analytics identify patients who are at high risk for asthma exacerbations by continuously monitoring their health data. The doctor receives warnings from the system, promptly modifies the patient's medication schedule, and offers remote guidance.

Patients in rural or remote locations with limited access to healthcare services would considerably benefit from the suggested patient-centric framework for telemedicine and remote patient monitoring. This makes it possible to provide proactive, individualized care, remote monitoring, and patient empowerment, which improves health outcomes and makes healthcare delivery more effective.

17.5 CHALLENGES ASSOCIATED WITH USING BLOCKCHAIN AND AI IN TELEMEDICINE AND RPM

Though the potential benefits of using blockchain and AI in telemedicine and RPM are significant, still there are challenges associated with using blockchain and AI in telemedicine and RPM, which needs to be taken care of (Vyas, S., et al 2023). Major challenges and possible resolutions are as follows:

- Limited Internet Access: The viability of telemedicine consultations and real-time remote monitoring may be impacted by limited or unstable

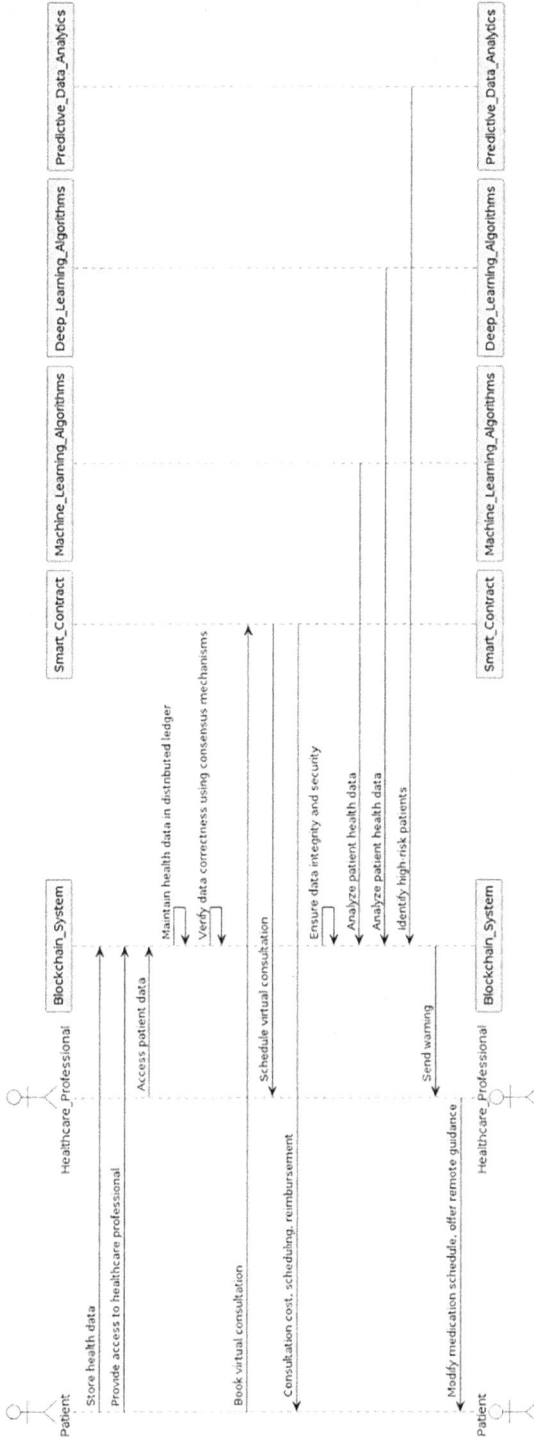

FIGURE 17.3 Use case scenario of patient-centric framework using blockchain and artificial intelligence.

internet access in remote or rural areas. Resolution: Utilizing other kinds of communication, such as phone calls or SMS-based systems, or using offline-capable telemedicine platforms are both potential alternatives, but will be limited to many aspects.

- Digital Gap: Some patients in remote regions may not have adequate access to technology or digital literacy, which may limit their capacity to engage fully in the suggested framework. Resolution: The digital gap can be closed by offering patient education and support to increase digital literacy as well as by looking into opportunities for easily accessible and affordable technology solutions.

- Legal and Regulatory Considerations: The legal and regulatory environment governing the use of blockchain technology and artificial intelligence in healthcare, including telemedicine and RPM, is still developing. Navigating the legal system and complying with data protection, privacy, and other requirements can be challenging. Liability, responsibility, and intellectual property rights are just a few of the legal and regulatory concerns that the use of blockchain, smart contracts, and machine intelligence in healthcare may bring up. Resolution: Legal and regulatory issues can be addressed by ensuring adherence to pertinent laws and regulations, seeking legal counsel, and establishing explicit agreements among players.

- Data Privacy and Security Issues: Using smart contracts and storing health data in a distributed ledger raises issues with data privacy and security. Resolution: You may protect patient data and ease privacy concerns by putting strong security measures in place, like encryption, authentication, and permission-based access controls. Additionally, adherence to pertinent data protection laws like HIPAA should be ensured.

- Ethical Concerns: The application of predictive data analytics and machine learning algorithms may give rise to ethical issues with bias, fairness, transparency, and consent. Resolution: Ethical concerns can be resolved by setting up ethical frameworks, doing routine audits, and ensuring algorithmic usage is transparent. Another crucial step is getting the patients' informed consent before using or disclosing their data.

- Integration and Implementation Challenges: It can be difficult to integrate a decentralized framework with current healthcare systems and implement distributed ledger technologies, smart contracts, and machine learning algorithms. Resolution: Implementation and integration issues can be resolved by collaboration between healthcare providers, technology suppliers, and other stakeholders. The creation of standardized interfaces and protocols for interoperability can promote easy integration with current healthcare systems.

- Adopt and Acceptance: Broad use of blockchain and AI in telemedicine and RPM would call for adaptations to organizational culture, processes, and outlook. It might be difficult to guarantee the acceptance, adoption, and engagement of healthcare professionals, patients, and other stakeholders.

- Human Interaction and Personal Touch: Patients may not receive the same level of personal attention and human connection during telemedicine consultations as they would during conventional in-person appointments, which can have an influence on patient engagement and satisfaction. Resolution: Including patient-centric approaches can assist in closing the gap and deliver a pleasant experience for patients. These methods include virtual patient interaction devices, personalized communication, and caring practices (Vyas, S., 2023).

Overall, overcoming the obstacles to the use of blockchain and AI in telemedicine and RPM requires a holistic approach that includes ethical concerns, regulatory compliance, technological proficiency, interoperability, adoption, and ongoing advancement. The potential advantages of these technologies in telemedicine and RPM can be maximized while minimizing the accompanying obstacles by adopting a proactive and comprehensive approach.

17.6 FUTURE AVENUES FOR INTEGRATION AND COLLABORATION OF HEALTHCARE WITH BLOCKCHAIN AND AI

The suggested patient-centric architecture for telemedicine and remote patient monitoring utilizing distributed ledger, consensus mechanisms, smart contracts, immutability, machine learning, deep learning, and predictive data analytics opens up a number of opportunities for further development. These possible areas for additional research are as follows:

- Enhancements of Privacy and Security Mechanisms: The privacy and security mechanisms in the suggested framework may be further improved through research. This could involve researching advanced encryption methods, blockchain-based privacy solutions, and innovative approaches to guarantee the availability, confidentiality, and integrity of patient data in a decentralized environment.
- Ethical Matters and Bias Reduction: Addressing ethical issues including bias, fairness, openness, and permission is vital as machine learning algorithms are employed for predictive data analytics. The development of approaches to reduce bias in machine learning models, the continual improvement of clear and comprehensible algorithms, and the establishment of moral guidelines for data usage and sharing in telemedicine and remote patient monitoring can be the subject of future research.
- Economic and Policy Impacts: Additional research can be done to determine how the suggested framework would affect economics and policy. Research can concentrate on examining the potential advantages for healthcare providers, payers, and patients, as well as the cost-effectiveness, reimbursement models, and policy implications of telemedicine and remote patient monitoring implementation utilizing a decentralized framework.

- Machine Learning Model Validation and Enhancements: Additional research can be done to validate and improve machine learning models utilized for predictive data analytics in telemedicine and remote patient monitoring. To achieve reliable and precise predictions, this may involve optimizing algorithms, including more varied knowledge bases, and assessing their accuracy, precision, sensitivity, and other performance measures.
- Real-World Implementation and Evaluation: Putting the suggested framework into practice in rural healthcare settings can shed light on its efficacy, viability, and influence on patient satisfaction. It is possible to find areas for development and confirm the framework's advantages by evaluating it in various healthcare settings with a variety of patient demographics.
- Integration with Current Healthcare Systems: It can be challenging to integrate the suggested framework with current healthcare systems, health information exchanges, and electronic health records (EHRs). The development of standardized protocols, interfaces, and interoperability solutions can be the subject of future study to enable seamless data interchange and integration amongst various healthcare stakeholders and systems.
- User Experience and Patient Participation: The proposed framework can be further optimized by improving the user experience and patient participation during telemedicine consultations and remote patient monitoring. It is possible to do research to create user-friendly user interfaces, virtual patient engagement tools, and personalized communication strategies that emphasize patient comfort, satisfaction, and engagement in their medical care.

Further research and innovation in these areas can help optimize the framework, enhance patient care, and improve healthcare outcomes in rural or remote areas.

17.7 CONCLUSION

Telemedicine and remote patient monitoring using blockchain and artificial intelligence (AI) have significant potential for revolutionizing the way healthcare is offered. Integrating AI's superior analytics, machine learning, and predictive capabilities with blockchain's inherent security, transparency, and immutability can lead to creative solutions that improve patient care, data privacy, and operational effectiveness.

To address concerns about data privacy, security, and interoperability, blockchain technology can facilitate safe and transparent data interchange, patient-controlled data sharing, and auditable records of medical information. With the use of AI, this data can be used to generate insights, develop personalized treatment plans, and enable telemedicine consultations, remote monitoring, and predictive analytics for early disease diagnosis. These innovations could lead to better patient outcomes, lower healthcare costs, greater access to care, and more patient-centered procedures. Further enhancing trust, accountability, and automation in telemedicine and remote patient monitoring is possible with the use of consensus mechanisms and smart contracts in blockchain-based solutions.

The use of blockchain and AI in healthcare also comes with challenges. These include issues with standardization, interoperability, regulatory compliance, and ethical issues. The responsible and successful adoption of these technologies in healthcare depends on strong data governance frameworks, transparent regulatory rules, and stakeholder collaboration.

In conclusion, telemedicine and remote patient monitoring using blockchain and AI have enormous potential to change the way healthcare is delivered, improve patient care, and advance the area of telemedicine.

REFERENCES

Ahmad, R.W., Salah, K., Jayaraman, R., Yaqoob, I., Ellahham, S., & Omar, M. (2021). The role of blockchain technology in telehealth and telemedicine. *International Journal of Medical Informatics*, *148*, 104399.

Bansal, H., Gupta, D., & Anand, D. (2022, October). Analysis of Consensus Algorithms in Context of the Blockchain based Applications. In *2022 10th International Conference on Reliability, Infocom Technologies and Optimization (Trends and Future Directions) (ICRITO)* (pp. 1–7). IEEE.

Bohr, A., & Memarzadeh, K. (2020). The rise of artificial intelligence in healthcare applications. In *Artificial intelligence in healthcare* (pp. 25–60). Academic Press.

Burstiq. (2017). Burstiq Bringing Health to Life. https://www.burstiq.com/wp-content/uploads/2017/09/BurstIQ-whitepaper_07Sep2017.pdf

Chen, J., Amaize, A., & Barath, D. (2021). Evaluating telehealth adoption and related barriers among hospitals located in rural and urban areas. *The Journal of Rural Health*, *37*(4), 801–811.

Coral Health. (2021). Reclaiming Your Medical Records. Medium. https://mycoralhealth.medium.com/reclaiming-your-medical-records-ce5f17777011

DigiDyl. (2021). AI Doctors - DigiDyl. Medium. https://medium.com/@digidyl/ai-doctors-ee6ae9a3a64d

Doc.Ai. (2018). The Future of Healthcare - doc.ai. Medium. https://medium.com/@_doc_ai/the-future-of-healthcare-d0971fff20b7

Edlich, A., Phalin, G., Jogani, R., & Kaniyar, S. (2019). Driving impact at scale from automation and AI. *McKinsey Global Institute*, *100*. https://www.mckinsey.com/~/media/McKinsey/Business%20Functions/McKinsey%20Digital/Our%20Insights/Driving%20impact%20at%20scale%20from%20automation%20and%20AI/Driving-impact-at-scale-from-automation-and-AI.ashx

El-Rashidy, N., El-Sappagh, S., Islam, S.R., M. El-Bakry, H., & Abdelrazek, S. (2021). Mobile health in remote patient monitoring for chronic diseases: principles, trends, and challenges. *Diagnostics*, *11*(4), 607.

Ghosh, P.K., Chakraborty, A., Hasan, M., Rashid, K., & Siddique, A.H. (2023). Blockchain application in healthcare systems: A review. *Systems*, *11*(1), 38.

Gupta, S., Sharma, H.K., & Kapoor, M. (2023). Digital medical records (DMR) security and privacy challenges in smart healthcare system. In *Blockchain for Secure Healthcare Using Internet of Medical Things (IoMT)*. Cham; Springer. 10.1007/978-3-031-18896-1_6

Haleem, A., Javaid, M., Singh, R.P., & Suman, R. (2021). Telemedicine for healthcare: Capabilities, features, barriers, and applications. *Sensors International*, *2*, 100117.

Hilty, D.M., Armstrong, C.M., Edwards-Stewart, A., Gentry, M.T., Luxton, D.D., & Krupinski, E.A. (2021). Sensor, wearable, and remote patient monitoring competencies for clinical care and training: scoping review. *Journal of Technology in Behavioral Science*, 6, 252–277.

Jabarulla, M.Y., & Lee, H.N. (2021, August). A blockchain and artificial intelligence-based, patient-centric healthcare system for combating the COVID-19 pandemic: Opportunities and applications. In *Healthcare* (Vol. 9, No. 8, p. 1019). MDPI.

Jiang, F., Jiang, Y., Zhi, H., Dong, Y., Li, H., Ma, S., … & Wang, Y. (2017). Artificial intelligence in healthcare: Past, present and future. *Stroke and Vascular Neurology*, 2(4).

Malasinghe, L.P., Ramzan, N., & Dahal, K. (2019). Remote patient monitoring: A comprehensive study. *Journal of Ambient Intelligence and Humanized Computing*, 10, 57–76.

Medicalchain. (2018). Medicalchain Whitepaper 2.1. https://medicalchain.com/Medicalchain-Whitepaper-EN.pdf

Medicohealth. (2018). The Biggest Doctor-Patient Environment Based on Blockchain. https://medicohealth.io/supporters/documents/wp_beta.pdf

Mc Cord, K.A., & Hemkens, L.G. (2019). Using electronic health records for clinical trials: Where do we stand and where can we go? *CMAJ*, 191(5), E128–E133.

Priya, R.L., & Jinny, S.V. (2020). Research Issues and Solution in Blockchain Healthcare. *Blockchain Technology*, 75–91.

Solve.Care. (2018). A Revolution in HealthCare. https://solve.care/wp-content/uploads/2018/09/lightpaper.pdf

Shi, S., He, D., Li, L., Kumar, N., Khan, M.K., & Choo, K.K.R. (2020). Applications of blockchain in ensuring the security and privacy of electronic health record systems: A survey. *Computers & Security*, 97, 101966.

Tagde, P., Tagde, S., Bhattacharya, T., Tagde, P., Chopra, H., Akter, R., … & Rahman, M.H.. (2021). Blockchain and artificial intelligence technology in e-Health. *Environmental Science and Pollution Research*, 28, 52810–52831.

Tech, S. (2021). Healthereum Using Blockchain Technology to Vastly Improve the Patient Experience. Medium. https://medium.com/ktrade/healthereum-using-blockchain-technology-to-vastly-improve-the-patient-experience-6cb11f9ebc7f

Todaro, J. (2018). Overview of MedCredits. Medium. https://medium.com/medxprotocol/introduction-to-medcredits-287f9fac03e6

Vyas, S. (2023). Extended reality and edge AI for healthcare 4.0: Systematic study. In *Extended Reality for Healthcare Systems* (pp. 229–240). Academic Press.

Vyas, S., Bhargava, D., & Khan, S. (2023). Healthcare 4.0: A Systematic Review and Its Impact Over Conventional Healthcare System. *Artificial Intelligence for Health 4.0: Challenges and Applications*, 1–17.

Wootton, R. (1996). Telemedicine: A cautious welcome. *BMJ*, 313(7069), 1375–1377.

18 The Prediction of Critical Health Diseases Using Artificial Intelligence with Lung Cancer as a Case Study

Harish Tiwari, Darpan Anand, and Mukesh Kalla

18.1 INTRODUCTION

The goal of artificial intelligence, a branch of computer science, is to build machines that are capable of learning, solving problems, and making decisions—all activities that are traditionally only performed by intelligent people. Machine learning, deep learning, computer vision, and natural language processing are examples of AI approaches (Aggarwal et al., 2022). These methods have been effectively used in a variety of industries, including healthcare and early medical diagnosis to save patient lives. The prediction of critical health diseases is a crucial task in the medical field. Early detection and treatment can significantly improve patient outcomes and reduce healthcare costs (Barrett et al., 2019). However, traditional methods for predicting diseases have limitations, such as the reliance on a few clinical parameters and subjective interpretation of medical imaging (Bi et al., 2019). AI is utilized as a powerful tool to overcome these limitations and improve disease prediction accuracy. In this chapter, we will review the latest developments in AI-based predictive modeling for critical health diseases.

AI algorithms are used to build predictive models for critical health diseases by analyzing large data sets of patient information. These data sets can include electronic health records, medical imaging, genomic data, and other sources. AI algorithms can be used to analyze these data sets and to identify patterns and relationships that may be associated with the development of critical health diseases (Tang et al., 2018). The models developed by AI algorithms can then be used to predict the likelihood of developing these diseases in individual patients.

The World Health Organization (WHO) estimates that non-communicable diseases (NCDs) would account for over 71% of all fatalities globally in 2021. The major causes of death worldwide include serious health conditions such cardiovascular disease, cancer, diabetes, and chronic respiratory illnesses (Bray et al., 2021).

DOI: 10.1201/9781003346289-18

The United Nations has a dedicated agency for worldwide public health called the World Health Organization (WHO). The group releases statistics on major illnesses, including information on world health. Diabetes, cancer, respiratory illnesses, and cardiovascular disorders are some of the main causes of death worldwide. According to the WHO, cardiovascular disorders, which include heart attacks and strokes, are the leading cause of mortality worldwide, accounting for an estimated 17.9 million deaths annually. Additionally, the COVID-19 pandemic has caused a significant increase in fatal diseases, particularly respiratory conditions. As of the start of 2022, the WHO estimates that COVID-19 has caused over 5 million mortalities worldwide (Murray et al., 1994).

Early detection and treatment can significantly improve patient outcomes and reduce healthcare costs. However, traditional methods for predicting diseases have limitations, such as the reliance on a few clinical parameters and subjective interpretation of medical imaging. AI has emerged as a powerful tool to overcome these limitations and improve disease prediction accuracy. There are various crucial diseases that impact a person and their family and cancer is one of them. There are various types of cancers identified by medical science and lung cancer is one of them. Lung cancer is a severe form of cancer, which has been responsible for many deaths worldwide. Early detection of lung cancer can lead to effective treatment and increased chances of survival. The conventional diagnosis of lung cancer involves a biopsy, which is an invasive procedure that can be painful and expensive. In recent years, several studies have focused on developing machine learning algorithms for the prediction of lung cancer. These algorithms use various data sources such as medical imaging, patient demographics, and clinical data. In this chapter, we will review the latest developments in AI-based predictive modeling for critical health diseases. The ability to predict critical health diseases using artificial intelligence (AI) has become a game changer in the medical field. AI is being used to analyze vast amounts of data from electronic health records (EHRs), medical imaging, genomics, and other sources to develop models that can predict the likelihood of developing critical health diseases, such as cancer, heart disease, and diabetes. In this chapter, the recent developments in lung cancer prediction using machine learning algorithms. Lung cancer is one of the leading causes of death worldwide, with a low survival rate if not detected in its early stages. Therefore, there is a critical need for accurate and reliable lung cancer prediction models to help diagnose the disease at an early stage.

The complexity and multidimensional nature of critical health disorders, as well as the wide range of risk variables involved, such as age, genetics, lifestyle, and environmental factors, should be noted. Prevention and management of critical health diseases require a multidisciplinary approach that includes public health interventions, healthcare policies, and individual behavior changes (Liu et al., 2022).

Here are some specific statistics about critical health diseases from WHO:

- Cardiovascular Diseases: Cardiovascular disorders are responsible for 17.9 million annual deaths worldwide, which account for 31% of all global deaths.

- Cancer: In 2020, an estimated 9.6 million people died from cancer worldwide, making it the second-leading cause of death globally.
- Diabetes: In 2021, diabetes is estimated to have caused 1.6 million deaths worldwide (Liu et al., 2022).
- Chronic Respiratory Diseases: Chronic obstructive pulmonary disease (COPD) alone caused an estimated 3.0 million deaths globally in 2019.

It's worth noting that these statistics may have changed since my knowledge cut-off date, and it's always best to refer to the most up-to-date data from reliable sources like WHO (Wensing et al., 2020).

18.2 LITERATURE SURVEY

The utilization of AI to predict critical health diseases has been an active and prominent area of research in past few years. Numerous studies have shown the potential of AI-based predictive models to improve patient outcomes by identifying diseases early and providing personalized treatment plans. A study conducted by Rajkomar et al. (2018) showed that an AI-based predictive model could accurately predict patient mortality, hospital readmission, and prolonged length of stay (Rajkomar et al., 2018).

Another study by Miotto et al. (2018) demonstrated the potential of AI-based predictive models to accurately diagnose diseases such as pneumonia and skin cancer (Kapoor et al., 2019). These studies highlight the promising potential of AI in predicting critical health diseases and suggest that further research in this area is warranted. An emerging area that has the potential to completely transform healthcare is the application of AI in the prediction of serious illnesses. AI algorithms have a powerful capability to analyze vast amounts of data from multiple sources to identify patterns and detect disease symptoms in patients before they become severe. This can lead to early discovery and well-timed treatment, which can improve patient outcomes (Ghazal et al., 2021). This literature survey aims to provide an overview of the current trends of AI in predicting critical diseases, its applications, benefits, and potential limitations.

"Deep learning for predicting hospitalization events in patients with COPD using electronic health records" by Zhang et al. (2021) used deep learning algorithms to predict hospitalization experiences in patients with chronic obstructive pulmonary disease (COPD). The research showed that deep learning models performed better than conventional prediction models (Si et al., 2021).

Patel et al. (2021) reviewed recent studies on the application of AI and machine learning in predicting prostate cancer diagnosis and prognosis. The authors highlighted the potential of AI in improving diagnosis and treatment decision making (Ma et al., 2022).

"Predicting ICU Mortality Risk Using Machine Learning Techniques: A Systematic Review" by Adeli et al. (2021) carried out a comprehensive study that predicted ICU mortality risk using machine learning techniques. The study found that machine learning algorithms were effective in predicting ICU mortality risk, but more studies are needed to assess their generalizability and clinical utility (Murugappan et al., 2021).

"A systematic review of artificial intelligence and deep learning in diabetic retinopathy screening" by Zhang et al. (2021) reviewed studies on the utilization of AI and deep learning in diabetic retinopathy screening. The study noticed that deep learning models performed better than traditional screening methods in detecting diabetic retinopathy (Wang et al., 2020a).

"Deep learning for predicting cardiovascular disease risk: A systematic review and meta-analysis" by Yang et al. (2021) has published a systematic review and meta-analysis of studies that utilized deep learning models to predict cardiovascular disease hazard. The study found that deep learning models were more accurate in predicting cardiovascular disease risk than traditional prediction models (Lan et al., 2021).

AI has also been applied in predicting the risk of diabetes. For instance, Ebrahimzadeh et al. (2022) developed an AI-based system that could predict type 2 diabetes by analyzing clinical data. The algorithm could identify patients who were at a high risk of acquiring type 2 diabetes and had a high accuracy rate (Ebrahimzadeh et al., 2022).

Similarly, Malo et al. (2019) demonstrated a machine learning algorithm to predict the risk of diabetic retinopathy. The authors reported that the algorithm had a high accuracy rate and could recognize patients with a high probability of developing diabetic retinopathy.

One of the early works in this field was investigated by Ng et al. (2011), which utilized machine learning algorithms to predict the likelihood of a patient having a heart attack within ten years. The study found that the algorithm could accurately predict the risk of heart attack and outperformed traditional risk prediction models (Parthiban & Srivatsa, 2012).

Machine learning algorithms were employed in a study by Rajkomar et al. (2018) to forecast in-hospital death rates for patients admitted to the hospital. The authors reported that the algorithm had a high accuracy rate and could predict mortality rates earlier than traditional models (Hsu et al., 2021).

In the field of cancer prediction, Esteva et al. (2017) used deep learning procedures to predict the likelihood of melanoma based on images of skin lesions. The authors reported that the algorithm had a high accuracy rate and could outperform human dermatologists in some cases (Vocaturo et al., 2019).

Liu et al. (2020) used ML algorithms to predict the danger of lung cancer in cigarette smokers. The authors reported that the algorithm could recognize patients at extreme risk of developing lung cancer earlier than traditional models (Wu et al., 2020).

In the field of diabetes prediction, a study by Mordi et al. (2022) used ML systems to predict the risk of type 2 diabetes in a population-based cohort. The authors reported that the algorithm had a high accuracy rate and could discover patients at great probability of developing type 2 diabetes (Mordi et al., 2022).

18.3 IMPORTANCE OF PREDICTING CRITICAL HEALTH DISEASES

Predicting critical health diseases is crucial in the medical field because it allows for early detection and intervention, which can significantly improve patient outcomes.

For example, predicting critical health diseases such as cancer, heart disease, and diabetes early can help healthcare providers develop treatment plans that are more effective and less invasive. Additionally, early detection can help reduce healthcare costs associated with treating advanced-stage diseases (Wang et al., 2021).

AI has demonstrated considerable promises in the prediction of serious illnesses like cancer, heart disease, and neurological conditions. An analysis can be done using AI algorithms on massive amounts of data from multiple sources to identify patterns and detect disease symptoms in patients before they become severe. Early discovery and prompt treatment may result from this, improving patient outcomes (Subramanian et al., 2020).

AI has been utilized to analyze medical imaging like mammograms and CT scans in the field of cancer detection and treatment for tumor detection and measure the stage of the disease. AI algorithms have also been used to identify genetic mutations that may be associated with certain types of cancer, helping doctors create personalized treatment plans (Sotoudeh et al., 2019).

In the field of cardiology, electrocardiograms (ECGs) have been analyzed using AI to find irregular cardiac rhythms and to determine the likelihood of developing heart disease. AI algorithms can also analyze medical images such as cardiac MRIs to detect structural abnormalities in the heart that may be indicative of heart disease (Zhou et al., 2021).

The ability of AI to anticipate neurological conditions like Parkinson's and Alzheimer's disease has also shown promise. AI algorithms can analyze brain images like MRI scans to detect structural changes in the brain that may be indicative of these diseases. AI can also be used to analyze speech patterns and other biomarkers to predict the onset of neurological disorders (Frizzell et al., 2022).

Despite the potential benefits of AI in predicting critical health diseases, there are also some limitations that need to be addressed. These limitations include the lack of transparency in AI algorithms, limited data quality, bias in data, limited generalizability, and legal and ethical concerns.

To address these limitations, researchers are exploring new directions for AI in predicting critical health diseases. These directions include the integration of multiple data sources, explainable AI, improved data quality, enhanced personalization, and the empowerment of patients.

Overall, the current state of research on AI to predict critical health diseases is promising and has the potential to revolutionize healthcare. However, it is crucial to address potential limitations to ensure that AI-based predictive models are accurate, reliable, and equitable.

Predicting critical health diseases is essential for several reasons:

- **Early Detection:** Before symptoms appear, predictive models can help identify people who are at a high risk of contracting serious illnesses. Early detection can lead to early intervention and management of the disease, which can improve health outcomes and reduce healthcare costs.
- **Personalized Medicine:** Predictive models can help identify highly efficient treatment options for persons with critical health diseases. This approach to

medicine is known as *precision medicine*, and it tailors treatments to a patient's specific characteristics, including their genetic makeup, lifestyle, and other factors.

- **Public Health Interventions:** Predictive models can help public health officials develop targeted interventions to prevent critical health diseases. For example, a predictive model could be utilized to identify societies or residents at high risk of developing heart illness, and public health officials could develop targeted interventions to reduce risk factors, such as promoting healthy lifestyle changes.

- **Resource Allocation:** Predictive models can help healthcare organizations allocate resources more efficiently. For example, A model could recognize patients with an extreme probability of developing critical health diseases and prioritize them for screening, testing, and treatment.

- **Prevention Efforts:** With this knowledge, individuals can make lifestyle changes, such as eating a healthier diet, exercising regularly, and reducing stress, which can lower their risk of developing critical health diseases.

- **Resource Allocation:** Predicting critical health diseases can help healthcare systems and policymakers allocate resources effectively, such as targeting screening efforts to high-risk populations or allocating funding to research efforts aimed at improving diagnosis and treatment.

- **Economic Impact:** Critical health diseases can have a significant economic impact on individuals, families, and society as a whole. Predicting critical health diseases can help with early detection and treatment, reducing the overall economic burden associated with managing critical health diseases.

18.4 METHODS OF USING AI TO PREDICT CRITICAL HEALTH DISEASES

There are several methods of using AI to predict critical health diseases. One of the most common methods is machine learning. Data is utilized by machine learning algorithms to discover trends and create forecasts. In healthcare, ML algorithms can be trained on high-volume data sets of patients to predict the likelihood of developing a critical health disease (Yan et al., 2020). Another method of using AI to predict critical health diseases is deep learning, which is the subdivision of machine learning that employs artificial neural networks to learn from data. Deep learning algorithms have been successfully applied to medical imaging data to predict critical health diseases such as cancer (Razzak et al., 2018).

There are several AI-based methods, shown in Figure 18.1, to predict critical health diseases.

18.4.1 Machine Learning

Data is used by machine learning algorithms to discover trends and create forecasts. Data is used by machine learning algorithms to discover trends and create forecasts. Machine learning models can analyze large data sets, including electronic health

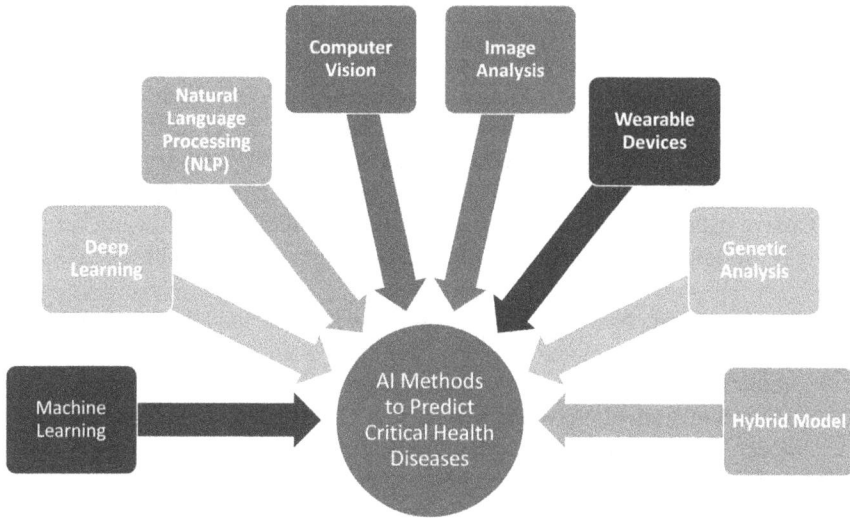

FIGURE 18.1 AI based methods of critical health diseases.

records and genomic data, to identify patterns and predict critical health diseases (Wahl et al., 2018).

A large data set of medical records, lifestyle factors, and genetic information of patients can be used for training the model using machine learning algorithms to identify patterns and risk factors associated with critical health diseases. These algorithms can then be used to predict the likelihood of an individual developing a critical health disease.

Using machine learning techniques, it is possible to predict serious illnesses like diabetes, cancer, and cardiovascular disease (Srivastava et al., 2020). Some of the most common machine learning techniques used in healthcare include:

- **Supervised Learning:** These types of algorithms are trained on labeled data sets, where the outcome variable is known. Once training completed successfully, new and unlabeled data can be analyzed to predict outcomes. For example, based on risk indicators like age, gender, blood pressure, and cholesterol levels, supervised learning algorithms can be used to forecast the possibility of acquiring cardiovascular disease.
- **Unsupervised Learning:** Unlike supervised, unlabeled data are utilized in unsupervised learning, algorithms to identify patterns or clusters in data set. Unsupervised learning algorithms can be utilized to analyze patient data and identify subgroups of patients with similar risk factors, symptoms, or disease progression.
- **Reinforcement Learning:** A type of machine learning called reinforcement learning trains algorithms to base their decisions on feedback from the outside environment. Reinforcement learning algorithms can be used to improve treatment strategies, like determining the optimal medication doses for a patient based on their response.

18.4.2 Deep Learning

A branch of machine learning that uses artificial neural networks to simulate the function of the human brain is known as deep learning. This method can identify complex patterns and relationships in large data sets, which can be useful for predicting critical health diseases (Harerimana et al., 2019). Deep learning methods can be used to predict critical health diseases, since it is a type of AI. These methods use massive data to train models to make accurate predictions. Following are some examples of deep learning methods utilized to predict critical health diseases:

- **Convolutional Neural Networks (CNNs):** These models are generally used in image recognition. They have also been used to diagnose and predict illnesses cancer, cardiovascular disease, stroke, etc.
- **Recurrent Neural Networks (RNNs):** RNNs are one more deep learning model type that can be used to forecast serious illnesses. They work particularly effectively for jobs involving sequential data, such forecasting the course of a disease or the effectiveness of a treatment.
- **Long Short-Term Memory (LSTM) Networks:** LSTM networks, a class of RNN that can manage long-term dependencies, are frequently employed in jobs involving natural language processing. Additionally, they have been used to forecast serious illnesses, such as predicting the likelihood of developing Alzheimer's disease based on genetic information.
- **Autoencoder Networks:** Autoencoder networks are used for unsupervised learning tasks and can be utilized to extract relevant attributes from unprocessed raw data, such as medical images or genetic data. These extracted features can then be used to predict critical health diseases.

18.4.3 Natural Language Processing (NLP)

NLP is a specialized branch of AI that focuses on how computers and human languages interact. NLP algorithms can investigate massive medical documents and texts, which are generally in unstructured form, to identify patterns and make predictions about critical health diseases. Natural language processing (NLP) algorithms can be used to obtain knowledgeable data from medical records, electronic health records, and patient surveys for the intention of predicting the likelihood of an individual developing a critical health disease (Ivanov et al., 2021). NLP primarily aims on understanding and processing human language. NLP techniques can be applied to predicting critical health diseases by analyzing non-attributed text from various sources (Jayaraman et al., 2020). Here are some methods of using NLP to predict critical health diseases:

- **Text Classification:** Text classification is a technique that involves categorizing text data into different categories. It used to classify electronic health records data, clinical transcripts, and social media into different categories based on depending on the existence or absence of particular phrases, symptoms, or risk factors linked to serious health conditions. This can help in identifying high-risk individuals and providing targeted interventions and treatments (Kumar et al., 2022).

- **Sentiment Analysis:** For analyzing text data to determine the sentiment, emotion, and opinion in the text, sentiment analysis can be utilized. NLP can be used to analyze text data from online health forums, social media, and other sources to identify individuals expressing symptoms or concerns related to critical health diseases. This can help in early detection and intervention (Gupta et al., 2020).
- **Topic Modeling:** Topic modeling is a technique that involves identifying topics or themes in text data. It is to analyze electronic health records texts, clinical transcripts, and social media to identify common topics related to critical health diseases. This can help in identifying risk factors, common symptoms, and potential interventions (Boon-Itt & Skunkan, 2020).
- **Text Summarization:** Text summarization is a technique that involves summarizing large volumes of text data into concise and meaningful summaries. NLP can be used to summarize electronic health records, clinical notes, and social media data related to critical health diseases. This can help in identifying important insights, trends, and patterns associated with critical health diseases (Chen et al., 2020).

18.4.4 COMPUTER VISION

The goal of the AI discipline of computer vision is to make it possible for machines to comprehend and interpret visual data from their surroundings. By analyzing medical pictures like X-rays and MRIs, computer vision can be utilized in the healthcare industry to spot trends and forecast serious diseases. Computer vision can be used in a variety of ways to predict critical health diseases (Garg & Mago, 2021). Here are some methods of using AI to predict critical health diseases using computer vision:

- **Medical Imaging Analysis:** Medical images can be used to detect various critical illness, including cancer, cardiovascular diseases, and neurological disorders (Liu et al., 2019). AI algorithms can analyze medical images to detect patterns that are invisible to the human. For example, AI can be trained to detect lung nodules in chest X-rays, which can be indicative of lung cancer.
- **Early Disease Detection:** AI can be used to identify early signs of diseases before they manifest in physical symptoms. For example, AI systems can examine retinal scans to find early indications of diabetic retinopathy, a condition that, if unchecked, can result in blindness (Khemasuwan et al., 2020).
- **Risk Assessment:** Based on elements including a patient's genetic makeup, lifestyle, and family history, AI can determine their likelihood of developing a specific disease (Yoon et al., 2003). For example, based on medical history and lifestyle decisions, AI systems can be trained to predict a person's risk of acquiring heart disease.

- **Monitoring Disease Progression:** It can be used to track the progression of an illness over time. For example, brain MRI scans can be examined by AI algorithms to track the progression of Alzheimer's disease (Greenberg et al., 2009).
- **Personalized Treatment Plans:** Personalized treatment plans can be created for patients using AI based on their individual medical histories and genetic makeup. AI, for instance, can be used to analyze a patient's genetic information to determine the most effective course of cancer treatment (Diamandis et al., 2010).

18.4.5   IMAGE ANALYSIS

Different types of critical disease diagnostic medical images like CT scans images, MRIs, and X-rays can be examined to identify important hidden patterns and features associated with critical health diseases. Once images are analyzed and hidden patterns are discovered using AI, the discovered data can be utilized to predict the likelihood of an individual developing a critical health disease. The image analysis based on AI algorithms has shown promising results in predicting critical health diseases (Koul et al., 2022). Here are some methods that can be used:

- **Radiology Image Analysis:** It is possible to train AI systems to recognize anomalies in radiology pictures including X-rays, CT scans, and MRI scans. For instance, radiological scans can be used to train AI-based systems to recognize lung nodules, breast cancer, and brain tumors (Debelee et al., 2020).
- **Dermatology Image Analysis:** Some algorithms can be trained to detect skin diseases by analyzing images of skin lesions. For example, AI-based algorithms can be trained to detect melanoma, psoriasis, and eczema (Hogarty et al., 2020).
- **Ophthalmology Image Analysis:** To detect eye diseases by analyzing images of the retina and optic nerve, AI algorithms can be used. For example, AI-based algorithms can be trained to detect glaucoma and diabetic retinopathy (Sarhan et al., 2020).
- **Endoscopy Image Analysis:** Models can be utilized to detect abnormalities in endoscopy images such as colonoscopy and gastroscopy. For example, AI-based algorithms can be trained to detect polyps and early signs of colorectal cancer (Viscaino et al., 2021).
- **Histopathology Image Analysis:** In order to identify cancer and other disorders, AI systems can be trained to analyze photos of tissue samples. For example, AI-based algorithms can be trained to detect breast cancer, lung cancer, and prostate cancer in histopathology images (Komura & Ishikawa, 2018).

18.4.6   WEARABLE DEVICES

Another promising area of AI to analyze data from wearable devices, such as smartwatches and/or fitness trackers, to identify patterns and risk factors associated with critical health diseases. This research can be used to forecast a person's risk of serious illness. As they provide a convenient and non-invasive way to monitor various health parameters, wearable devices are growing in popularity in the healthcare sector. With the advancements in AI technology, it is now possible to use data collected from these devices to predict critical health diseases (Nahavandi et al., 2022). Here are some methods of using AI to predict critical health diseases using wearable devices:

- **Machine Learning Algorithms:** The data gathered from wearable devices can be trained to identify trends and abnormalities using machine learning algorithms. These algorithms can be used to analyze the data and spot pattern changes that could portend the start of serious illnesses including cancer, diabetes, and heart disease (Awotunde et al., 2021).
- **Predictive Modeling:** Predictive modeling is another method of using AI to predict critical health diseases. It involves developing models that make use of historical data to forecast the probability of a future event, such as the development of a serious illness. Wearables can gather information that can be utilized to train these models and increase their precision (Qureshi et al., 2020).
- **Natural Language Processing (NLP):** Interactions using voice- or text-based wearable technology can be used to collect data. This data can be analyzed using NLP algorithms to spot trends or warning signs of serious illnesses (Le Glaz et al., 2021).
- **Deep Learning:** It can be used to analyze the huge volumes of data obtained via mobile devices and find patterns or anomalies that might signal the beginning of serious illnesses (Muthu et al., 2020).
- **Sensor Fusion:** it includes combining data from several sensors to improve the accuracy and reliability of predictions. Wearable devices can collect data from multiple sensors. By combining these data, AI algorithms can more accurately forecast serious illnesses (Muzammal et al., 2020).

18.4.7 GENETIC ANALYSIS

Artificial intelligence (AI) can be used to analyze genetic data and find risk factors for serious diseases. The results of this analysis can be used to forecast a person's risk of contracting a serious illness (Dias & Torkamani, 2019).

18.4.8 HYBRID MODELS

Hybrid models increase prediction accuracy by combining many AI techniques into a single more accurate technique. A hybrid approach, for instance, may integrate machine learning and natural language processing to analyze both structured and unstructured data and forecast serious diseases (Kavitha et al., 2021).

18.5   APPLICATION OF AI TECHNIQUE TO PREDICT LUNG CANCER: A CASE STUDY

The most severe and common type of cancer, lung cancer is now the main reason for cancer-related deaths globally. Lung cancer causes around 25% of all cancer deaths in the United States (Greenlee et al., 2000). Lung cancer causes 8.1% of all cancer-related deaths and 5.9% of all cancers in India (Singh et al., 2021). The conventional diagnosis of lung cancer involves a biopsy, which is an invasive procedure that can be painful and expensive. Therefore, non-invasive methods are required for the early identification of lung cancer (Marrugo-Ramírez et al., 2018).

It begins in the lungs and spread throughout the body. There are two primary types of lung cancer classification. Non-small cell lung cancer and small cell lung cancer are two subtypes of lung cancer. The malignant cells of small cell lung cancer form in the pulmonary tissues of the lungs. Lung cancers that are neither tiny nor small cells are together referred to as small cell carcinomas. Cigarette smoking is the main cause of small cell lung cancer. Non-small cell lung cancer is a condition in which the tissues of the lungs develop cancerous cells (Ambrosini et al., 2012).

Chest discomfort, shortness of breath, loss of weight, etc. are only some of the patient-specific symptoms that may be present (Chowienczyk et al., 2020). Detection in an early stage is crucial, as it increases the likelihood that mortality may be reduced by vigilant monitoring and treatment. Lung cancer survival rates have been improving thanks to better diagnosis and treatment options. Using machine learning for both illness diagnosis and prognosis has already shown to be very successful. The higher degree of accuracy in the diagnosis is a result of machine learning. Today's standard of medical treatment is based mostly on machine learning. There are healthcare programs all throughout the world that are using machine learning algorithms. It's possible that disagreements over medical diagnoses may arise with the advent of AI in computers.

Lung cancer can be substantially more successfully treated and has a higher survival probability with early identification. Machine learning techniques have shown great potential in predicting the occurrence of lung cancer by analyzing various risk factors (Joshua et al., 2020a). We give a thorough analysis of recent research on machine learning–based lung cancer prediction. By examining the clinical and demographic characteristics of patients, we also suggest a machine learning–based method for the early diagnosis of lung cancer.

The purpose of machine learning (ML) is to use data analysis to discover a condition's symptoms. Because of this, doctors will have an easier time pinpointing where illnesses first began. Since the data set was investigated using several ML techniques, the research has been validated. It facilitates a more succinct analysis of the circumstances, leading to better allocation of resources and, ultimately, a higher profit margin. ML keeps an eye on disease outbreaks so that prompt measures may be implemented.

ML algorithms have been increasingly common in the field of medical research in recent years. Large data sets can be analyzed by ML algorithms to find patterns that human specialists might not see. The development of machine learning algorithms for the prediction of lung cancer has so been the focus of various investigations (Radhika et al., 2019).

The lung cancer begins in the lungs, which are vital organs necessary for breathing, and it has the potential to spread to other body areas. It is one of the most prevalent diseases in the world and causes a significant percentage of fatalities due to cancer each year (Rudin et al., 2021). Patients with lung cancer have a low chance of survival, and the disease's prognosis greatly depends on the stage of the cancer when it is discovered. To improve patient outcomes and raise survival rates for lung cancer patients, early detection and correct diagnosis are crucial. It raises the likelihood that a therapy will be successful (Shin et al., 2020). As a result, there is a critical need for precise and trustworthy techniques for lung cancer early diagnosis.

By examining multiple risk factors, ML approaches have demonstrated considerable promise in predicting the development of lung cancer. Large data sets can be used to train ML algorithms to find patterns and relationships that are difficult for people to spot (Cruz et al., 2011). ML techniques have been employed in various research recently for the early diagnosis of lung cancer. We provide a thorough analysis of these research in this work and suggest a ML-based strategy for lung cancer early detection.

Machine learning algorithms have demonstrated significant promise in properly detecting lung cancer in recent years. These algorithms discover patterns and forecast the probability that a patient has lung cancer using information from patient records, medical pictures, and other sources (Joshua et al., 2020b). Additionally, ML models can assist in identifying the most important variables that affect the likelihood of developing lung cancer, giving healthcare professionals important information.

In this case study, we explore various machine learning algorithms that can predict lung cancer accurately. We also evaluate the performance of these algorithms using publicly available lung cancer data sets. Finally, we propose an ensemble model that combines multiple algorithms to achieve even better performance.

ML methods have been applied in numerous research for the early identification of lung cancer. In a study by Yang et al. (2020), based on demographic and clinical data, a ML model was created to forecast the likelihood of developing lung cancer. The model's area under the curve (AUC) was 0.83 and its accuracy was 85.6% (Wang et al., 2020b). Another study by Pan et al. (2020) used a deep learning algorithm with chest computed tomography (CT) scans to forecast the likelihood of developing lung cancer. The model's accuracy was 90.6% and its AUC was 0.91 (Pan et al., 2020). In another study by Wang et al. (2020), they suggested another prediction model the risk of lung cancer on demographic, clinical, and radiological features. An accuracy of 87.9% and an AUC of 0.85 are reported form this approach (Xia et al., 2020). In a recent study by Liu et al. (2021), a machine learning model was developed to predict the survival rate of lung cancer patients based on clinical and genomic features. The model's accuracy was 78.7% and its AUC was 0.76 (Xie et al., 2021).

FIGURE 18.2   Methodology to predict lung cancer.

18.5.1 Methodology

We conducted a systematic review of the literature, in which several electronic databases like PubMed, Scopus, and Web of Science were utilized. Different good-quality lung cancer–related articles published between 2015 and 2022 were analyzed and summarized. The following search terms were used: "lung cancer prediction," "machine learning," "deep learning," "artificial intelligence," and "medical imaging."

The proposed approach shown in Figure 18.2 consists of the following steps:

- **Data Collection:** Collection of good quality related clinical and demographic data of lung cancer patients from publicly available data sets.
- **Data Pre-Processing:** To remove missing values, handling categorical variables, and scaling numerical features.
- **Feature Selection:** Selection of the most appropriate features using various techniques used for feature selection.
- **Machine Learning Model Development:** We will develop a machine learning model using various ML algorithms.
- **Model Evaluation:** Various performance evaluation metrics are used to evaluate the performance of the developed model.

18.5.2 Introduction to WEKA Tool

Many of the AI researchers use WEKA, a free and open-source Java tool distributed under the GNU General Public License. It can run on practically any current computing platform because it is fully implemented in the Java programming language, making it portable and platform independent. Data pre-processing, clustering, classification, association, visualization, and feature selection are just a few of the common data mining activities that WEKA can perform.

The primary goal of the WEKA tool is to provide scientists and engineers with a comprehensive suite of automated learning algorithms and data processing capabilities.

We can examine the data using the visualization tool it offers. The same data set can be used to apply several models. Then, we can evaluate the results of various models and choose the one that best serves our needs. Consequently, using WEKA, we can speed up the overall development of ML models.

The office space of the WEKA is equipped with tools for regression analysis, classification, compilation, code extraction, and attribute selection. Numerous pre-processing techniques and a clear presentation of data allow for a thorough analysis of preliminary results. WEKA's various GUIs make even the most fundamental operations easy to do. Navigating Explorer's primary user interfaces. It has panel-based interface that can perform a variety of data mining activities for a wide range of panel types. The first panel, dubbed "pre-processing," is where you'll find the tools you'll need to import and transform data in preparation for applying WEKA data filters. Information may be retrieved from a wide variety of repositories, including files, URLs, and databases. WEKA supports a variety of file formats, including the ARFF format, CSV, LibSVM, and C4.5.

18.5.3 DATA MINING WITH WEKA

The first user interface of WEKA may be described as seen in the image below. The following six buttons on the WEKA GUI chooser activate the WEKA's graphical environment:

- **Explorer:** An environment for data exploration called Explorer enables the processing of unprocessed data, character selection, the use of learning algorithms, and presenting discoveries in various visual formats.
- **Experiment:** A platform for conducting controlled experiments and statistically evaluating different machine learning methods.
- **Knowledge Flow:** Explorer, provides a graphical representation of the KDD procedure and has a similar drag-and-drop interface.
- **Simple CLI:** It has a user-friendly command line interface (CLI) for running WEKA scripts.
- **Load:** A data set from the library and explores the interface's initial displays. This study makes reference to the data set "lung cancer 'tr' data set.csv."

There are different panels in the Explorer interface to provide access to the main components of the workbench:

- **The Pre-Process Panel:** It pre-processes the data by importing it from a database, a CSV file, an ARFF, etc., and employing a filtering technique that can be used to change the data's format, such as turning numerical properties into discrete ones. On the pre-process screen, it is also possible to delete instances and attributes in accordance with particular criteria. You may also view the graph for a specific characteristic.
- **The Classify Panel:** It enables the user to apply classification and regression algorithms to the data set and measure the accuracy of the resultant model. Additionally, incorrect predictions, ROC curves can also be seen. The output region of the classifier contains the results of classification.
- **The Cluster Panel:** It is used to access WEKA's clustering tools, such as the straightforward k-means, EM, DBScan, and XMeans algorithms. The omit Attribute button makes it feasible to omit certain attributes while utilizing the clustering process.
- **The Associate Panel:** It provides access to algorithms like Apriori and Predictive Apriori. Once the proper parameter for the association rule has been selected, the result list enables viewing or saving of the result set.

- **The Select Attributes Panel:** It provides for the search of the best subset of attributes for making predictions across all possible combinations of attributes in the data set.
- **The Visualize Panel:** it is used to visualize 2D plots of current relations.

18.5.4 EXPERIMENTAL DATA

Experimental data are collected from Cancer Genome Atlas (TCGA), a trusted data repository. Images from 422 non-small cell lung cancer (NSCLC) patients are included in this collection. Clinical outcome information and manual delineation of the 3D volume of the gross tumor volume by a radiation oncologist for these patients' pre-treatment CT scans are available.

WEKA is used for data analysis. The data set is pre-processed to eliminate missing and faulty values using WEKA. Various data mining classification techniques are utilized, such as Bayes, functional, lazy, and tree classifiers, to analyze the data set and to find out the best classifier having maximum accuracy (other performance parameters too). The effect of classes Balancer and Resampling are also analyzed to observe the effect on accuracy and other performance parameters. After achieving the goal of finding the best classifier for analysis of lung cancer, various performance tuning approaches are utilized to maximize the accuracy and other performance parameters.

18.5.5 RESULT AND SIMULATION

The results of practice and inspection are shown on the right panel in WEKA. It also displays the percentage of samples that were correctly categorized and the number of samples that were incorrectly labeled. The next step for the expert, once the model has been built and tested, is to put a numerical value on the display of the model. In this study, the researcher used four measures of accuracy: precision, recall, accuracy, and F-Measure. The formulas for these measures are as follows:

$$Accuracy = (TP + TN)/(TP + TN + FP + FN)$$

$$Recall = TP/(TP + FN)$$

$$Precision = TP/(TP + FP)$$

$$Accuracy(A) = (tp + tn)/Total \; \# \; sample$$

$$F - Measure = (2 * Recall * Precision)/(Recall + Precision)$$

In this context, the symbols "TP," "FP," "TN," and "FN" stand for "true positive," "false negative," "false negative," and "false positive."

TABLE 18.1

Performance Evaluation of Different Algorithms

Classifier	Accuracy	Precision	Recall	F-Mesure
SVM	98.98	0.958	0.965	0.988
Naïve-Bayes	97.98	0.935	0.939	0.936
Multilayer Perceptron	99.01	0.998	0.987	0.988
J48	97.12	0.945	0.950	0.966

Table 18.1 shows the results acquired from the experiment done on the data set using the WEKA tool and the results are analyzed through the graph representation in Figure 18.3. The accuracy of multilayer perceptron neural network is best, while the J48 technique is the least accurate for the same data set. Therefore, it is obvious that the MLP technique gives the best results in this case.

In this study, the algorithm of the decision tree is connected to information on lung cancer. Detecting new lung cancer treatments requires combining data from several patients. The data extraction tool WEKA is utilized in this research for its ability to classify data using different classification methods like SVM, Naïve-Bayes, Multilayer Perceptron, and J48. Among the suggested algorithms, Multiplayer Perceptron is found to be the best classifier that can be used to predict lung cancer data, with the highest degree of accuracy (Figure 18.4).

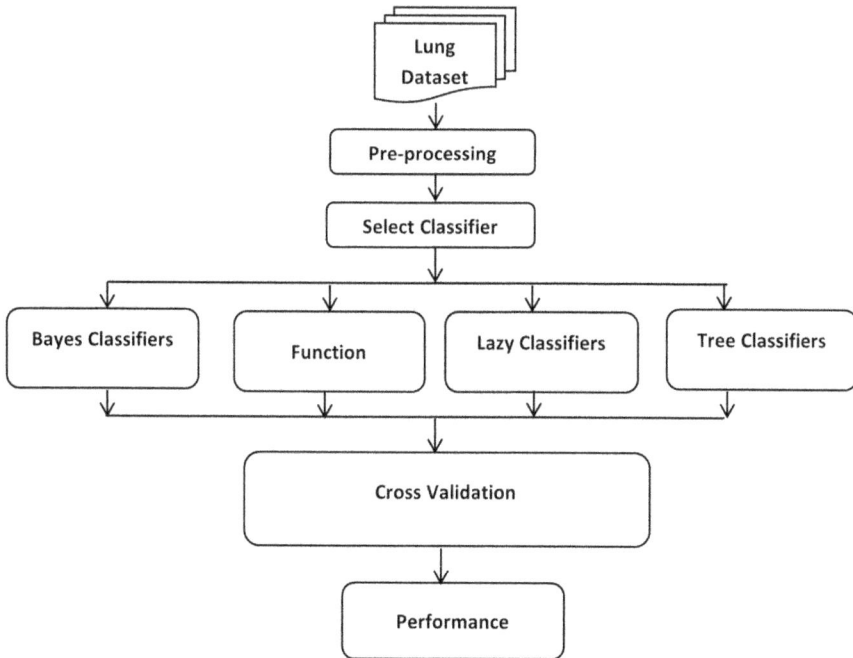

FIGURE 18.3 Methodology to predict lung cancer.

Accuracy of proposed algorithms

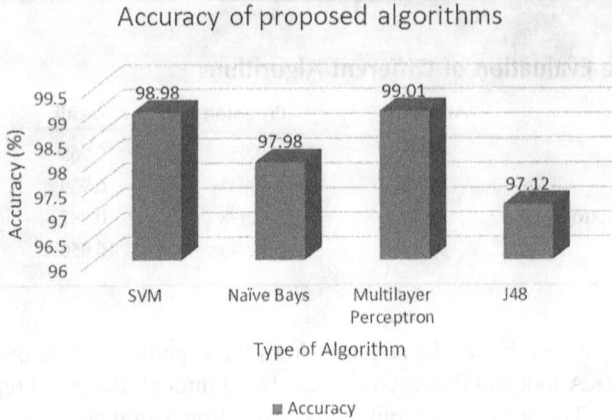

FIGURE 18.4   Graph representation of accuracy of all different classifiers.

18.6   BENEFITS OF USING AI TO PREDICT CRITICAL HEALTH DISEASES

The benefits of using AI to predict critical health diseases are numerous. Firstly, AI can assist healthcare professionals in identifying individuals who are at a high risk of contracting serious illnesses, enabling early intervention and treatment. Secondly, AI can help reduce healthcare costs associated with treating advanced-stage diseases. Thirdly, AI can improve patient outcomes by developing personalized treatment plans that are more effective and less invasive (Dlamini et al., 2020).

Using AI to predict critical health diseases can provide several benefits, including:

- Early Detection: AI algorithms can identify patterns and anomalies in large data sets that is not easily possible for humans to notice. By analyzing patient data, AI can identify early warning signs of critical health diseases, which can help healthcare professionals intervene earlier and potentially save lives.
- Improved Accuracy: AI algorithms can analyze massive data sets with incredible speed and accuracy, reducing the risk of human error. This can improve the accuracy of critical health disease diagnoses, leading to better patient outcomes.
- Personalized Treatment: To create individualized treatment regimens, AI systems can examine patient data including medical history, way of life, and genetic information. By doing so, the likelihood of problems is decreased, and treatment outcomes may be more successful.
- Reduced Healthcare Costs: Early detection and personalized treatment can help reduce healthcare costs by preventing expensive emergency room visits and hospitalizations.

- Better Healthcare Planning: AI algorithms can analyze population-level health data to identify trends and predict disease outbreaks. This information can help healthcare professionals plan and prepare for potential health crises.

18.7 POTENTIAL LIMITATIONS OF AI IN PREDICTING CRITICAL HEALTH DISEASES

Despite the many benefits of using AI to predict critical health diseases, there are potential limitations that need to be considered. One of the main limitations is the quality of data used for the purpose of training AI algorithms. If these training data is biased or incomplete, the predictions made by AI algorithms may not be accurate. Additionally, the interpretability of AI algorithms can be challenging, making it difficult for healthcare providers to understand how predictions are made (Battineni et al., 2020). While AI has significant potential to improve healthcare outcomes, there are also some potential limitations to consider when using AI to predict critical health diseases. Some of these limitations include:

- **Limited Data Availability:** The high quality and massive amount of data is required for the training and testing of the AI models effectively. In some cases, there may be a limited amount of data available, which can limit the performance of the AI model.
- **Bias in Data:** The AI model may be biased if the data used to train it was biased. Particularly in populations that are marginalized or under-represented, this can result in erroneous projections and errors in diagnosis or treatment.
- **Lack of Transparency:** AI models can be intricate and challenging to understand. Healthcare providers may find it difficult to trust and use these predictions due to the lack of transparency that can make it difficult to understand how the model arrived at a certain forecast.
- **Dependence on Technology:** The use of AI for critical health disease prediction is dependent on technology. If the technology fails or malfunctions, the predictions generated by the model can be compromised.
- **Legal and Ethical Issues:** The application of AI in healthcare raises a number of moral and legal concerns, such as data privacy, liability, and informed permission. These issues must be addressed to ensure the ethical use of AI in healthcare.
- **Limited Data Quality:** AI requires massive high-quality data to function effectively. However, in some cases, there may be limited or poor-quality data available, which can lead to inaccurate predictions.
- **Limited Generalizability:** AI-based predictive models may perform well in one population but may not be as accurate in other populations. This limited generalizability can make it challenging to apply AI algorithms to diverse patient populations.
- **Legal and Ethical Concerns:** The application of AI algorithms in healthcare may raise legal and ethical issues. Who is liable for the

repercussions, for instance, if a diagnostic or treatment error is generated by an AI algorithm? There can also be issues with patient privacy and data security (Graham et al., 2020).

18.8 FUTURE DIRECTIONS OF AI IN PREDICTING CRITICAL HEALTH DISEASES

The potential of AI in forecasting serious illnesses is great. As AI technology continues to evolve, healthcare providers can expect to see improved accuracy in predictions, more personalized treatment plans, and reduced healthcare costs (Hashimoto et al., 2018). Additionally, there is potential for AI to be used in combination with other technologies, such as wearable devices, to monitor patient health and predict critical health diseases. The future directions of AI in predicting critical health diseases are promising and can significantly improve healthcare (Chen et al., 2019). Some potential directions include:

- **Integration of Multiple Data Sources:** Currently, AI algorithms mostly rely on structured data, such as medical records or lab results. In the future, wearable technology and social media will be examples of more varied and unstructured data sources that AI can be linked with to provide a more thorough knowledge of a patient's health (Daneshjou et al., 2021).
- **Explainable AI:** To address the lack of transparency in AI algorithms, there is a growing need for explainable AI. This type of AI would provide clinicians with clear explanations of how predictions are made and allow them to understand the reasoning behind the predictions.
- **Improved Data Quality:** As more data becomes available, there will be a growing need to ensure that the data is of high quality, which requires better data collection tools and data cleaning methods to ensure that AI algorithms are trained on accurate and reliable data (Roh et al., 2019).
- **Enhanced Personalization:** Personalized treatment plans for individuals based on particular attributes, such as their genetic profile, lifestyle, and health history, can be developed using AI algorithms. In the future, AI can be integrated with other advanced technologies such as robotics and 3D printing to create personalized medical devices or implants (Goetz & Schork, 2018).
- **Empowerment of Patients:** AI can also be used to empower patients by providing them with real-time feedback and personalized recommendations for managing their health. Patients can use wearable devices and mobile apps that are integrated with AI algorithms to monitor their health and receive guidance on managing their conditions (Albahri et al., 2018; Vyas, 2023).

18.9 CONCLUSION

In conclusion, utilizing AI to forecast serious illnesses has the potential to greatly enhance patient outcomes, lower healthcare expenses, and advance medical science. The advantages of utilizing AI exceed the dangers, despite any potential drawbacks. Healthcare professionals should keep looking into using AI to predict serious illnesses.

REFERENCES

Aggarwal, K., Mijwil, M.M., Al-Mistarehi, A.H., Alomari, S., Gök, M., Alaabdin, A.M.Z., & Abdulrhman, S.H. (2022). Has the future started? The current growth of artificial intelligence, machine learning, and deep learning. *Iraqi Journal for Computer Science and Mathematics*, *3*(1), 115–123.

Albahri, O.S., Albahri, A.S., Mohammed, K.I., Zaidan, A.A., Zaidan, B.B., Hashim, M., & Salman, O.H. (2018). Systematic review of real-time remote health monitoring system in triage and priority-based sensor technology: Taxonomy, open challenges, motivation and recommendations. *Journal of Medical Systems*, *42*, 1–27.

Ambrosini, V., Nicolini, S., Caroli, P., Nanni, C., Massaro, A., Marzola, M.C., … & Fanti, S. (2012). PET/CT imaging in different types of lung cancer: an overview. *European Journal of Radiology*, *81*(5), 988–1001.

Awotunde, J.B., Folorunso, S.O., Bhoi, A.K., Adebayo, P.O., & Ijaz, M.F. (2021). Disease Diagnosis System for IoT-based Wearable Body Sensors with Machine Learning Algorithm. *Hybrid Artificial Intelligence and IoT in Healthcare*, 201–222.

Barrett, M., Boyne, J., Brandts, J., Brunner-La Rocca, H.P., De Maesschalck, L., De Wit, K., … & Zippel-Schultz, B. (2019). Artificial intelligence supported patient self-care in chronic heart failure: a paradigm shift from reactive to predictive, preventive and personalised care. *EPMA Journal*, *10*, 445–464.

Battineni, G., Sagaro, G.G., Chinatalapudi, N., & Amenta, F. (2020). Applications of machine learning predictive models in the chronic disease diagnosis. *Journal of Personalized Medicine*, *10*(2), 21.

Bi, W.L., Hosny, A., Schabath, M.B., Giger, M.L., Birkbak, N.J., Mehrtash, A., … & Aerts, H.J. (2019). Artificial intelligence in cancer imaging: clinical challenges and applications. *CA: A Cancer Journal for Clinicians*, *69*(2), 127–157.

Boon-Itt, S., & Skunkan, Y. (2020). Public perception of the COVID-19 pandemic on Twitter: sentiment analysis and topic modeling study. *JMIR Public Health and Surveillance*, *6*(4), e21978.

Bray, F., Laversanne, M., Cao, B., Varghese, C., Mikkelsen, B., Weiderpass, E., & Soerjomataram, I. (2021). Comparing cancer and cardiovascular disease trends in 20 middle-or high-income countries 2000–19: A pointer to national trajectories towards achieving Sustainable Development goal target 3.4. *Cancer Treatment Reviews*, *100*, 102290.

Chen, P.H.A., Leibrand, A., Vasko, J., & Gauthier, M. (2020). Ontology-based and User-focused Automatic Text Summarization (OATS): Using COVID-19 Risk Factors as an Example. *arXiv preprint arXiv:2012.02028*.

Chen, P.H.C., Gadepalli, K., MacDonald, R., Liu, Y., Kadowaki, S., Nagpal, K., … & Stumpe, M.C. (2019). An augmented reality microscope with real-time artificial intelligence integration for cancer diagnosis. *Nature Medicine*, *25*(9), 1453–1457.

Chowienczyk, S., Price, S., & Hamilton, W. (2020). Changes in the presenting symptoms of lung cancer from 2000–2017: a serial cross-sectional study of observational records in UK primary care. *British Journal of General Practice*, *70*(692), e193–e199.

Cruz, C.S.D., Tanoue, L.T., Matthay, R.A. (2011). Lung cancer: Epidemiology, etiology, and prevention. *Clinics in Chest Medicine*, *32*(4), 605–644.

Daneshjou, R., Smith, M.P., Sun, M.D., Rotemberg, V., & Zou, J. (2021). Lack of transparency and potential bias in artificial intelligence data sets and algorithms: a scoping review. *JAMA Dermatology*, *157*(11), 1362–1369.

Debelee, T.G., Kebede, S.R., Schwenker, F., & Shewarega, Z.M. (2020). Deep learning in selected cancers' image analysis—a survey. *Journal of Imaging*, *6*(11), 121.

Diamandis, M., White, N.M., & Yousef, G.M. (2010). Personalized medicine: marking a new epoch in cancer patient management. *Molecular Cancer Research*, *8*(9), 1175–1187.

Dias, R., & Torkamani, A. (2019). Artificial intelligence in clinical and genomic diagnostics. *Genome Medicine*, *11*(1), 1–12.

Dlamini, Z., Francies, F.Z., Hull, R., & Marima, R. (2020). Artificial intelligence (AI) and big data in cancer and precision oncology. *Computational and Structural Biotechnology Journal*, *18*, 2300–2311.

Ebrahimzadeh, A., Ebrahimzadeh, A., Mirghazanfari, S.M., Hazrati, E., Hadi, S., & Milajerdi, A. (2022). The Effect of Ginger Supplementation on Metabolic Profiles in Patients with Type 2 Diabetes Mellitus: A Systematic Review and Meta-Analysis of Randomized Controlled Trials. *Complementary Therapies in Medicine*, 102802.

Frizzell, T.O., Glashutter, M., Liu, C.C., Zeng, A., Pan, D., Hajra, S.G., ... & Song, X. (2022). Artificial Intelligence in Brain MRI Analysis of Alzheimer's Disease Over the Past 12 Years: A Systematic Review. *Ageing Research Reviews*, 101614.

Garg, A., & Mago, V. (2021). Role of machine learning in medical research: A survey. *Computer Science Review*, *40*, 100370.

Ghazal, T.M., Hasan, M.K., Alshurideh, M.T., Alzoubi, H.M., Ahmad, M., Akbar, S.S., ... & Akour, I.A. (2021). IoT for smart cities: Machine learning approaches in smart healthcare—A review. *Future Internet*, *13*(8), 218.

Goetz, L.H., & Schork, N.J. (2018). Personalized medicine: Motivation, challenges, and progress. *Fertility and Sterility*, *109*(6), 952–963.

Graham, S.A., Lee, E.E., Jeste, D.V., Van Patten, R., Twamley, E.W., Nebeker, C., ... & Depp, C.A. (2020). Artificial intelligence approaches to predicting and detecting cognitive decline in older adults: A conceptual review. *Psychiatry Research*, *284*, 112732.

Greenberg, N., Grassano, A., Thambisetty, M., Lovestone, S., & Legido-Quigley, C. (2009). A proposed metabolic strategy for monitoring disease progression in Alzheimer's disease. *Electrophoresis*, *30*(7), 1235–1239.

Greenlee, R.T., Murray, T., Bolden, S., & Wingo, P.A. (2000). Cancer statistics, 2000. *CA: A Cancer Journal for Clinicians*, *50*(1), 7–33.

Gupta, P., Kumar, S., Suman, R.R., & Kumar, V. (2020). Sentiment analysis of lockdown in India during covid-19: A case study on twitter. *IEEE Transactions on Computational Social Systems*, *8*(4), 992–1002.

Harerimana, G., Kim, J.W., Yoo, H., & Jang, B. (2019). Deep learning for electronic health records analytics. *IEEE Access*, *7*, 101245–101259.

Hashimoto, D.A., Rosman, G., Rus, D., & Meireles, O.R. (2018). Artificial intelligence in surgery: Promises and perils. *Annals of Surgery*, *268*(1), 70–76.

Hogarty, D.T., Su, J.C., Phan, K., Attia, M., Hossny, M., Nahavandi, S., ... & Yazdabadi, A. (2020). Artificial intelligence in dermatology—where we are and the way to the future: a review. *American Journal of Clinical Dermatology*, *21*, 41–47.

Hsu, S.D., Chao, E., Chen, S.J., Hueng, D.Y., Lan, H.Y., & Chiang, H.H. (2021). Machine learning algorithms to predict in-hospital mortality in patients with traumatic brain injury. *Journal of Personalized Medicine*, *11*(11), 1144.

Ivanov, O., Wolf, L., Brecher, D., Lewis, E., Masek, K., Montgomery, K., ... & Reilly, C. (2021). Improving ED emergency severity index acuity assignment using machine learning and clinical natural language processing. *Journal of Emergency Nursing*, *47*(2), 265–278.

Jayaraman, P.P., Forkan, A.R.M., Morshed, A., Haghighi, P.D., & Kang, Y.B. (2020). Healthcare 4.0: A review of frontiers in digital health. *Wiley Interdisciplinary Reviews: Data Mining and Knowledge Discovery*, *10*(2), e1350.

Joshua, E.S.N., Chakkravarthy, M., & Bhattacharyya, D. (2020a). An extensive review on lung cancer detection using machine learning techniques: A systematic study. *Rev. d'Intelligence Artif.*, *34*(3), 351–359.

Joshua, E.S.N., Chakkravarthy, M., & Bhattacharyya, D. (2020b). An extensive review on lung cancer detection using machine learning techniques: A systematic study. *Rev. d'Intelligence Artif.*, *34*(3), 351–359.

Kapoor, R., Walters, S.P., & Al-Aswad, L.A. (2019). The current state of artificial intelligence in ophthalmology. *Survey of ophthalmology*, *64*(2), 233–240.

Kavitha, M., Gnaneswar, G., Dinesh, R., Sai, Y.R., & Suraj, R.S. (2021, January). Heart Disease Prediction Using Hybrid Machine Learning Model. In *2021 6th International Conference on Inventive Computation Technologies (ICICT)* (pp. 1329–1333). IEEE.

Khemasuwan, D., Sorensen, J.S., & Colt, H.G. (2020). Artificial intelligence in pulmonary medicine: Computer vision, predictive model and COVID-19. *European Respiratory Review*, *29*(157).

Komura, D., & Ishikawa, S. (2018). Machine learning methods for histopathological image analysis. *Computational and Structural Biotechnology Journal*, *16*, 34–42.

Koul, A., Bawa, R.K., & Kumar, Y. (2022). Artificial intelligence in medical image processing for airway diseases. In *Connected e-Health: Integrated IoT and cloud computing* (pp. 217–254). Cham: Springer International Publishing.

Kumar, Y., Koul, A., & Mahajan, S. (2022). A deep learning approaches and fastai text classification to predict 25 medical diseases from medical speech utterances, transcription and intent. *Soft Computing*, *26*(17), 8253–8272.

Lan, T., Liao, Y.H., Zhang, J., Yang, Z.P., Xu, G.S., Zhu, L., & Fan, D.M. (2021). Mortality and Readmission Rates After Heart Failure: A Systematic Review and Meta-Analysis. *Therapeutics and Clinical Risk Management*, 1307–1320.

Le Glaz, A., Haralambous, Y., Kim-Dufor, D.H., Lenca, P., Billot, R., Ryan, T.C., … & Lemey, C. (2021). Machine learning and natural language processing in mental health: systematic review. *Journal of Medical Internet Research*, *23*(5), e15708.

Liu, J., Bai, R., Chai, Z., Cooper, M.E., Zimmet, P.Z., & Zhang, L. (2022). Low-and middle-income countries demonstrate rapid growth of type 2 diabetes: An analysis based on Global Burden of Disease 1990–2019 data. *Diabetologia*, *65*(8), 1339–1352.

Liu, X., Faes, L., Kale, A.U., Wagner, S.K., Fu, D.J., Bruynseels, A., … & Denniston, A.K. (2019). A comparison of deep learning performance against health-care professionals in detecting diseases from medical imaging: a systematic review and meta-analysis. *The Lancet Digital Health*, *1*(6), e271–e297.

Ma, K., Harmon, S.A., Klyuzhin, I.S., Rahmim, A., & Turkbey, B. (2022). Clinical application of artificial intelligence in positron emission tomography: Imaging of prostate cancer. *PET Clinics*, *17*(1), 137–143.

Marrugo-Ramírez, J., Mir, M., & Samitier, J. (2018). Blood-based cancer biomarkers in liquid biopsy: A promising non-invasive alternative to tissue biopsy. *International Journal of Molecular Sciences*, *19*(10), 2877.

Mordi, I.R., Trucco, E., Syed, M.G., MacGillivray, T., Nar, A., Huang, Y., … & Doney, A.S. (2022). Prediction of major adverse cardiovascular events from retinal, clinical, and genomic data in individuals with type 2 diabetes: a population cohort study. *Diabetes Care*, *45*(3), 710–716.

Murray, C.J., Lopez, A.D., & World Health Organization. (1994). Global Comparative Assessments in the Health Sector: Disease Burden, Expenditures and Intervention Packages.

Murugappan, M., Murugesan, L., Jerritta, S., & Adeli, H. (2021). Sudden Cardiac Arrest (SCA) prediction using ECG morphological features. *Arabian Journal for Science and Engineering*, *46*, 947–961.

Muthu, B., Sivaparthipan, C.B., Manogaran, G., Sundarasekar, R., Kadry, S., Shanthini, A., & Dasel, A. (2020). IOT based wearable sensor for diseases prediction and symptom analysis in healthcare sector. *Peer-to-peer Networking and Applications*, *13*, 2123–2134.

Muzammal, M., Talat, R., Sodhro, A.H., & Pirbhulal, S. (2020). A multi-sensor data fusion enabled ensemble approach for medical data from body sensor networks. *Information Fusion*, *53*, 155–164.

Nahavandi, D., Alizadehsani, R., Khosravi, A., & Acharya, U.R. (2022). Application of artificial intelligence in wearable devices: Opportunities and challenges. *Computer Methods and Programs in Biomedicine*, *213*, 106541.

Pan, Y., Shi, D., Wang, H., Chen, T., Cui, D., Cheng, X., & Lu, Y. (2020). Automatic opportunistic osteoporosis screening using low-dose chest computed tomography scans obtained for lung cancer screening. *European radiology*, *30*, 4107–4116.

Parthiban, G., & Srivatsa, S.K. (2012). Applying machine learning methods in diagnosing heart disease for diabetic patients. *International Journal of Applied Information Systems*, *3*(7), 25–30.

Qureshi, K.N., Din, S., Jeon, G., & Piccialli, F. (2020). An accurate and dynamic predictive model for a smart M-Health system using machine learning. *Information Sciences*, *538*, 486–502.

Radhika, P.R., Nair, R.A., & Veena, G. (2019, February). A Comparative Study of Lung Cancer Detection Using Machine Learning Algorithms. In *2019 IEEE International Conference on Electrical, Computer and Communication Technologies (ICECCT)* (pp. 1–4). IEEE.

Rajkomar, A., Oren, E., Chen, K., Dai, A.M., Hajaj, N., Hardt, M., ... & Dean, J. (2018). Scalable and accurate deep learning with electronic health records. *NPJ Digital Medicine*, *1*(1), 18.

Razzak, M.I., Naz, S., & Zaib, A. (2018). Deep learning for medical image processing: Overview, challenges and the future. *Classification in BioApps: Automation of Decision Making*, 323–350.

Roh, Y., Heo, G., & Whang, S.E. (2019). A survey on data collection for machine learning: A big data-ai integration perspective. *IEEE Transactions on Knowledge and Data Engineering*, *33*(4), 1328–1347.

Rudin, C.M., Brambilla, E., Faivre-Finn, C., & Sage, J. (2021). Small-cell lung cancer. *Nature Reviews Disease Primers*, *7*(1), 3.

Sarhan, M.H., Nasseri, M.A., Zapp, D., Maier, M., Lohmann, C.P., Navab, N., & Eslami, A. (2020). Machine learning techniques for ophthalmic data processing: A review. *IEEE Journal of Biomedical and Health Informatics*, *24*(12), 3338–3350.

Shin, H., Oh, S., Hong, S., Kang, M., Kang, D., Ji, Y.G., ... & Choi, Y. (2020). Early-stage lung cancer diagnosis by deep learning-based spectroscopic analysis of circulating exosomes. *ACS Nano*, *14*(5), 5435–5444.

Si, Y., Du, J., Li, Z., Jiang, X., Miller, T., Wang, F., ... & Roberts, K. (2021). Deep representation learning of patient data from Electronic Health Records (EHR): A systematic review. *Journal of Biomedical Informatics*, *115*, 103671.

Singh, N., Agrawal, S., Jiwnani, S., Khosla, D., Malik, P.S., Mohan, A., ... & Prasad, K.T. (2021). Lung cancer in India. *Journal of Thoracic Oncology*, *16*(8), 1250–1266.

Sotoudeh, H., Shafaat, O., Bernstock, J.D., Brooks, M.D., Elsayed, G.A., Chen, J.A., ... & Friedman, G.K. (2019). Artificial intelligence in the management of glioma: era of personalized medicine. *Frontiers in Oncology*, *9*, 768.

Srivastava, S., Pant, M., & Agarwal, R. (2020). Role of AI techniques and deep learning in analyzing the critical health conditions. *International Journal of System Assurance Engineering and Management*, *11*, 350–365.

Subramanian, M., Wojtusciszyn, A., Favre, L., Boughorbel, S., Shan, J., Letaief, K.B., ... & Chouchane, L. (2020). Precision medicine in the era of artificial intelligence: implications in chronic disease management. *Journal of Translational Medicine*, *18*(1), 1–12.

Tang, A., Tam, R., Cadrin-Chênevert, A., Guest, W., Chong, J., Barfett, J., ... & Canadian Association of Radiologists (CAR) Artificial Intelligence Working Group. (2018). Canadian Association of Radiologists white paper on artificial intelligence in radiology. *Canadian Association of Radiologists Journal*, *69*(2), 120–135.

Viscaino, M., Bustos, J.T., Muñoz, P., Cheein, C.A., & Cheein, F.A. (2021). Artificial intelligence for the early detection of colorectal cancer: A comprehensive review of its advantages and misconceptions. *World Journal of Gastroenterology*, *27*(38), 6399.

Vocaturo, E., Perna, D., & Zumpano, E. (2019, November). Machine Learning Techniques for Automated Melanoma Detection. In *2019 IEEE International Conference on Bioinformatics and Biomedicine (BIBM)* (pp. 2310–2317). IEEE.

Vyas, S. (2023). Extended reality and edge AI for healthcare 4.0: Systematic study. In *Extended Reality for Healthcare Systems* (pp. 229–240). Academic Press.

Wahl, B., Cossy-Gantner, A., Germann, S., & Schwalbe, N.R. (2018). Artificial intelligence (AI) and global health: How can AI contribute to health in resource-poor settings? *BMJ Global Health*, *3*(4), e000798.

Wang, S., Zhang, Y., Lei, S., Zhu, H., Li, J., Wang, Q., ... & Pan, H. (2020a). Performance of deep neural network-based artificial intelligence method in diabetic retinopathy screening: A systematic review and meta-analysis of diagnostic test accuracy. *European Journal of Endocrinology*, *183*(1), 41–49.

Wang, S., Rong, R., Yang, D.M., Fujimoto, J., Yan, S., Cai, L., ... & Xiao, G. (2020b). Computational staining of pathology images to study the tumor microenvironment in lung cancer. *Cancer Research*, *80*(10), 2056–2066.

Wang, W., Yan, Y., Guo, Z., Hou, H., Garcia, M., Tan, X., ... & Suboptimal Health Study Consortium and European Association for Predictive, Preventive and Personalised Medicine. (2021). All around suboptimal health—A joint position paper of the Suboptimal Health Study Consortium and European Association for Predictive, Preventive and Personalised Medicine. *EPMA Journal*, *12*, 403–433.

Wensing, M., Sales, A., Armstrong, R., & Wilson, P. (2020). Implementation science in times of Covid-19. *Implementation Science*, *15*, 1–4.

Wu, Y., Liu, J., Han, C., Liu, X., Chong, Y., Wang, Z., ... & Li, S. (2020). Preoperative prediction of lymph node metastasis in patients with early-T-stage non-small cell lung cancer by machine learning algorithms. *Frontiers in Oncology*, *10*, 743.

Xia, X., Gong, J., Hao, W., Yang, T., Lin, Y., Wang, S., & Peng, W. (2020). Comparison and fusion of deep learning and radiomics features of ground-glass nodules to predict the invasiveness risk of stage-I lung adenocarcinomas in CT scan. *Frontiers in Oncology*, *10*, 418.

Xie, Y., Meng, W.Y., Li, R.Z., Wang, Y.W., Qian, X., Chan, C., ... & Leung, E.L.H. (2021). Early lung cancer diagnostic biomarker discovery by machine learning methods. *Translational Oncology*, *14*(1), 100907.

Yan, L., Zhang, H.T., Xiao, Y., Wang, M., Guo, Y., Sun, C., ... & Yuan, Y. (2020). Prediction of criticality in patients with severe Covid-19 infection using three clinical features: A machine learning-based prognostic model with clinical data in Wuhan. *MedRxiv*, 2020-02.

Yoon, P.W., Scheuner, M.T., & Khoury, M.J. (2003). Research priorities for evaluating family history in the prevention of common chronic diseases. *American Journal of Preventive Medicine*, *24*(2), 128–135.

Zhou, J., Du, M., Chang, S., & Chen, Z. (2021). Artificial intelligence in echocardiography: Detection, functional evaluation, and disease diagnosis. *Cardiovascular Ultrasound*, *19*(1), 1–11.

19 Revolutionizing Healthcare

The Impact of Augmented and Virtual Reality

Pawan Whig, Ashima Bhatnagar Bhatia, Radhika Mahajan, and Ankit Sharma

19.1 INTRODUCTION

The healthcare industry is in a state of rapid change, with new technologies being introduced at an unprecedented rate. One of the most exciting and rapidly evolving areas of healthcare technology is augmented reality (AR) and virtual reality (VR). These technologies have the potential to transform the way healthcare is delivered and received, offering new and innovative solutions to age-old problems. From patient education to telemedicine and surgical planning, AR and VR are being used to improve outcomes and enhance the overall healthcare experience.

In this chapter, we will delve into the world of AR and VR in healthcare, exploring the many applications of these technologies and their impact on the industry. We will examine the benefits and challenges of implementing AR and VR in healthcare and discuss the potential for further growth and development in this field. By the end of this chapter, readers will have a comprehensive understanding of the role of AR and VR in healthcare and the potential they have to revolutionize the way we think about healthcare delivery and patient experiences.

The healthcare augmented and virtual reality market was valued at USD\$2.3 billion in 2021 and is expected to grow at a compound annual growth rate (CAGR) of 26.88% from 2022 to 2030 to reach USD\$19.6 billion. According to recent estimates, the market is projected to reach around \$9.7 billion in the next five years, and by 2027, it is expected to grow almost 3.5 times its current value of \$2.7 billion, as depicted in Figure 19.1.

The causes of this occurrence are many. Most importantly, AR and VR have the potential to allow a range of new healthcare modalities, from enhancing the training of doctors and other medical professionals to expanding their capacity to practice medicine through telehealth and telemedicine (Whig et al., 2022). Technology has opened up a completely new way to look at medical practice. One can understand the difference between VR, AR, and MR, as shown in Figure 19.2.

DOI: 10.1201/9781003346289-19

Augmented Reality & Virtual Reality in Healthcare Market Size, By Region, 2016 - 2028 (USD Billion)

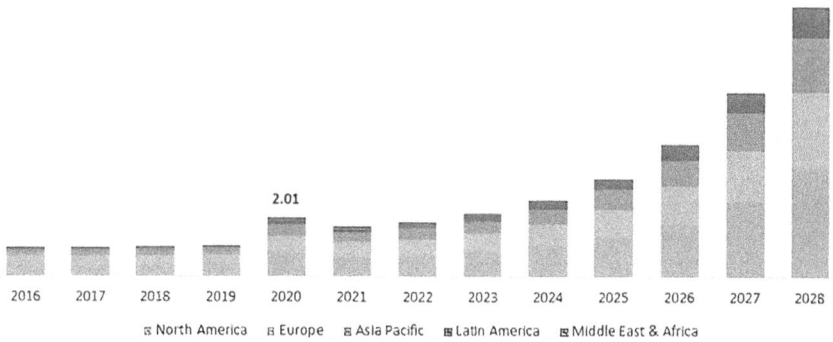

FIGURE 19.1 AR and VR market size by 2028.

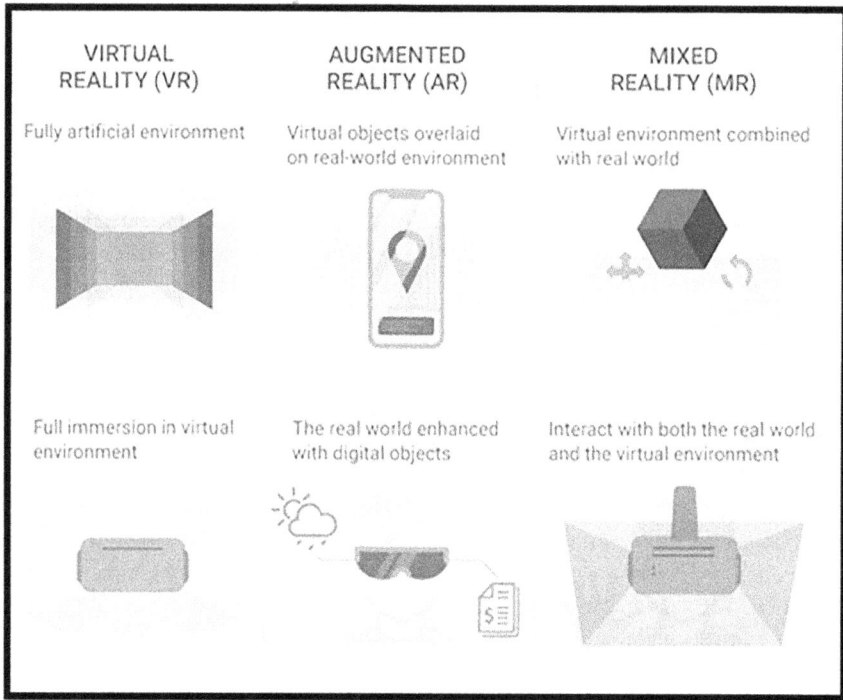

FIGURE 19.2 Difference between AR, VR, and MR.

Advancements in AR/VR technology offer numerous benefits, such as providing quality education and enhancing collaboration in the healthcare industry. However, relying heavily on technology can also pose a significant risk, particularly in terms of security and privacy concerns with digital transmission of patient data.

Global VR Device Shipment Share by Vendor

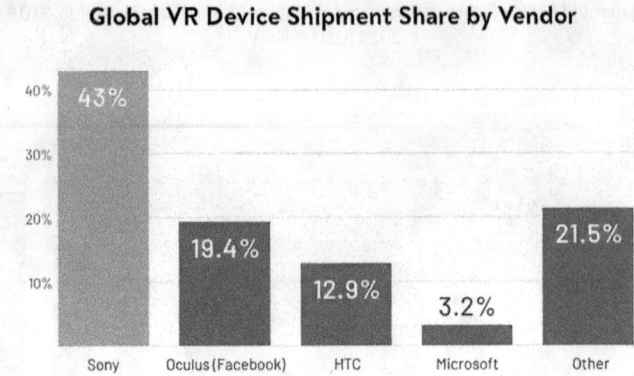

FIGURE 19.3   Global VR device shipment share.

Moreover, healthcare is a people-centric profession, and there are concerns about whether the increased use of AR/VR could negatively impact the patient-physician relationship by reducing the human element. Despite the potential advantages, it is essential to consider the potential drawbacks before implementing AR/VR technology in healthcare (Vyas, S., & Gupta, S., 2023).

The hardware segment, which accounted for about 67.8% of the market's revenue in the past and is predicted to grow well in the years ahead, is expected to dominate the market in terms of components (Whig et al., 2022). Rising demand for various devices, including 3D sensors, smart glasses, and head-mounted displays, will propel market expansion in the years ahead. The healthcare industry uses each of these tools extensively. These gadgets are used for training and simulation and are also widely utilized for carrying out different types of procedures (Jupalle et al., 2022). The application of these tools for diagnostics will also be important. The global VR device shipment share is shown in Figure 19.3.

The surgery application category is anticipated to command the biggest market share over the next several years. They are employed in the performance of minimally invasive operations. The technology will eventually be used in operations due to ongoing improvements in the field. To bring innovative items to the market, many businesses are forming partnerships and collaborating (Tomar et al., 2021).

Based on the technology being employed, the AR category is anticipated to outpace the VR segment in terms of market share. In terms of revenue, augmented reality previously held a market share of roughly 59% and will increase over the next several years (Whig et al., 2022). Various areas in which AR/ VR is very popular nowadays with its share are shown in Figure 19.4.

19.2   LATEST MARKET UPDATE

The market for AR and VR in healthcare is expected to continue to grow in the coming years. According to recent market research, the global AR and VR healthcare market is projected to reach $5.1 billion by 2025, growing at a compound annual growth rate (CAGR) of 47.5% from 2020 to 2025. This growth is largely due

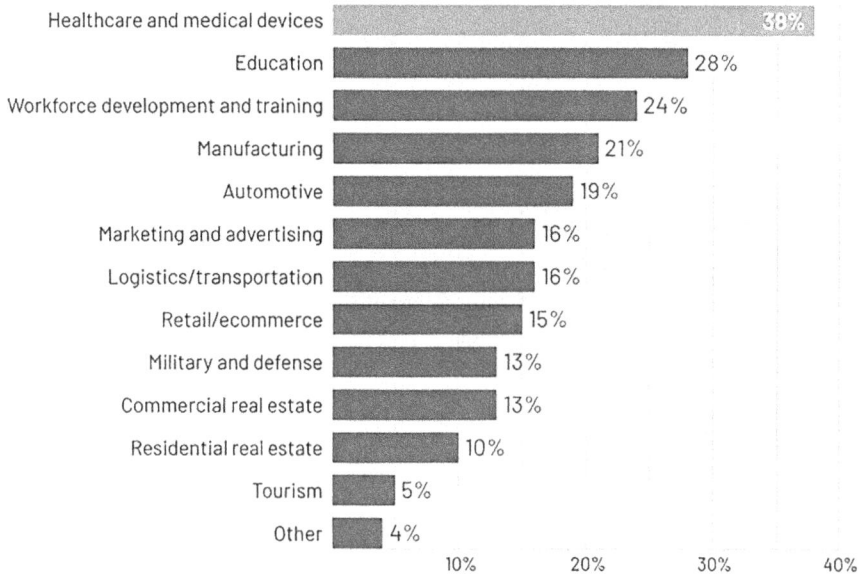

FIGURE 19.4　Various areas in which AR/VR are popular, apart from gaming.

to the increasing adoption of these technologies in various healthcare applications such as patient education, telemedicine, and surgical planning.

The COVID-19 pandemic has further accelerated the adoption of AR and VR in healthcare as remote consultations and virtual care become more important. With the rise of remote work and telemedicine, AR and VR are being used to provide virtual tours of healthcare facilities, remote consultations, and remote training for healthcare professionals.

Moreover, advancements in AR and VR technology are making these solutions more accessible and cost-effective, allowing a wider range of healthcare providers to implement these technologies in their practices. This, combined with increasing investment in the development of AR and VR solutions, is expected to drive the continued growth of the AR and VR healthcare market.

In recent years, there has been an upsurge in demand for wearable technology. The rising usage of VR and AR has also improved how customers experience exercise. The adoption of this technology has produced more affordable and easily accessible healthcare facilities (Anand et al., 2022; Chopra and Whig, 2022). In the upcoming years, it is anticipated that the adoption of various wearable devices, including rings, fit bands, and goggles, will drive market expansion.

In the upcoming years, the market expansion will be driven by the usage of these devices to treat patients' mental health. The industry is anticipated to expand well in the next years due to how well it treats depression and makes patients' surroundings safer. The usage of these devices also improves doctor-patient communication and aids in a better understanding of the course of therapy, both of which will spur market expansion in the years to come. In the upcoming years, the adoption of AR and VR by surgeons and psychiatrists is anticipated to propel market expansion (Rupani et al., 2017).

19.2.1 RESTRAINTS

Due to several technological constraints in the market, the use of virtual reality and augmented reality technologies will be constrained. VR may be useless for treating a few medical disorders, which will prevent the market from expanding in the upcoming years. Other factors affecting market expansion include the specs of the computer being used and its resolution.

19.2.2 OPPORTUNITIES

Inpatient psychiatrists are heavily utilizing these technologies to treat anxiety and depression because there has been an increase in the use of augmented reality and virtual reality in the diagnosis of various diseases and it also aids in planning a proper path for treatment. As a result, the demand for these platforms and technologies is expected to grow in the coming years. Because these technologies have shown to be very helpful in executing difficult procedures, the demand for these platforms is anticipated to increase significantly in the healthcare sector over the course of the projected year.

19.2.3 CHALLENGES

The high cost of augmented and virtual reality systems is one of the main obstacles preventing the sector from expanding. The platforms for AR that are employed in the healthcare industry are quite costly and intricate. Because of the numerous security and privacy issues surrounding the usage of augmented reality, industry development will be constrained in the years to come.

19.3 IMPACT OF COVID-19 ON THE HEALTHCARE MARKET FOR AUGMENTED REALITY AND VIRTUAL REALITY

AR and VR in the worldwide healthcare market were worth USD\$2.0 billion in 2020, and it is predicted to increase at a compound annual growth rate (CAGR) of 27.2% between 2021 and 2028. Among the key determinants expected to drive the growth and adoption of AR and VR innovations in the healthcare sector include technological advancements and digitization in health coverage, favorable regulatory proposals, soaring healthcare spending, increasing usage in surgeries, and clinical experience (Whig & Ahmad, 2012). Pre- and post-pandemic impacts on AR and VR are shown in Table 19.1. These technologies have a wide range of uses in medicine, such as surgery, diagnosis, rehab, education, and learning. The impact analysis of COVID-19 is concisely presented in Figure 19.5.

TABLE 19.1
Pre- and Post-COVID Impact on AR/VR

Pandemic Impact	Post-COVID Outlook
According to prior predictions, the market for augmented reality and virtual reality in healthcare was predicted to reach more than USD$900 million in 2020, but instead, it increased by 27.7% from 2019 to 2020.	Due to the shift in trends toward AR and VR applications and acceptance in the healthcare sector, the market will grow by 85% from 2020 to 2021.
The primary drivers of market expansion are the rising demand for services including telemonitoring, medical training and education, and patient management.	The advancement of technology in software and hardware systems is lowering costs and increasing the user and developer experience. This, in turn, will accelerate the use of augmented and virtual reality.
Furthermore, during a pandemic emergency, there is a quick spike in the need for rapid digitization, training of healthcare specialists, and an increase in usage by the healthcare sector.	Key healthcare firms are increasing their investment in this industry to improve patient outcomes, education, and communication.

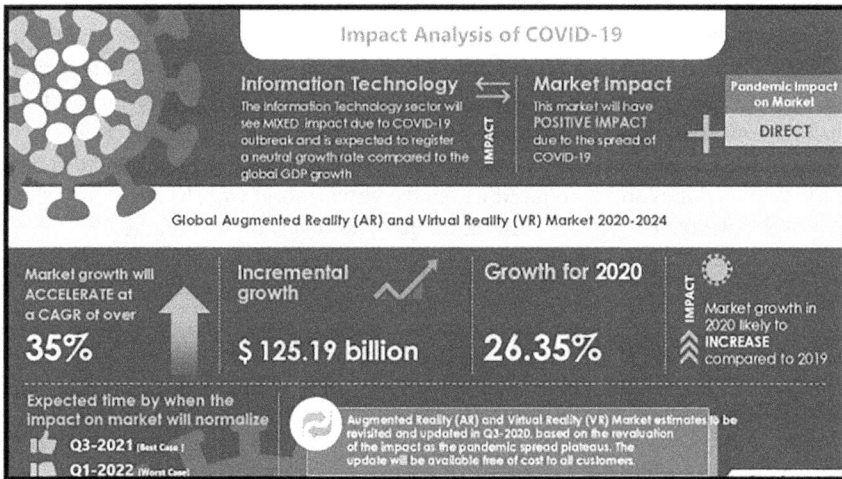

FIGURE 19.5 Impact analysis of COVID-19.

19.4 HEALTHCARE INDUSTRY AND FRESH OPPORTUNITIES

The following section discusses some of the key ways that virtual reality is transforming the healthcare industry and opening up fresh opportunities to enhance various medical treatment modalities.

19.4.1 TREATMENT OF PAIN

When a patient is engaged in virtual reality, the somatosensory cortex and the insula, which are responsible for pain, are less active. By relieving pain in this way, VR pain management can help patients bear uncomfortable medical procedures.

19.4.2 THERAPY

Physicians and medical experts are using virtual reality to treat patients who have phobias. Patients are forced to confront their concerns in VR-controlled surroundings, which trains them to overcome their fears. Virtual simulation is also being used in clinics and hospitals in India to assist people in coping with terrible occurrences in the past.

They are designed to encounter genuine situations within immersive surroundings to help patients cope with worries. The VR-enabled solution is causing quite a stir in the Indian healthcare industry, where it is effectively assisting those suffering from stress and depression in their recovery.

19.4.3 COGNITIVE RETRAINING

Doctors frequently utilize virtual reality chronic pain therapy to see patients executing complicated real-world tasks after they have suffered from a chronic stroke or brain injury.

Certain duties are reproduced within the virtual world via virtual reality, allowing patients to recover quickly. Cancer rehabilitation suits were developed in collaboration with AIIMS, one of India's premier healthcare facilities, to provide an all-in-one rehabilitation solution. Doctors frequently utilize virtual reality to watch patients execute complicated real-world tasks after they have suffered from a chronic stroke or brain injury. Certain duties are reproduced within the virtual world via VR, allowing patients to recover quickly. The AR/VR healthcare ecosystem is shown in Figure 19.6.

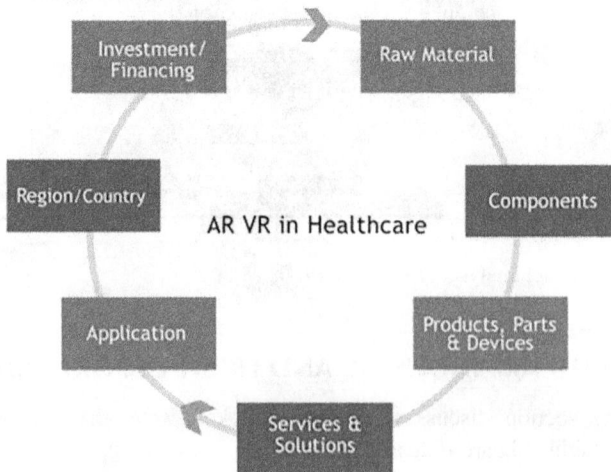

FIGURE 19.6   AR/VR healthcare ecosystem.

19.4.4   REHABILITATION OF THE PHYSICAL

Virtual reality is benefiting patients suffering from phantom limb discomfort tremendously in the Indian healthcare industry. VR has made caretakers' jobs simpler by allowing patients to accomplish tasks with their missing limbs, relieving pain and stress.

Virtual reality is considerably benefiting individuals suffering from phantom limb discomfort in the healthcare industry. VR has made caretakers' jobs simpler by allowing patients to accomplish tasks with their missing limbs, relieving pain and stress.

19.4.5   IMPROVED SURGICAL TECHNIQUES

With the inclusion of sophisticated technology solutions, augmented reality in medical education has been elevated and enhanced as India advances with better communication and education facilities.

Surgeons employ VR-created 3D models to plan procedures, improve accuracy, and reduce the risks involved with difficult surgery. With the inclusion of sophisticated technology solutions, medical education and augmented reality aided surgery have risen and enhanced as India advances with greater communication and education facilities. Surgeons employ VR-created 3D models to plan procedures ahead of time, improve precision, and reduce the risks involved with difficult surgery.

19.4.6   AUTISM AND ALZHEIMER'S TREATMENT

The use of virtual reality in treating conditions like autism and Alzheimer's The first immersive, all-in-one autism suite in the world, co-created by Lady Harding Hospital, including virtual reality games for autism. It includes cognitive exercises, games, excursions, and training.

19.4.7   BETTER MEDICAL TREATMENT TECHNIQUES

By enhancing the actual world in a three-dimensional paradigm, virtual reality is not only helping doctors identify issues more accurately, but is also giving medical students the chance to gain firsthand experience in an immersive setting.

Healthcare practitioners can safely practice, develop, and evaluate their medical abilities through the use of artificial representations of virtual environments that mirror real-world scenarios.

Due to its ability to lower human mistake rates and enable learners to produce more correct output, it has become crucial for our society. To better manage patients with serious illnesses including cancer, speech disorders, spinal injuries, and multiple sclerosis, Indian hospitals have largely used this strategy.

19.4.8   VR IN THE MEDICAL SECTOR

The healthcare industry has taken note of virtual reality since it offers so many advantages. Over the past ten years, it has seen exponential development and is now steadily rising to redefine its potential and effect in the healthcare business. The advancement of virtual reality will enhance the healthcare sector and provide a brighter future for the field.

19.5 FUTURE OF AR/VR IN THE HEALTHCARE SECTOR

Virtual reality has made strides and is predicted to have an extremely promising future. The Indian healthcare sector is incorporating it to improve and become a more practical answer for both healthcare professionals and patients. Key technologies used as a human interface in industry 4.0 are shown in Figure 19.7.

19.6 VIRTUAL REALITY VERSUS AUGMENTED REALITY

VR and AR are two closely related but distinct technologies that have found applications in the healthcare industry.

VR refers to a simulated experience that replaces the real world with a digital environment. In VR, users wear a headset that completely immerses them in a digital environment, blocking out the physical world around them. VR is typically used for entertainment, gaming, and training purposes. In healthcare, VR is being used for pain management, rehabilitation, and phobia treatment. AR is being used in healthcare for patient education, telemedicine, and surgical planning.

Despite being a technique that emerged decades ago, the AR paradigm is still new to many individuals. Confusing the term VR with augmented reality is still very popular.

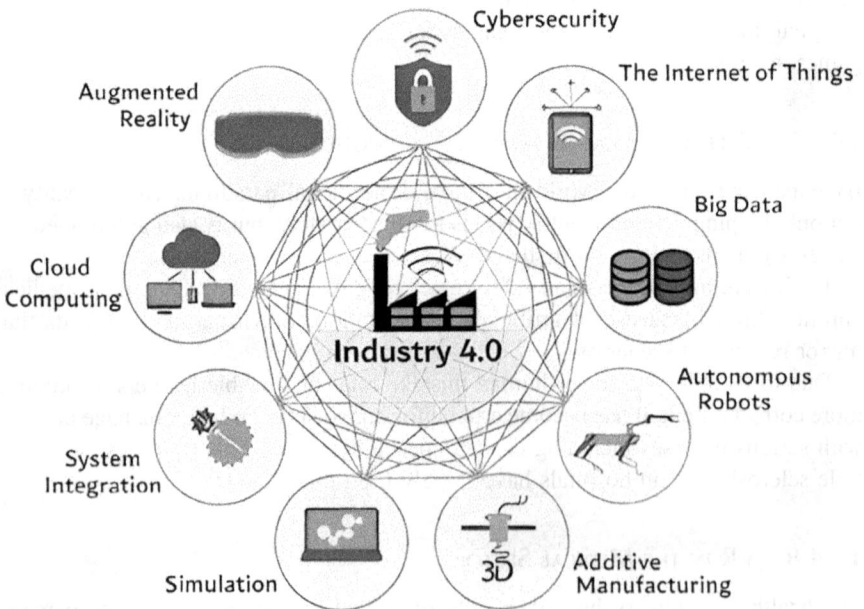

FIGURE 19.7 Key technologies used as human interface.

The biggest distinction between the two is that, with a different headset, VR creates the world in which we immerse ourselves. It is interactive and all we see is part of a digitally created world through pictures, sounds, etc. In AR, on the other hand, our universe becomes the context through which objects, pictures, and the like are put. All we see is in the real world, and wearing a headset might not be specifically required. Pokémon Go is the clearest and most mainstream example of this notion.

There is, though, a mixture of these realities called blended or mixed reality as well. For example, this hybrid technology makes it possible to see virtual objects in the real world and to create an environment where the tangible and the digital are essentially indistinguishable.

19.7   HISTORY OF VR

The concept of VR dates back to the 1930s and 1940s, when scientists and engineers first began exploring the idea of creating a simulated environment. However, it wasn't until the 1960s and 1970s that the technology advanced enough to allow for the creation of primitive VR systems.

One of the first VR systems was the Sword of Damocles, created by Ivan Sutherland in 1968. This system used a head-mounted display (HMD) and provided a simple wireframe environment that could be interacted with using a joystick. The technology continued to advance throughout the 1970s and 1980s, with the development of more advanced VR systems such as the VPL DataGlove and the Super Cockpit.

The 1990s saw the commercialization of VR with the release of consumer VR products such as the Virtuality arcade system and the Nintendo Virtual Boy. However, these products were not successful due to technical limitations and a lack of consumer interest.

In the 2010s, VR experienced a resurgence with the development of new, more advanced VR systems such as the Oculus Rift, HTC Vive, and PlayStation VR. These systems were more sophisticated and provided a more immersive VR experience, which led to a surge in interest in VR and its potential applications, including in the healthcare industry.

Today, VR technology continues to advance and is being used in a wide range of industries, including healthcare, entertainment, gaming, and education. The future of VR holds much promise, with ongoing developments in areas such as haptic feedback, improved displays, and increased processing power expected to further enhance the VR experience (Vyas, S., 2023).

It's a bit fuzzier as true VR takes hold of our brains as an all-encompassing simulacrum.

As for other scientific breakthroughs, the vision undoubtedly originated with science fiction, primarily the short story "Pygmalion's Spectacles" by Stanley G. Weinbaum in 1935, in which a physicist creates a pair of glasses that will "make it so that you are in the story, you speak to the shadows, and the shadows respond, and the story is all about you and you are in it instead of being on a screen."

19.8 BEGINNING OF VIRTUAL

1838

The stereoscope was invented by Charles Wheatstone. He noticed that you'll see a 3D vision if you draw something from two subtly different viewpoints and then see each image from a different lens. The stereoscope gave such a chance to the spectator. For the tools we use today, this gadget paved the way: cinematography, lighting, and others.

1849

David Brewster upgrades the stereoscope and makes a "lenticular stereoscope" in this manner. To create the first portable 3D viewer, he used the effects of his experiments and physical optics.

1901

L. Frank Baum publishes a novel and, for the first time in history, references AR-like technology. *The Master Key: An Electrical Fairy Tale* is a book featuring a young boy who is passionate about electricity and electronics. When the Demon of Electricity was summoned, he gave him a gift-the "character marker," a pair of glasses that exposed the latent character defects of humans.

1929

Ed Link produces the "Link Trainer" flight simulator. Thanks to the use of pumps, valves, and other equipment, this simulator helped pilots to obtain an accurate depiction of how it feels to fly an aircraft. This was a successful effort to introduce a VR prototype.

1935

In his novel *Pygmalion Spectacles*, Stanley G. Weinbaum describes a pair of glasses that allow the user to explore simulated worlds with the assistance of holographic pictures, smell, touch, and taste. Science fiction anticipated, as it always was, what we now have.

1939

The View-Master was created by William Gruber. It is a simple stereoscopic viewer that makes you look at two similar cross-eyed images to see a single 3D image. The View-Master was affordable to the consumer market and could be seen in the bedroom of almost any child.

19.9 VIRTUAL REALITY IN THE 1950S AND 1960S

In 1952

Sensorama, the first VR-like device of immersive multimodal technology that included a stereoscopic color monitor, odor emitters, a sound system, and fans, was invented by Morton Heilig. Thanks to the capacity of Sensorama to involve users in a wide-angle stereoscopic image, Morton succeeded in attracting the full attention of the audience.

In 1960

Morton Heilig invented the first head-mounted monitor once again. The customer was given a stereoscopic 3D image and stereo sound by the proprietary Telesphere Mask. Can it share a VR gear resemblance?

In 1961

The headlight is rendered by Comeau and Bryan. This wearable system monitored the movement of the head and projected each eye on a screen. It had options for magnetic tracking and a remote camera that matched the movement of the head. No computer simulation was done, but the device was partially identical to current VR helmets.

In 1968

The Sword of Damocles, another VR head-mounted show, was invented by Ivan Sutherland and his pupil, Bob Sproull. Ivan was already known for his successes in the development of computer graphics, and his expertise helped him create this gadget. It showed computer-generated wireframe spaces, and the picture perspective relied on the data for head tracking.

In 1969

Myron Krueger creates a series of computer-generated worlds. They referred to the individuals within the community that behaved. The technologies developed, for instance, allowed individuals to interact directly with each other. He called this "artificial reality."

1974

Myron Krueger is designing the Video Place Augmented Reality Lab. From his experimentation with simulated surroundings, his concept was created. The lab was developed by Myron to enable people to interact without gloves in "artificial reality" that would monitor gestures and other equipment. They were all the required hardware components at the Video place that allowed the user to be placed in the virtual environment.

In 1982

On TV, Dan Reitan uses AR. AR gained mass acceptance thanks to this guy, and the following technology is being used even today. To add graphics to a weather broadcast, Dan used space and radar cameras. This digital weather chart was the first time that AR could be used publicly by people.

In 1987

Thanks to Jaron Lanier, the name "virtual reality" was officially born. He had previously created VPL Testing, and it was the first company to market VR products. Specialized software was also distributed that allowed VR applications to be created. The devices developed at VPL Research were fairly primitive: the EyePhone, the DataGlove for data entry, and even 3D image renderers and stereo sound simulators were just other head-mounted monitors. The DataGlove was later licensed by Mattel to create the Power Glove for Nintendo, but it was not a success.

19.10   VIRTUAL REALITY IN THE 1990S AND 2000S

1990

The word "augmented reality" was introduced by Tom Caudell. He worked at Boeing and came up with an alternative to the diagrams used to direct staff in the area. He recommended equipping staff with head-mounted wearables that would project the schematics of the aircraft on reusable screens. With the assistance of a computer machine, the displayed photographs may conveniently be edited.

1991

Virtuality Party produces VR arcade machines at video game stores that can be located. The computers had a fast reaction time and equipped players with multi-player games with stereoscopic vision cameras, game controls, and the ability to collaborate. A bunch of hardware devices were already powered by VR gaming in the 1990s, such as VR headsets, graphics rendering subsystems, 3D trackers, and exoskeleton-like wearable parts. This pattern is already underway.

1992

Lawnmower Man, the premiere of the show. The plot was based on a scientist's fictitious story that used VR on a mentally ill patient. This was another case of how VR became a natural part of the film business through the mainstream.

1993

A VR headset is being developed by SEGA. The version was intended to supplement the consoles and arcades of the Sega Genesis and Saturn, but only one arcade was released.

1994

Julie Martin delivers the first-ever "Dancing in Cyberspace" theatrical AR show. Digital artifacts and environments were manipulated by the performers, providing an immersive picture.

1995

Nintendo is introducing the patented VR-32 unit, which will later be known as the Virtual Boy. Nintendo reported at the Consumer Electronics Show that their new device would provide players with a beautiful experience of engaging with virtual reality. Virtual Boy was the first home VR system, and they released the first home VR product ever, which was a huge risk for the company.

1996

The first-ever AR device appears with CyberCode 2D markers. It was focused on the 2D barcodes that even low-cost cameras mounted on mobile devices could identify. The device was able to evaluate the tagged object's 2D location and has been a base for a variety of AR applications.

1998

For the NFL game localization, Sportvision applies AR. With the assistance of an AR overlay, the perspective of the spectator was improved. Smoothly drawn on the ground was a yellow first-down marker. It gave a guide to the state of play to the audience.

1999

The first AR wearable equipment for BARS soldiers was published. The Battlefield Augmented Reality System aimed to help soldiers enhance battlefield vision, communications, location-identification of enemies, and overall situational awareness.

AR was used by NASA to guide the X-38 spacecraft. The car was fitted with an AR-powered navigation dashboard.

2000

Released by ARToolKit. This was a groundbreaking library tracking computer that allowed AR applications to be developed. It's open-source and hosted on GitHub right now.

A wearable EyeTap system was developed. EyeTap perceived the eye of the user as both a display and a camera and, with computer-generated data, and enhanced the environment the user saw.

19.11 CASE STUDY

The assessment of the brain-computer interface model is a crucial step in creating engineering systems to guarantee the effectiveness and quality of the system. The suggested concept seeks to replace heavy machinery-filled traditional mechanical laboratories with virtual technologies that would aid students' comprehension of and engagement with machines. How to assess the system such that student knowledge could be shown was a difficulty. After extensive investigation, it was discovered that a brain-computer interface (BCI), which is frequently used in applications like the one our team was working on, is the best method to accomplish that.

Brain-computer interface (BCI) technology allows a direct connection between the brain and an external device like a computer or robotic arm using electrodes. BCI systems can be established invasively or non-invasively. In the invasive method, electrodes are surgically implanted under the skin, while the non-invasive method involves attaching electrodes to the head. Electroencephalography (EEG) is a popular non-invasive method for developing a BCI system that records brain activity due to its ease of use, portability, affordability, and high temporal resolution. BCI has various applications such as stress detection, motor imagery exercises, and spinal cord rehabilitation. In a typical BCI system, information is extracted from the brain and sent to an external device for specific tasks. The proposed BCI system using EEG aims to extract signals from the brain, categorize them, and obtain specific information, as shown in Figure 19.8; the distribution of the information is displayed in Figure 19.9.

```
import mne
# Importing numpy
import NumPy as np
# Importing Scipy
import scipy as sp
# Importing Pandas Library
```

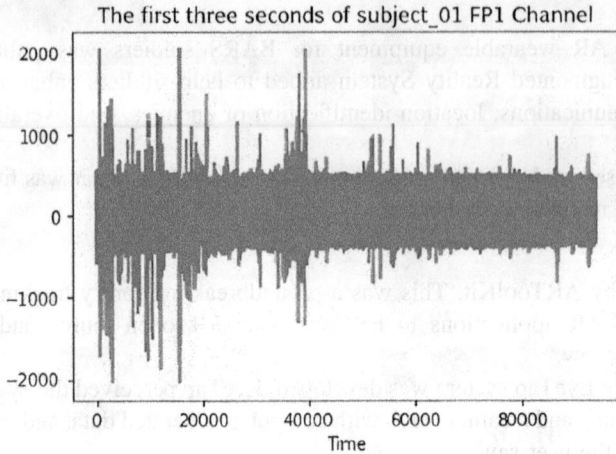

FIGURE 19.8 BCI with EEG to extract the signals.

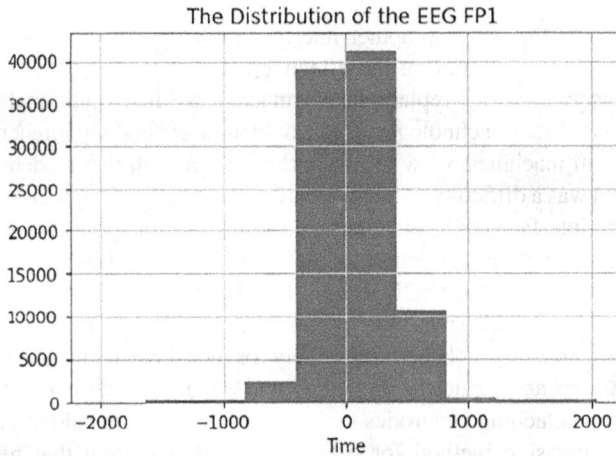

FIGURE 19.9 Distribution of the EEG.

```
import pandas as PD
# import glob function to scrap files path
from glob import glob
# import display() for better visualizations of DataFrames and arrays
from IPython.display import display
# import plot for plotting
import matplotlib.pyplot as plt
import math
from skimage.restoration import denoise_wavelet
from scipy.signal import savgol_filter
from scipy.signal import medfilt
import seaborn as sns
import yet
```

```
d_frame=pd.read_csv('../input/eeg-dataset-collected-from-students-using-vr/EEG_
Dataset/Subject00_0.csv')
d_frame.drop(columns='TimeDate',inplace=True) # converting the raw file to data
frame
plt.plot(d_frame['RAW'][0:]) # visualizing the first 10000 values of channel
ECG ECG
plot.label('Time')
plt. title ('The first three seconds of subject_01 FP1 Channel')
d_frame.head()


d_frame['RAW'].hist()
plt.xlabel('Time')
plt. title ('The Distribution of the EEG FP1 ')
```

19.11.1 PROCESSING OF SIGNALS

The system proposed in the study uses signal processing algorithms similar to those used by the authors. To remove artifacts and reduce background noise, the raw signals are filtered using a third-order median filter and low-pass and high-pass Butterworth filters with cutoff frequencies of 0.5 Hz and 50 Hz, respectively. The filtered signals are then segmented into four-second windows with a 50% overlap, which is later optimized to a three-second overlap for better accuracy. These segments are then transformed from the time domain to the frequency domain using the Fast Fourier transform.

19.12 CONCLUSION AND RESULTS

Everything in the world will be digitalized as a result of the power of computers and the current technological advancements. Thousands of articles are produced each year that employ technology in many aspects of life, including education. Because the coronavirus pandemic halted normal daily activities on Earth, the team came up with the notion of employing technology to provide an effective manner of continuing the educational process as it was before the epidemic. The answer is illustrated by the construction of a heavy machine lab in virtual reality to assist engineering students in maintaining their practical expertise through remote learning.

The proposed system underwent three phases: developing the engine, integrating it with the Unity VR engine to make it compatible with the virtual reality environment, and evaluating the system's performance by creating a brain-computer interface (BCI) system. The BCI utilized students' EEG signals to measure their interest levels during the use of the VR engine. As a result, it is possible to determine if the suggested system is as successful as the actual lab on campus.

In conclusion, the findings of our initial model were notable since students loved using it and showed a strong interest in it. Furthermore, the BCI system can anticipate a student's interest with an accuracy of up to 99% using EEG output.

19.12.1 FUTURE SCOPE

The future of VR in the healthcare industry holds much promise, and is expected to continue to grow and evolve in the coming years. Some of the areas where VR is expected to have the greatest impact include:

1. Telemedicine: VR is expected to play an increasingly important role in telemedicine, allowing healthcare professionals to remotely diagnose and treat patients using virtual environments.
2. Patient education: VR is being used to educate patients about their conditions and treatments, providing a more engaging and interactive experience. This is expected to become even more prevalent in the future.
3. Pain management: VR is being used to manage pain, providing patients with a distraction from their pain and helping to reduce their reliance on medications.
4. Rehabilitation: VR is being used to help patients recover from injuries and surgeries, providing them with an immersive and interactive rehabilitation experience.
5. Mental health: VR is being used to treat mental health conditions such as anxiety and phobias, providing patients with exposure therapy in a controlled virtual environment.

Overall, the future of VR in healthcare looks bright, with continued advancements in VR technology expected to provide new and innovative solutions to age-old healthcare problems. As VR becomes more widely adopted in the healthcare industry, it has the potential to transform the way healthcare is delivered and received, providing patients with a more engaging and personalized experience.

REFERENCES

Alkali, Y., Routray, I., & Whig, P. (2022). Strategy for reliable, efficient and secure IoT using artificial intelligence. *IUP Journal of Computer Sciences, 16*(2), 16–25.

Anand, M., Velu, A., & Whig, P. (2022). Prediction of loan behaviour with machine learning models for secure banking. *Journal of Computer Science and Engineering (JCSE), 3*(1), 1–13.

Antonio, J.M., Luis, S.R.J., & Faustino, S.P. (2013). Augmented and Virtual Reality Techniques for Footwear. *Computers in Industry.*

Azuma, R., Baillot, Y., Behringer, R., Feiner, S., Julier, S., & MacIntyre, B. (2001) Recent advances in augmented reality. *IEEE Computer Graphics and Applications, 21*(6), 34–47.

Azuma, R.T. (1997). A survey of augmented reality. *Presence, 6*(4), 355–385.

Benbelkacem, S., Belhocine, M., Bellarbi, A., Zenati-Henda, N., & Tadjine, M. (2013) Augmented reality for photovoltaic pumping systems maintenance tasks. *Renewable Energy, 55*, 428–437.

Chi, H.L., Kang, S.C., & Wang, X. (2013). Research trends and opportunities of augmented reality applications in architecture, engineering, and construction. *Automation in Construction, 33*, 116–122.

Chopra, G., & Whig, P. (2022). Smart agriculture system using AI. *International Journal of Sustainable Development in Computing Science*, *4*(1).

Ciampa, F. & Meo, M. (2010). Acoustic Emission Source Localization and Velocity Determination of the Fundamental Mode A0 Using Wavelet Analysis and a Newton-based Optimization Technique. *Smart Materials and Structures*, *19*(4), Article ID 045027.

Debernardi'ss, S., Fiorentino, M., Gattullo, M., Monno, G., & Uva, A.E. (2013). Text Readability in Head-worn Displays: Color and Style Optimization in Video vs. Optical See-through Devices. *IEEE Transactions on Visualization and Computer Graphics*, (99), 1.

Farhidzadeh, A., Dehghan-Niri, E., Moustafa, A., Salamone, S., & Whittaker, A. (2012). Damage Assessment of Reinforced Concrete Structures Using Fractal Analysis of Residual Crack Patterns. *Experimental Mechanics*, 1–13.

Jiwani, N., Gupta, K., & Whig, P. (2021). Novel HealthCare Framework for Cardiac Arrest With the Application of AI Using ANN. In *2021 5th International Conference on Information Systems and Computer Networks (ISCON)*, 1–5. IEEE.

Junaid, M., Shaikh, A., Hassan, M.U., Alghamdi, A., Rajab, K., Al Reshan, M. S., & Alkinani, M. (2021). Smart Agriculture Cloud Using AI Based Techniques. *Energies*, *14*(16), 5129.

Jupalle, H., Kouser, S., Bhatia, A.B., Alam, N., Nadikattu, R.R., & Whig, P. (2022). Automation of Human Behaviors and Its Prediction Using Machine Learning. *Microsystem Technologies*, 1–9.

Khera, Y., Whig, P., & Velu, A. (2021a). Efficient effective and secured electronic billing system using AI. *Vivekananda Journal of Research*, *10*, 53–60.

Koo, B., Choi, H., & Shon, T. (2009). Viva: Win monitoring framework based on 3D visualization and augmented reality in mobile devices. In M. Mokhtari, I. Khalil, J. Bauchet, D. Zhang, & C. Nugent, (Eds), *Ambient assistive health and wellness management in the heart of the city* (pp. 158–165), vol. 5597 of Lecture Notes in Computer Science. Berlin, Germany: Springer.

Krevelen, D. & Poleman, R. (2010). A survey of augmented reality technologies, applications and limitations. *The International Journal of Virtual Reality*, *9*(2), 1–20.

Kundu, T., Das, S., & Jata, K.V. (2007). Point of impact prediction in isotropic and anisotropic plates from the acoustic emission data. *Journal of the Acoustical Society of America*, *122*(4), 2057–2066.

Kundu, T., Das, S., & Jata, K.V. (2009). Detection of the point of impact on a stiffened plate by the acoustic emission technique. *Smart Materials and Structures*, *18*(3), Article ID 035006.

Liverani, A. & Ceruttii, A. (2010). Interactive GT code management for mechanical part similarity search and cost prediction. *Computer-Aided Design and Applications*, *7*(1), 1–15.

Liverani, A., Ceruttii, A., & Caligiana, G. (2013). Tablet-based 3D sketching and curve reverse modeling. *International Journal of Computer Aided Engineering and Technology*, 8, *5*(2–3), 188–215.

Niri, E.D., Farhidzadeh, A., & Salamone, S. (2013). Nonlinear Karmann Filtering for Acoustic Emission Source Localization in Anisotropic Panels, Ultrasonics.

Perelli, A., de Marchi, L., Marzani, A., & Speciale, N. (2012). Acoustic emission localization in plates with dispersion and reverberations using sparse PZT sensors in passive mode. *Smart Materials and Structures*, *21*(2), Article ID 025010.

Rose, J.L. (1999). *Ultrasonic waves in solid media*, Cambridge University Press.

Rupani, A., Whig, P., Sujediya, G., & Vyas, P. (2017). A Robust Technique for Image Processing Based on Interfacing of Raspberry-Pi and FPGA Using IoT. In *2017 International Conference on Computer, Communications and Electronics (Comptelix)*, 350–353. IEEE.

Salamone, S., Bartoli, I., di Leo, P. et al. (2010). High-velocity impact location on aircraft panels using macro-fiber composite piezoelectric rosettes. *Journal of Intelligent Material Systems and Structures, 21*(9), 887–896.

Seydel, R. & Chang, F.-K. (2001). Impact identification of stiffened composite panels: I. System development. *Smart Materials and Structures, 10*(2), 354–369.

Sharma, A., Kumar, A., & Whig, P. (2015). On the performance of CDTA based novel analog inverse low pass filter using 0.35 mm CMOS parameter. *International Journal of Science, Technology & Management, 4*(1), 594–601.

Su, Z., Ye, L., & Lu, Y. (2006). Guided lamb waves for identification of damage in composite structures: A review. *Journal of Sound and Vibration, 295*(3–5), 753–780.

Tomar, U., Chakroborty, N., Sharma, H., & Whig, P. (2021). AI-Based smart agriculturee system. *Transactions on Latest Trends in Artificial Intelligence, 2*(2).

Vyas, S. (2023). Extended reality and edge AI for healthcare 4.0: Systematic study. In *Extended reality for healthcare systems* (pp. 229–240). Academic Press.

Vyas, S., & Gupta, S. (2023). WBAN-based remote monitoring system utilising machine learning for healthcare services. *International Journal of System of Systems Engineering, 13*(1), 100–108.

Vyas, S., Bhargava, D., & Khan, S. (2023). Healthcare 4.0: A Systematic Review and Its Impact Over Conventional Healthcare System. *Artificial Intelligence for Health 4.0: Challenges and Applications,* 1–17.

Wang, X., Kim, M.J., Love, P.E., & Kang, S.C. (2013). Augmented reality in the built environment: classification and implications for future research. *Automation in Construction, 32,* 1–13.

Wang, B., Takatsubo, J., Akimune, Y., & Tsuda, H. (2005). Development of a remote impact damage identification system. *Structural Control and Health Monitoring, 12*(3-4), 301–314.

Whig, P.. (2019). Exploration of viral diseases mortality risk using machine learning. *International Journal of Machine Learning for Sustainable Development, 1*(1), 11–20.

Whig, P., & Ahmad, S.N. (2012). DVCC based readout circuitry for water quality monitoring system. *International Journal of Computer Applications, 49*(22), 1–7.

Whig, P., Nadikattu, R.R., & Velu, A. (2022). COVID-19 pandemic analysis using application of AI. *Healthcare Monitoring and Data Analysis Using IoT: Technologies and Applications, 1,* 1–25.

Whig, P., Velu, A., & Nadikattu, R.R. (2022). Blockchain platform to resolve security issues in IoT and smart networks. In *AI-Enabled Agile Internet of Things for Sustainable FinTech Ecosystems* (pp. 46–65). IGI Global.

Whig, P., Velu, A., & Ready, R. (2022). Demystifying federated learning in artificial intelligence with human-computer interaction. In *Demystifying Federated Learning for Blockchain and Industrial Internet of Things* (pp. 94–122). IGI Global.

Whig, P., Velu, A., & Sharma, P. (2022). Demystifying federated learning for blockchain: A case study. In *Demystifying Federated Learning for Blockchain and Industrial Internet of Things* (pp. 143–165). IGI Global.

Whig, P., Velu, A., & Naddikatu, R.R. (2022). The economic impact of AI-enabled blockchain in 6G-based industry. In *AI and Blockchain Technology in 6G Wireless Network* (pp. 205–224). Singapore: Springer.

Whig, P., Kouser, S., Velu, A., & Nadikattu, R.R. (2022). Fog-IoT-assisted-based smart agriculture application. In *Demystifying Federated Learning for Blockchain and Industrial Internet of Things* (pp. 74–93). IGI Global.

20 Augmented and Virtual Reality-Based Interventions for Learning Disabilities
Current Practices and Future Prospects

Samiya Khan

20.1 INTRODUCTION

A neurological disorder that affects the information processing and responding abilities of the brain is classified as a learning disability. Consequently, a learning disability can affect the reading, writing, problem solving, and speaking abilities of the individual, who may be an adult or child. A learning disability can affect one or more of these abilities. Collaterally, individuals with disabilities may also possess difficulties in managing time, socializing, and organizing things or tasks. The cause of learning disabilities in individuals is largely unknown. However, research indicates that injuries or developmental disorders may have a substantial contribution in the manifestation of one or multiple forms of learning disabilities.

Interventions and management strategies are focused on specific disabilities. Some of the commonly used therapeutic interventions include cognitive behavioral therapy or CBT (Kroese, 1998). Several interventions are implemented at the school-level to help children struggling with learning disabilities, which include tutoring, individualized instruction, and specialized curriculum (Fuchs &Fuchs, 1998). With a focus on simplifying the process of learning, appropriate and adequate modifications are made to the curriculum and classroom environment. As a result of its impactful nature, the use of assistive technology for intervention has recently gained popularity in therapeutic and restorative interventions for management of learning disabilities in children and adults (Bryant et al., 1998; Devi & Sarkar, 2019). Tools such as voice recognition software, software programs to help with reading, writing, and math, are increasingly being put to practice.

Technologies such as AR and VR have also found varied applications in management of learning disabilities for children and adults (Bailey et al., 2022),

DOI: 10.1201/9781003346289-20

FIGURE 20.1 Student exploring different VR-based learning applications.

including educational applications for schoolchildren (Figure 20.1) and therapeutic applications for adults such as drawing and games (Figure 20.2). Some of the popular applications include the use of AR for providing reminders in the form of visual cues to individuals with learning disabilities. These applications allow individuals to stay on their tasks and complete them as required (Bryant & Hemsley, 2022). Social anxiety is a significant manifestation of disorders or traumas that typically lead to learning disabilities. VR applications allow individuals struggling with social anxiety to practice skills that they require to lead a healthy life in a controlled and safe environment (Bryant et al., 2020). A combination of AR and VR applications are actively used for providing personalized instruction to individuals with learning disabilities (Donally, 2022). This allows them to receive instruction or training that is specific to their needs and abilities. Active usage of AR and VR applications for cognitive development and restoration is also recorded in existing literature (Papanastasiou et al., 2019). In these applications, there is increased focus on accessibility so that individuals with learning disabilities can use immersive environments at their own pace and to suit their specific needs.

This chapter explores the different aspects of the use of AR and VR for the management of learning disabilities in children and adults. The rest of the chapter is organized in the following manner: Section 20.2 provides a brief background on the prevalence and causes of learning disabilities in children and adults. Section 20.3 describes the different types of disabilities and existing AR/VR applications available for the same. Section 20.4 elaborates on the future scope of research in this field. Lastly, Section 20.5 synopsizes the conclusions of this work.

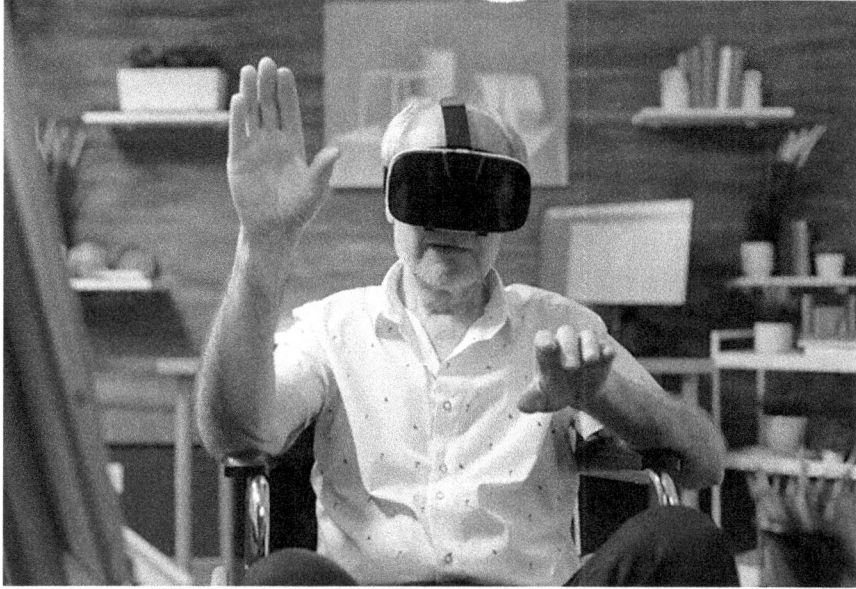

FIGURE 20.2 Adult doing free-hand drawing on a VR application.

20.2 BACKGROUND

Learning disabilities are known to affect as many as 5–15% children and adults around the world (Kesherim, 2023). It is worth mentioning that this estimate is based on the data available for countries such as Canada, United Kingdom, and the United States of America. According to Mencap (2020), there are around 1.5 million people with a learning disability in the United Kingdom. Country-wise variability in prevalence is expected to occur based on educational system procedures and policies, cultural attitudes of people towards learning disabilities, and access to available resources.

The actual cause of learning disability manifestation is not fully understood. However, the cause leading to a disability may be diagnosed from family and medical history. Moreover, these causes may vary from one individual to another. Typically, it is believed that learning disabilities are caused because of environmental and genetic causes. Furthermore, there is evidence that supports the idea that injuries or specific medical conditions may also result in the development of one or multiple learning disabilities (Pullen et al., 2017).

Injuries and development disorders can impact an individual's life in multi-faceted ways. Some of these facets include the individual's learning, behavior, and social interaction patterns. Injuries can potentially cause emotional, cognitive, and physical impairments. For instance, traumatic brain injuries or TBIs can alter the information processing capabilities of the affected individual's brain (Taylor et al., 2002). Consequently, the individual may face issues with writing, reading, and performing mathematical calculations. Moreover, some injuries may affect specific parts of an individual, which may directly or indirectly impact the individual's abilities. Some of the development disorders that are known for these issues include attention deficit hyperactivity disorder (ADHD) and autism spectrum disorder (Efstratopoulou et al., 2012).

20.3   LEARNING DISABILITIES: TYPES, CAUSES, AND MANAGEMENT

There are different types of learning disabilities, which are identified based on the skill or ability that is affected. This section provides details of some common learning disabilities, their causes, possible interventions, and existing AR/VR applications available for them.

20.3.1   DYSLEXIA

Dyslexia is one of the most known learning disabilities that affects an individual's ability to process language, read, and write. In the United Kingdom, around 10% of the population is known to be affected by dyslexia with more males affected than females (NHS n.d.). The causes for its manifestation are typically environmental and genetic in nature. Treatment and management strategies employed for dyslexia involve the use of specialized teaching methods that make use of assistive technology and psychological interventions such as cognitive behavioral therapy (CBT) (Reid & Peer, 2016). In some cases, interventions such as vision therapy, occupational therapy, and speech and language therapy may also be used (Srinithya, 2017).

Many VR applications have been developed and successfully deployed for dyslexia management. These applications typically focus on improving the writing and reading skills of the affected individual. Besides this, AR and VR applications also allow individuals to get an immersive experience, which allows them to explore and develop the concerned skills in a safe and controlled environment. Examples of AR and VR applications available for dyslexia management include Augmentally (Oswald Labs, 2023) and Dyslexia Quest (Nessy Learning, 2023), respectively.

20.3.2   DYSCALCULIA

The class of learning disability that affects an individual's ability to use and comprehend mathematics is referred to as dyscalculia. According to an estimate, around 6% of the people in the United Kingdom have dyscalculia (British Dyslexia Association n.d.). Like dyslexia, the causes that lead to dyscalculia are unknown. However, this condition is usually associated with environmental and genetic factors. According to Mahmud et al. (2020), one of the possible causes of dyscalculia may be a lack of exposure to mathematics during early childhood. A combination of interventions may be used to treat dyscalculia, which include educational methods such as one-on-one tutoring and psychological interventions such as CBT (Soares et al., 2018). Several software tools may also be used to help individuals with dyscalculia improve their mathematical skills (Kohn et al., 2020).

Several VR applications that can assist individuals with dyscalculia exist. Examples of VR applications include Math Reality (Math Reality, 2023) and Math World VR (Meta, 2023). Besides this, existing AR applications for dyscalculia include Maths AR (Apple Inc., 2023) and Math Worlds AR (Google Play, 2023). Other AR applications to assist individuals with dyscalculia are under development and focus on development of mathematical skills (Cai et al., 2021). As a rule, the applications developed for management of dyscalculia are centered around creating interactive elements and 3D visuals to practice mathematics in an engaging, immersive environment.

20.3.3 DYSGRAPHIA

Dysgraphia is a specific type of learning disability that is known to affect the writing ability of the affected individual. It is typically characterized by bad handwriting, poor spelling, and inability to organize work. Such individuals also struggle in forming alphabets, words, and sentences. Besides this, individuals with dysgraphia also find it difficult to express their ideas and thoughts in writing.

According to a study, dysgraphia is estimated to be prevalent in around 5–20% of people around the world (Structural Learning n.d.). The exact cause of manifestations that are classified as dysgraphia are unknown. However, there is some literary evidence to support the fact that neurological issues and developmental delay may act as a contributing factor (Kushki et al., 2011). Other plausible causes include physical impairments such as hearing loss and poor vision (Mogasale et al., 2012).

Management and treatment of dysgraphia typically involves the use of speech and language therapy along with occupational therapy (Chu, 1997). While speech and language therapy aids in improving language and writing skills, occupational therapy can play a profound role in helping with organizing written work and improving handwriting. VR applications are popularly used for assisting individuals with dysgraphia with letter formation and word recognition. One of the applications used for this purpose is Spell Bound (Gilbert et al., 2020).

20.3.4 AUDITORY PROCESSING DISORDER

When an individual finds it difficult or is unable to process and interpret auditory information, this is referred to as auditory processing disorder (APD). It is worth mentioning that APD is not the same as a hearing impairment as the former is characterized by difficulty in understanding the heard information. This learning disability can exist in children as well as adults and shall lead to issues related to communication, language, and learning. Although the causes of APD are not exactly known, it is believed to be associated with an impairment of the way the human brain processes sound (Obuchi et al., 2017). Genetic factors and environmental factors such as exposure to excessively loud noise and head trauma can also contribute to a manifestation of this learning disability.

The exact statistics around prevalence of APD in the United Kingdom are not known. However, a study suggests that 3–5% of schoolchildren in the

United Kingdom have some form of APD. Treatment planning for APD typically involves the use of language and speech therapy, in addition to interventions that can potentially help the individual in improving listening skills (Bamiou, 2006). Other approaches, such as CBT, use of assistive listening devices, and educational interventions such as special education approaches and one-on-one tutoring, may also be employed. AR and VR are actively used for management of APD with the help of applications such as Soundscape VR (Groove Science Studios, 2017), which uses virtual reality simulations to help improve sound localization, and AR-based augmented hearing provided by IQBuds (Vyas, S., & Gupta, S., 2022).

20.3.5 VISUAL PROCESSING DISORDER

In contrast to APD, which is characterized by the inability to process audio information, visual processing disorder (VPD) is characterized by an individual's inability to process visual information. VPD is also known as visual information processing disorder and visual perception disorder. Therefore, individuals with VPD find it difficult to interpret the visuals they see, which also affects their learning and day-to-day functioning. The causes behind VPS are not exactly known. However, factors such as genetic and environmental circumstances are believed to be contributory (Robinson et al., 1999).

The actual statistics around the prevalence of VPD in the United Kingdom are unknown. However, the study suggested that the prevalence of VPD in school-aged children in the USA is approximately 5%. The typical treatment plan for VPD includes CBT, vision therapy, and occupational therapy (Kurtz, 2006). Vision therapy assists the individual with improving eye coordination and focusing. On the other hand, occupational therapy aids in developing visual tracking, scanning, and focusing on objects. CBT can help in improving the individual's ability to interpret visual information. There are several VR applications used for managing VPD in the United Kingdom and worldwide. These applications include NeuroVR2 (Informer Technologies Inc., 2023), which uses virtual reality simulations to help improve visual attention.

20.3.6 NONVERBAL LEARNING DISABILITY

The form of learning disability that deals with an individual's inability to comprehend non-verbal cues such as body language and facial expressions is referred to as nonverbal learning disability (NLD). This neurological disorder can also affect an individual's ability to socialize and interact with others. Typically, individuals with NLD may find it difficult to organize information, solve problems, and comprehend abstract concepts. Although the exact cause of NLD is unknown, it is usually associated with the brain's inability to process specific kinds of information (Brue & Wilmshurst, 2016). Other causal factors include genetic issues and environmental reasons such as exposure to specific toxins and trauma during development.

Statistics around the prevalence of NLD in the United Kingdom and worldwide is not available. The treatment methodology adopted for individuals with

NLD includes the use of occupational therapy, speech and language therapy, CBT, and educational support (Burkhardt, 2005). Some of the VR applications that are used for NLD management include Mind Maze (Mind Maze Labs, 2023), which is an immersive VR experience designed to help people with NLD improve their social skills, and Virtually Better (Virtually Better Inc., 2023), which is an interactive VR program designed to help people with NLD improve their communication skills.

20.4 DISCUSSION

There are two main aspects of exploring AR/VR potential for managing learning disabilities in children and adults. These include development of innovative applications for dealing with specific issues and challenges associated with management of learning disabilities in individuals belonging to different age groups. There is immense scope of innovation in development of applications that support self-regulation strategies and improve academic performance in children with learning disabilities. Besides this, it is equally important to investigate the effectiveness and impact of AR/VR-based interventions in developing and restoring specific areas such as problem solving, attention, memory, executive functioning, and cognitive skills.

One of the lateral skills that is typically affected by the prevalence of learning disabilities in adults as well as children is social anxiety. Therefore, applications and feasibility or impact studies to investigate the influence of learning disabilities on an individual's social skills is a potential area of work. Consequently, development of applications that can assist in social skills development and restoration can also be explored.

Evidently, there are several stakeholders with respect to delivery and impact assessment of any therapeutic or restorative intervention. Thus, AR/VR applications will need to play an instrumental role in facilitating communication between therapists and/or teachers, parents or carers, and other professionals involved in providing support to individuals with learning disabilities. To steer this development towards a pathway of self-management and regulation, individuals or patients also need to play an important part in designing, developing, and delivering an intervention. All these facets must be included in future AR/VR interventions for individuals with learning disabilities to make them future proof.

Some of the other aspects that require attention include accessibility and personalization. It is crucial to investigate and explore ways in which AR and VR can be used to make resources more accessible for individuals with learning disabilities. Besides this, improving the accessibility of existing AR/VR applications and understanding means to improve their usability is also crucial for long-term adoption and continued use of AR/VR solutions for this class of therapeutic interventions. Personalized healthcare is expected to carve the path for next-generation healthcare. Therefore, an analysis of the potential efficacy of AR/VR in providing personalized healthcare to individuals with learning disability shall be considered for future research in this field (Vyas, S., 2023).

20.5 CONCLUSION

The use of AR and VR for management of learning disabilities in children and adults offers a huge potential for innovation and impactful research. Although, the causes leading to learning disabilities in adults and children vary from individual to individual and may largely be unknown, the manifestation of learning disabilities include issues with basic skills such as reading, writing, math, problem solving, and performing other cognitive tasks. In entirety, individuals with learning disabilities are unable to process and comprehend information in one way or the other.

AR and VR are impactful technologies that can be used to create immersive environments or simulations of real-world scenarios for individuals with learning disabilities to practice specific skills to provide personalized therapy in a safe and controlled manner. Finally, these technologies can also be used to provide feedback to the stakeholders. Therefore, it can be used by performing holistic assessment of an individual's therapeutic needs for therapy planning purposes. With development and increased adoption of these technologies for management of learning disabilities in children and adults, the possibilities are expected to increase across diverse lines of enquiry.

REFERENCES

Apple Inc (2023). AR Math, *Apple Store*, Available at: https://apps.apple.com/us/app/ar-math-arithmetic/id1276691203

Bailey, B., Bryant, L., & Hemsley, B. (2022). Virtual reality and augmented reality for children, adolescents, and adults with communication disability and neurodevelopmental disorders: A systematic review. *Review Journal of Autism and Developmental Disorders*, 9.2, 160–183.

Bamiou, D.-E., Campbell, N., & Sirimanna, T. (2006). Management of auditory processing disorders. *Audiological Medicine*, 4.1, 46–56.

British Dyslexia Society. About Dyscalculia. British Dyslexia Society. n.d. Web. 30 Apr. 2023. Available at: https://www.bdadyslexia.org.uk/dyscalculia/how-can-i-identify-dyscalculia

Brue, A.W., & Wilmshurst, L. (2016). *Essentials of intellectual disability assessment and identification*. John Wiley & Sons.

Bryant, D.P., Bryant, B.R., & Raskind, M.H. (1998). Using assistive technology to enhance the skills of students with learning disabilities. *Intervention in School and Clinic*, 34.1, 53–58.

Bryant, L., & Hemsley, B. (2022) Augmented reality: a view to future visual supports for people with disability. *Disability and Rehabilitation: Assistive Technology*, 1–14.

Bryant, L., Brunner, M., & Hemsley, B. (2020). A review of virtual reality technologies in the field of communication disability: implications for practice and research. *Disability and Rehabilitation: Assistive Technology*, 15.4, 365–372.

Burkhardt, S. (2005). Non-verbal learning disabilities. *Current perspectives on learning disabilities* (Vol. 16, pp. 21–33). Emerald Group Publishing Limited.

Cai, M., Akcayir, G., & Epp, C.D. (2021). Exploring Augmented Reality Games in Accessible Learning: A Systematic Review. *arXiv preprint arXiv:2111.08214*.

Chu, S. (1997). Occupational therapy for children with handwriting difficulties: A framework for evaluation and treatment. *British Journal of Occupational Therapy*, 60.12, 514–520.

Devi, C.R., & Sarkar, R. (2019). Assistive Technology for Educating Persons with Intellectual Disability. *European Journal of Special Education Research.*

Donally, J. (2022). *The immersive classroom: Create customized learning experiences with AR/VR.* International Society for Technology in Education.

Efstratopoulou, M., Janssen, R., & Simons, J. (2012). Differentiating children with attention-deficit/hyperactivity disorder, conduct disorder, learning disabilities and autistic spectrum disorders by means of their motor behavior characteristics. *Research in Developmental Disabilities, 33.1,* 196–204.

Fuchs, L.S., & Fuchs, D. (1998). General educators' instructional adaptation for students with learning disabilities. *Learning Disability Quarterly, 21.1,* 23–33.

Gilbert, T. et al. (2020). Spell Bound, *DevPost,* Available at: https://devpost.com/software/spell-bound

Google Play (2023). Math Worlds AR. *Google Play Store,* Available at: https://play.google.com/store/apps/details?id=com.hmhco.arbookcovers&hl=en&gl=US

Groove Science Studios (2017). Sound Scape VR: 2017. *STEAM,* Available at: https://store.steampowered.com/app/636930/Soundscape_VR/

Informer Technologies Inc. (2023). NeuroVR2, *Informer Technologies,* Available at: https://neurovr2.software.informer.com/

Kesherim, R. (2023). 31 Learning Disabilities Statistics: Facts. *Supportive Care - ABA Therapy.* Web. 30 Apr. 2023. Available at: https://www.supportivecareaba.com/statistics/learning-disabilities

Kohn, J., et al. (2020). Efficacy of a computer-based learning program in children with developmental dyscalculia. What influences individual responsiveness? *Frontiers in Psychology, 11,* 1115.

Kroese, B.S. (1998). Cognitive-behavioural therapy for people with learning disabilities. *Behavioural and Cognitive Psychotherapy, 26.4,* 315–322.

Kurtz, E.A. (2006). *Visual perception problems in children with AD/HD, autism, and other learning disabilities: A guide for parents and professionals.* Jessica Kingsley Publishers.

Kushki, A., et al. (2011). Changes in kinetics and kinematics of handwriting during a prolonged writing task in children with and without dysgraphia. *Research in Developmental Disabilities, 32.3,* 1058–1064.

Mahmud, M.S., et al. (2020). Dyscalculia: What we must know about students' learning disability in mathematics? *Universal Journal of Educational Research, 8.12B,* 8214–8222.

Math Reality (2023). About the Project, *Math Reality,* Available at: https://math-reality.eu/about/

Mencap (2020). How common is learning disability? *Mencap.* Web. 30 Apr. 2023. Available at: https://www.mencap.org.uk/learning-disability-explained/research-and-statistics/how-common-learning-disability

Meta (2023). Math World VR. *Meta Oculus Store,* Available at: https://www.oculus.com/experiences/quest/4923914040997217/

Mogasale, V.V., et al. (2012). Prevalence of specific learning disabilities among primary school children in a South Indian city. *The Indian Journal of Pediatrics, 79,* 342–347.

Mind Maze Labs (2023). Pioneering Digital Neurotherapeutics. *Mind Maze,* Available at: https://mindmaze.com/

NHS. Dyslexia. NHS. n.d. Web. 30 Apr. 2023. Available at: https://www.nhs.uk/conditions/dyslexia/

Obuchi, C., et al. (2017). Auditory symptoms and psychological characteristics in adults with auditory processing disorders. *Journal of Otology, 12.3,* 132–137.

Oswald Labs (2023). Augmenta11y, *Oswald Labs,* Available at: https://oswaldlabs.com/platform/shravan/apps/augmenta11y/

Papanastasiou, G., et al. (2019). Virtual and augmented reality effects on K-12, higher and tertiary education students' twenty-first century skills. *Virtual Reality, 23*, 425–436.

Pullen, P.C., et al. (2017). Specific learning disabilities. *Handbook of special education* (pp. 286–299) Routledge.

Reid, G., & Peer, L. (2016). Special Educational Needs: A Guide for Inclusive Practice. *Special Educational Needs*, 1–416.

Robinson, G.L., et al. (1999). Understanding the causal mechanisms of visual processing problems: A possible biochemical basis for Irlen syndrome? *Australian Journal of Learning Difficulties, 4.4*, 21–29.

Soares, N., Evans, T., & Patel, D.R. (2018). Specific learning disability in mathematics: a comprehensive review. *Translational Pediatrics, 7.1*, 48.

Srinithya, G. (2017). *Effectiveness of vision therapy as an adjunct to occupational therapy in improving visual motor skills in learning disabled children. Diss.* Komarapalayam: JKK Muniraja Medical Research Foundation.

Structural Learning. Dysgraphia. Structural Learning. n.d. Web. 30 Apr. 2023. Available at: https://www.structural-learning.com/post/dysgraphia

Taylor, H.G., et al. (2002). A prospective study of short-and long-term outcomes after traumatic brain injury in children: Behavior and achievement. *Neuropsychology, 16.1*, 15.

Virtually Better Inc (2023) Virtual Reality for a Healthier You. *Virtually Better*, Available at: https://www.virtuallybetter.com

Vyas, S. (2023). Extended reality and edge AI for healthcare 4.0: Systematic study. In *Extended reality for healthcare systems* (pp. 229–240). Academic Press.

Vyas, S., & Gupta, S. (2022). Case study on state-of-the-art wellness and health tracker devices. In *Handbook of research on lifestyle sustainability and management solutions using AI, big data analytics, and visualization* (pp. 325–337). IGI Global.

21 Employ Metrics in the Data Warehouse's Requirements Model for Hospitals

Tanu Singh and Bhavana Kaushik

21.1 INTRODUCTION

Bill Inmon was praised for coining the word "data warehouse" and it is defined as follows:

> "A Data Warehouse (DW) is a time-variant, integrated, subject-oriented and non-volatile data collection which supports in the various important decisions in the organizational" (Inmon, 1991); also depicted in Figure 21.1.

The managers must have access to historical information through the DW in order to make relevant decisions (Jeusfeld et al., 1998; Jarke et al., 2002). Additionally, because DW is becoming more complex, quality assessment must constantly be kept in mind throughout its design and development (Bouzeghoub & Kedad, 2002; Serrano et al., 2008).

For the development of DW, several approaches were recommended in literature (Ballard et al., 1998; Castro et al., 2002; Golfarelli et al., 1998a; Hüsemann et al., 2000; Inmon, 2005). These strategies can generally be classified into two categories: approaches that are requirements driven and those that are data driven (Winter & Strauch, 2003). The end user's needs are collected at the beginning of the process using the requirements-driven approach, and then the process is started by putting these demands into practise while leveraging the relevant data sources. However, the data-driven approach begins with a DW design and uses it from the final decision making to the thorough examination of data sources (Golfarelli et al., 1998b; Inmon, 2005). The development of DW should follow the same guidelines as traditional software development, which takes requirements analysis into account as well as the design and implementation phases (Busborg et al., 1998a; 1998b; Golfarelli & Rizzi, 1998). The phases of the development life cycle of DW are presented (Golfarelli, 2009) and shown in Figure 21.2.

As organizational managers use DW to make business decisions, the information quality of DW becomes crucial. As a result, it's critical to preserve high-quality information starting with DW design and development (Inmon, 1991). Poor DW

DOI: 10.1201/9781003346289-21

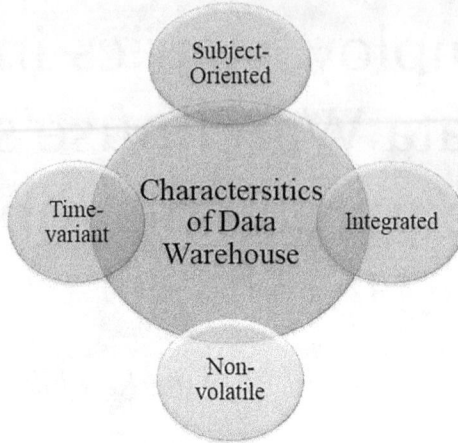

FIGURE 21.1   Characteristics of DW.

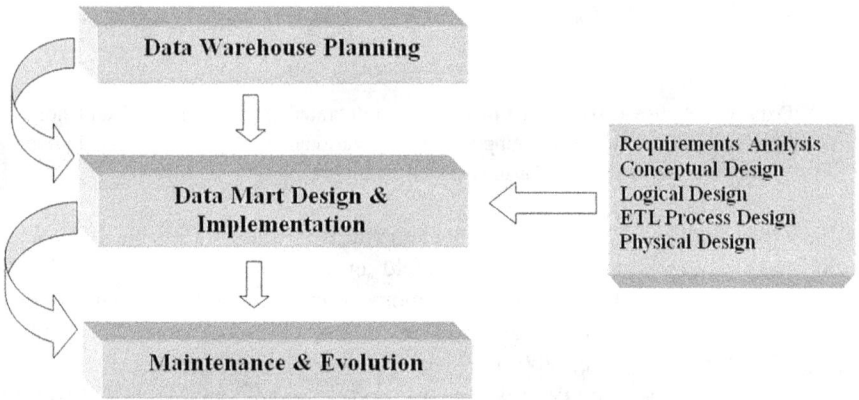

FIGURE 21.2   DW development phases.

quality can have severe effects from a technical and organizational standpoint, including customer loss, financial loss, and employee dissatisfaction (English, 1996). This inspired researchers to guarantee the DW's information quality so that it can be utilized in making relevant decisions.

The following factors impact the information quality of DW stated by (Serrano et al., 2007):

1. the DWs overall quality
2. the data presentation quality, as depicted in Figure 21.3.

The data model quality is of four types, which are presented by (Kumar et al., 2012; Serrano et al., 2007), depicted in Figure 21.3, which involves in the quality assessment of the overall DW. To begin with, a number of design methodologies

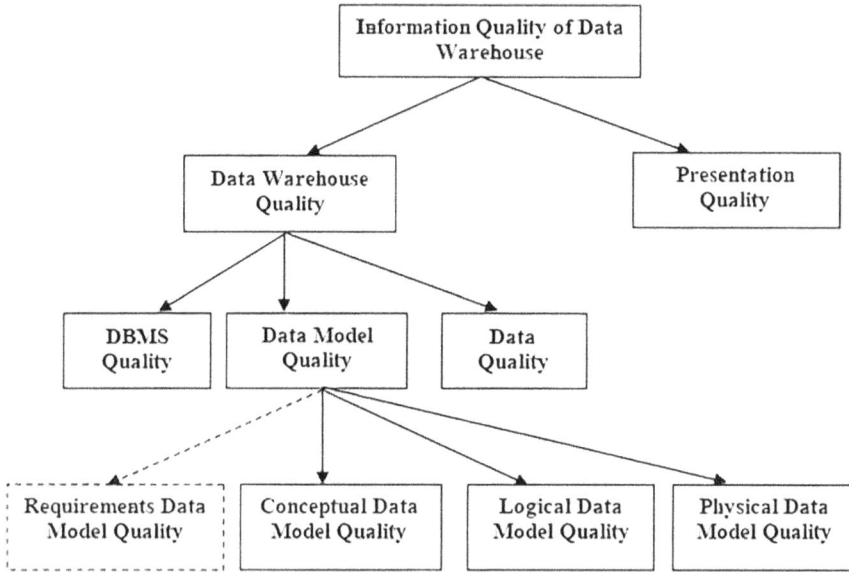

FIGURE 21.3   Information quality of DW.

(Debevoise, 1999; Frendi & Salinesi, 2003; Golfarelli & Rizzi, 1999; Giorgini et al., 2008; Husemann et al., 2000; Kumar et al., 2010; Mazón et al., 2007; Paim & Castro, 2003; Prakash & Gosain, 2008) have been suggested in the literature to create high-quality DW. The majority of their methodologies include conceptual, logical, and physical design of DW as their starting point during the requirements engineering (RE) phase. Although necessary, a methodology might not be enough to ensure the quality of DW (Serrano et al., 2008). They made the point that a precise process for evaluating the quality of a product utilizing metrics and methodologies is required.

Although data model quality has a higher impact on the final development than any other phase, it only makes up a small fraction of the overall system development. The data model's quality must be assessed because it is the main factor in determining the whole system's quality (Moody & Simsion, 1995). Data model quality is comprised of requirements, conceptual, logical, and physical data model quality, as shown in Figure 21.3. The next subsection of this section only discusses the quality of the requirements data model.

It is essential to pay attention and put forth effort in the early stages of the requirements phase to spot faults as they emerge or stop them from forming. Additionally, a study by the group of Standish found that 30% of DW projects fail because they only deal with physical, logical, and conceptual data models without explicitly incorporating the requirements data model (Frendi & Salinesi, 2003). Various DW methodologies were presented in the literature, emphasizing the significance of the RE phase for DW development (Mazón et al., 2007; Schiefer et al., 2002).

By presenting metrics to anticipate external quality characteristics, ongoing research in the field of software engineering (SE) can be seen in the context of quality assessment and software measurement (ISO/IEC, "ISO/IEC 9126-1, 2001; ISO/IEC 25010, 2010). Metrics are extremely important and play a big part in determining the level of SE quality. Given the enormous gap between empirical and formal validation in the software metrics, measurements still sag at the edges of SE (Fenton & Neil, 2000). Therefore, in order to make software metrics practically useful, it is crucial to implement detailed formal and empirical validation.

Kumar et al. (Kumar et al., 2013) suggested requirements completeness and traceability metrics based on the agent-goal-decision-information (AGDI) model (Kumar et al., 2010) of the RE approach as objective indicators to evaluate the DW requirements data model's quality. From the literature, there exist few DW requirements models based on AGDI model, i.e., university requirements schema (Kumar et al., 2010) and banking requirement schema [19, 25; 34].

In this chapter, the hospital requirements model is created to evaluate the DW requirements data model quality of DW that helps in the overall quality of hospital DW. As a result, the main objective of this chapter is to raise the standard of the hospital requirements data model that was created initially and the early stages of DW development.

The rest of the chapter is organized as follows: Section 21.2 explains the related work in the requirement data model of DW. Further, Section 21.3 depicts the importance of requirements engineering. In Section 21.4, RE approaches are illustrated in detail. Further, the AGDI model is explained, which is based on the RE approach in Section 21.5. Moreover, the existing RE approach for the hospital requirements model is discussed in Section 21.6. Also, the requirements completeness metrics is explained using a hospital requirements model of DW in Section 21.7. In addition, Section 21.8 revises the lesson learnt from the whole chapter, followed by the conclusion and the future scope in Section 21.9.

21.2 RELATED WORK

As many business decisions count on DW information, organizations are constantly concerned about its quality. Due to the high level of DW complexity in the first phases of DW design and development, it becomes vital to assure the evaluation of DW system and its quality (Inmon, 1991). It becomes crucial to draw attention to the quality problems at each data model level, including the physical, logical, conceptual, and needs data models.

At the level of physical, logical, and conceptual data models, a variety of methodologies and procedures have been investigated in the literature. However, there was very little research on the quality of DW requirements data models in the literature (Kumar et al., 2012; Kumar, 2015; Singh & Kumar, 2020; 2021a; 2021b; 2022a; 2022b), which could serve as the foundation for designing and creating a high-quality DW conceptual data model. Only requirements data model quality issues are deemed to be pertinent in this case and are examined in the sections that follow.

To understand the stakeholder intentions, Kumar et al. (Kumar et al., 2012) suggested a quality-oriented RE strategy for DW. Here, the model, which is enhanced with quality analysis, is termed an organizational, decision, and information model used to represent the banking requirements model of DW. Additionally, a number of rules and guidelines that the analyst must adhere to when performing modeling activities were presented on the ground of banking organizations.

Later, Kumar et al. (Kumar et al., 2013) proposed requirements metrics to assess and track the development of the requirements model quality of DW. The organization, decision, and information models were used to capture the DW requirements in this case, and the given requirements metrics were successfully applied to the requirements model/schema for the banking industry.

Based on the metrics proposed by (Kumar et al., 2013), this chapter will cover the DW requirements model for the hospital sector based on the AGDI model.

In the section that follows, we'll talk about the significance of RE of DW before talking about the hospital requirements model.

21.3 IMPORTANCE OF REQUIREMENTS ENGINEERING (RE)

A software system's success is mostly determined by how well it accomplishes the objectives it was designed to do. Finding out what those needs are and documenting them in a way that can be utilized for analysis, communication, and implementation is the process of RE (Nuseibeh & Easterbrook, 2000). Additionally, as can be seen from the literature, many researchers have expanded the definition of RE.

RE is the process of creating requirements that can be created by an iterative cooperative process of issue analysis, documentation of all outcomes in various formats, and verification of the accuracy of understanding obtained (Jackson, 1995). Later, (Zave, 1997) defined RE as a subfield of SE that is mostly focused on the constraints and practical objectives of software system functions. Additionally, recent definitions of RE by numerous researchers have broadened the definition of RE to encompass the entire organizational system as well as the process of developing software applications (Greenspan et al., 1994; Loucopoulos et al., 1996; Mylopoulos et al., 1994; Yu & Mylopoulos, 1994).

Therefore, RE deals with issues relating to organizational plans, processes, goals, etc., and develops systems to help the organization achieve its goals (Loucopoulos et al., 1996; Macaulay, 1996; Mylopoulos et al., 2001).

The main processes of the RE are requirements development and requirements management. There are various steps of requirements developments like requirements elicitation, requirements analysis, etc. It can be shown that the RE processes must capture the needs of the consumer. Early RE and late RE phases, which are both equally important and are separated into two categories (Yu, 1995), are discussed in the next sub-section to follow.

The requirements completeness, consistency, and automatic verification are the primary concerns of the late phase of RE (Bubenko, 1980; Dardenne et al., 1993; Dubois et al., 1986) whereas the goal of the early stages of RE is to model and analyze how different stakeholders' interests are served or harmed by various system and environmental alternatives (Yu, 1995; 1997). Early phases of the RE process place more emphasis on understanding the "whys" than on precisely defining "what" the system is supposed to achieve (Yu & Mylopoulos, 1994). In contrast to the late phase, the RE early phase is also crucial, according to (Yu, 1997).

In the literature, a number of RE approaches have been put forth (Dubois et al., 1986; Mylopoulos et al., 1994) to aid thinking and express knowledge in RE; however, these approaches made no distinction between the RE early and late phases. However, only a few RE approaches, such as agent and goal-oriented techniques, are utilized in the two RE phases i.e., early and late phase (Chung, 1993; Dubois, 1989; Feather, 1987; Fickas & Helm, 1992).

21.4   REQUIREMENTS ENGINEERING APPROACHES

Numerous RE strategies have been put forth in the literature for designing DW while highlighting the significance of the RE phase.

At first, the RE phase received no attention from the DW development. However, Inmon (Inmon, 1996) drew attention to the fact that, in contrast to the traditional software development life cycle (SDLC), the requirements of DW were the final ones to be looked into. Furthermore, Golfarelli and Rizzi (Golfarelli & Rizzi, 1999) presented the SDLC for DW, emphasizing RE and representing workloads as requirements analysis and specification. A model-based approach was also suggested by Williams et al. (Williams et al., 1999) to enhance the effectiveness of the RE process. Additionally, Bohnlein and Vom Ende (Böhnlein & Ulbrich-vom Ende, 2000) proposed a technique for DW requirements that were derived from the business process model and involved determining DW metrics and dimensions based on preliminary research on organizational goals and services. Later, the DW requirements definition (DWARF) technique was introduced by (Paim & Castro, 2003), which employed the conventional RE procedure for defining and managing information requirements, by the computer assistance SE (CASE) tool.

Contrary to transactional systems, practically all RE techniques (Berenbach & Borotto, 2006; Bresciani & Donzelli, 2003) not to be counted between the early and late RE for DW, according to Yu and Mylopoulos (Yu & Mylopoulos, 1998). Business process requirements and planned decision processes can also be used to meet DW requirements (Frendi & Salinesi, 2003). With this method, existing data models and DW requirements are combined to establish data models of DW, which can be further used to fulfill new requirements.

Additionally, Kimball and Ross (Kimball & Ross, 2002) presented the four-step approach based on business processes. In this case, the facts and the dimensions were selected as the main business process choice and described as the organizational operational process backed by the transactional system. Additionally, a different RE approach was proposed by (Winter & Strauch, 2003; 2004), i.e., a user-driven bottom-up approach, where it describes information requirements but does not depict decision requirements; which is basically not good practice in the early process of DW design. In order to tackle this issue, a goal-driven RE strategy was presented in the literature and is described below.

The goal-driven RE strategy developed by Bonifati et al. (Bonifati et al., 2001) is based on a top-down goal-question-metric (GQM) technique for establishing goals and is focused on removing the information system from an organization's data marts. Later, Schiefer et al (Schiefer et al., 2002). suggested a comprehensive strategy for managing the requirements for DW systems on the basis of goal modeling at various levels of abstraction. To establish the DW system needs for the DW design phases, however, no rules were mentioned in this technique. There are various RE strategies for DW that simply concentrate on the information requirements and ignore decision requirements. However, there exist many recent RE approaches for DW that make a difference between the two phases of RE, i.e., early and late (Giorgini et al., 2008; Mazón et al., 2007; Prakash & Gosain, 2008).

Later, Prakash and Gosain (Prakash & Gosain, 2008) introduced a novel RE technique that captured decision requirements. This method started with decisional goals (find the set of pertinent decisions), and at the end, information needs that were necessary to achieve these goals were modeled. Early requirements, however, were not recorded, which means that other stakeholders' dependencies were not modeled. There were no standards for converting decisional criteria into DW conceptual models using the GDI schema. Additionally, an agent goal decision information (AGDI) approach was presented by (Kumar et al., 2009) to model the early stage of RE, where the early requirements were recorded through organization and goal modeling activities. However, (Kumar et al., 2009) did not support the late RE approach. Many RE approaches have been documented in the literature, and they are discussed in the sections that follow.

By modifying the GDI model (Prakash & Gosain, 2008) into the AGDI model (Kumar et al., 2010), it was able to model both early (organization and decision model) and late (information) DW requirements. Additionally, the expansion of the AGDI model (Kumar et al., 2009) into a quality-oriented DW requirements model (Kumar et al., 2011) where the goals of the stakeholders were recorded. The two soft goals in this situation are quality and decisional. The stakeholders' quality soft targets are divided into those with clearly defined success criteria and those that categorize different restrictions including budget, time, and others. As a result, the stakeholders in the organization were able to make more important decisions because all of these soft goals were retained in the DW meta-data.

Additionally, to support the design of DW, a RE approach that comprises of three phases, early, late, and conceptual RE phase was presented by (Kumar et al., 2016). In addition, Amaral and Mernik (Amaral & Mernik, 2016) assess the quality of the models itself as well as the model driven engineering (MDE) strategy, which starts with requirements. Additionally, Gupta and Gupta (Gupta & Gupta, 2019) propose the original idea of requirements slicing and backtracking to deal with dependency and developers' preferences. They also control the dependence of technical limitations among the requirements. To enhance the DW quality, (Amalfitano et al., 2020) has studied a novel way of carrying out questionnaire-based gap analysis (QBGA) processes using the MDE approach, and application lifecycle management (ALM) methodology.

All of the above RE approaches are modeled using different techniques and methods. These approaches are further validated formally and empirically to validate the methods and metrics. In this chapter, the AGDI model is considered to depict the hospital requirements model. But before, explaining the requirements metrics, let's discuss the basic AGDI model based on RE approach of DW and the hospital requirements model.

21.5 AGDI MODEL BASED ON RE APPROACH

The AGDI model was announced as an expansion to the existing GDI model (Prakash & Gosain, 2003) in the prior work of (Kumar et al., 2010) to model both phases of the requirements (early and late) and are represented in Figure 21.4 (with dotted lines).

In the AGDI model, the concept of a goal was used to represent the stakeholders' goals, and agent dependencies were used to represent their relationships with one another. Following is a discussion of the notions of agent, goal, decision, and information.

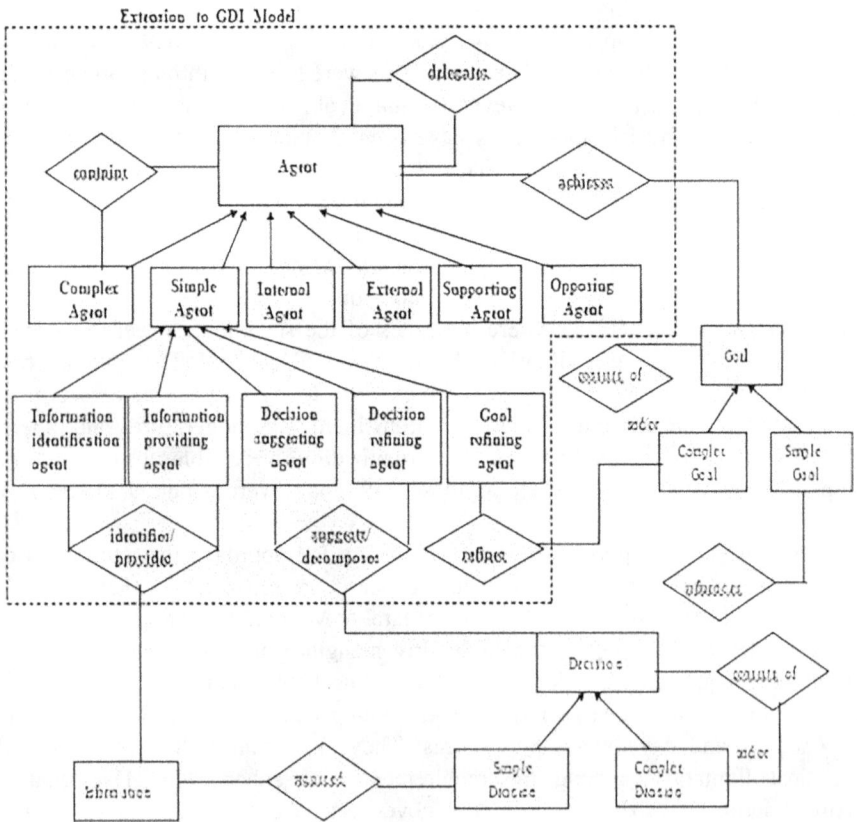

FIGURE 21.4 AGDI model for DW.

1. **Agent:** Kumar et al. (Kumar et al., 2010) claim that the idea of the agent simulates the stakeholders' diverse responsibilities in decision-making processes within an organization (see Figure 21.4). Additionally, he classified the agents in a variety of ways. The agent can be segregated into two sets: basic agents (which model various stakeholder roles) and complex agents (which model additional stakeholders inside it). Prior to achieving the simple goals, the complex goals are refined.

2. **Goal:** In an organization, there are two different sorts of goals, namely complex and simple goals, according to Kumar et al. (Kumar et al., 2010). An organization employing the "achieves" connection will refine the complicated goals until the simple goals are accomplished; see Figure 21.4. Goal dependency is the relationship between the agents; an agent may use this relationship to outsource its complex goals to a goal-refining agent. The AGDI paradigm (Kumar et al., 2010) also emphasizes achieving decisional goals. To achieve their decisional aims, the agents may recommend a set of pertinent decisions. These decisions are addressed in the next section.

3. **Decision:** Kumar et al. (Kumar et al., 2010) proposed simple or complex options, and to accomplish the objectives, complex decisions are broken into simple decisions. The agents are dependent on decisions; hence, an agent may use the "delegate" connection (see Figure 21.4) to outsource its straightforward aim to a decision-suggesting/refining agent. The agents make these decisions based on a set of pertinent information that is stated in the following manner.

4. **Information:** Kumar et al. (Kumar et al., 2010) claim that if higher-level agents are required to make decisions in an organization, they are given the information. Additionally, the agents' information dependency is represented by a "delegate" connection, which assigns simple decision making to an agent that provides information or performs identification (see Figure 21.4).

Now, we will implement the above AGDI model into the hospital requirements schema, which is discussed in the next section to follow.

21.6 HOSPITAL REQUIREMENTS MODEL OF DW BASED ON THE AGDI MODEL

As previously said, there aren't many research studies on DW requirements schemas in literature (Kumar et al., 2012; 2010; 2013; 2016), and the existing DW requirements schema isn't enough for further research. Hence, a hospital requirements model based on the AGDI model (Kumar et al., 2010) is created from different domains using the RE approach of DW. The domain is familiar to avoid every complexity. Hence, this section demonstrates an example of the hospital requirements model of DW.

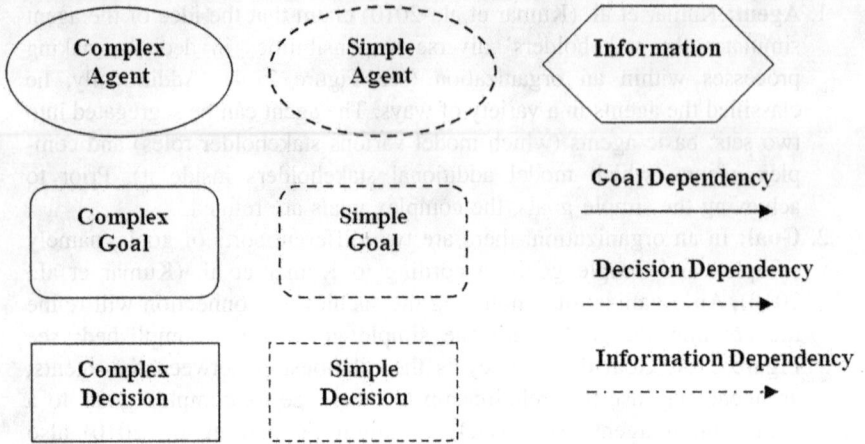

FIGURE 21.5 Legends of RE approach of DW.

The legends of the RE technique (Kumar et al., 2013) are all shown in Figure 21.5, which may be utilized to calculate the values of the existing DW needs metrics.

The early, late, and conceptual RE phases of the RE approach are described by (Kumar et al., 2016). Information modeling was obtained in the late RE phase, whereas organization and decision modeling were acquired in the early RE phase. Finally, in the conceptual design phase, the information model from the late RE phase and the dimensions, measures, and facts from the early RE phase are identified.

The hospital requirements model that contains all the models is shown in Figure 21.6.

One complex agent from the hospital schema is broken into two simple and one complex agent that is again further broken into simple agents (Figure 21.7).

FIGURE 21.6 Agents and their goals of the hospital requirements model.

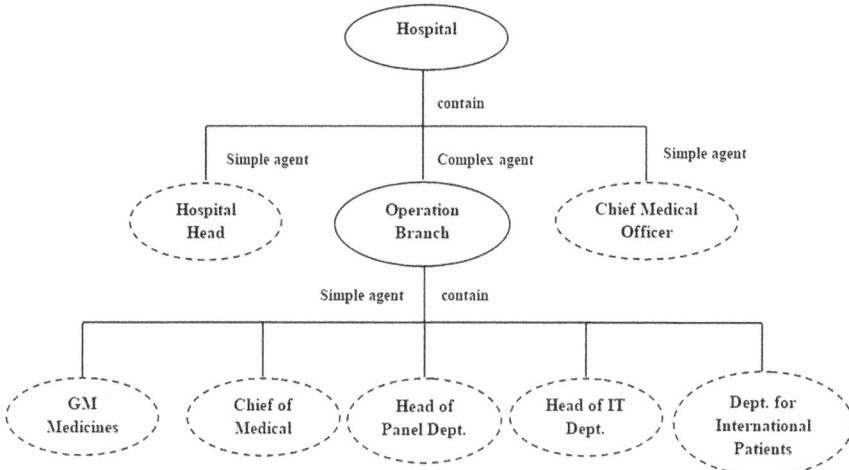

FIGURE 21.7   Refined complex agents into simple agents.

The complex goals are broken into simple ones. Once the organization's agents have suggested both complex and simple decisions, such simple goals are then achieved. Additionally, these complex decisions are streamlined and transformed into decision models as the interaction between the agents develops.

Also, the information model, where agents give unique and exclusive information, so as input these decision models, is used. This information is subsequently sent to higher-level agents to make important business decisions to achieve the objectives of the organization.

Figure 21.8 illustrates the hospital descriptive DW needs schema and divides the final model into three categories: organizational modeling, decisional modeling, and information modeling. This chapter does not concentrate on the conceptual design process and does not explain it here.

As a result, we employed the AGDI model to capture the full requirements of DW.

This section provides a detailed description of the DW hospital requirements model. Now in the next section, the requirements completeness metrics are explained based on the AGDI model of DW.

21.7   REQUIREMENTS COMPLETENESS METRICS OF THE HOSPITAL REQUIREMENT MODEL OF DW

Kumar et al. (Kumar et al., 2013) suggested requirements completeness and traceability metrics as the requirements metrics, while requirements traceability measures include metrics for coverage (COV) and depth and height coverage (DHCOV) and where requirements completeness metrics include metrics for specification to be completed (STBC) and requirements decomposition (RD).

In this chapter, we only include the requirements completeness metrics, and the detailed description of requirements metrics is depicted in Table 21.1. The values of

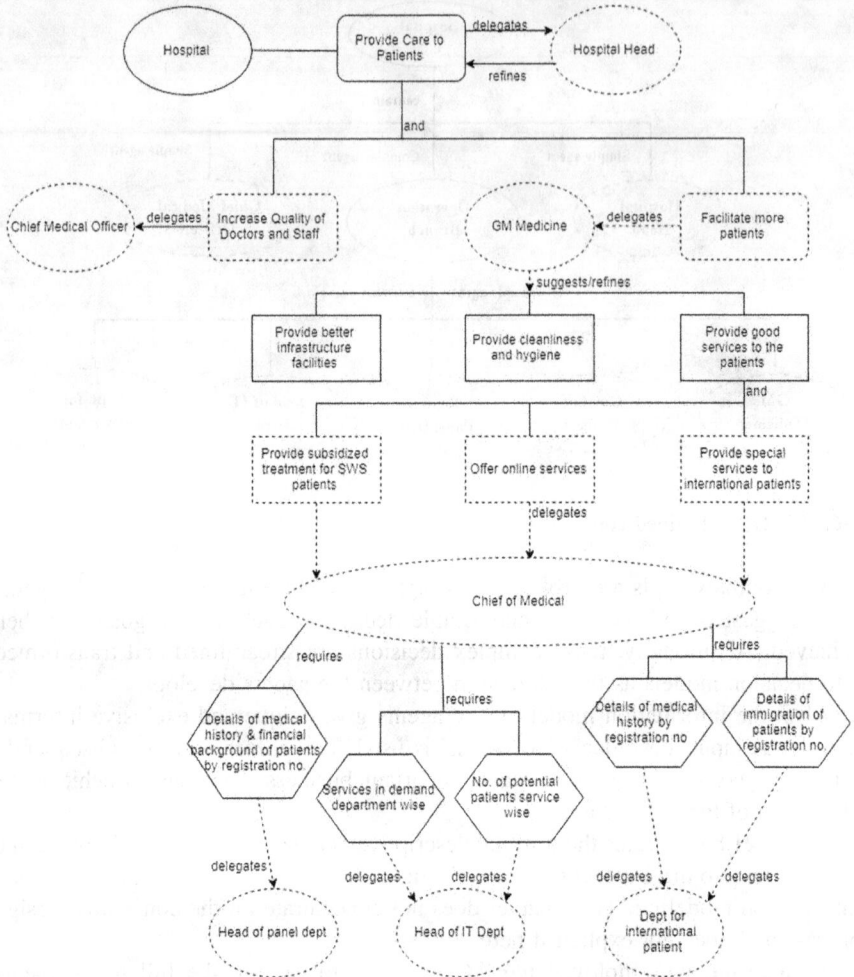

FIGURE 21.8   Hospital requirements schema of DW.

requirements metrics can be added by referring to Figures 21.6, 21.7, and 21.8 of the hospital requirements model and the calculated values are shown in Table 21.1.

The following is a detailed explanation of the hospital requirements model of ten completeness metrics, as proposed by (Kumar et al., 2013):

1. **Requirements Decomposition Metrics:** In this metrics, we have six metrics that contains all the goals, decisions, and information and the detailed explanation is as under:
   a. **NSA Metric:** The quantity of simple agents is counted. Therefore, based on Figure 21.8, NSA metric value is 7, i.e., Hospital head, GM Medicine, Chief Medical Officer, Chief of medical, Head of panel dept., Head of IT dept., Department of international affairs, which

TABLE 21.1

Calculated Values of Requirements Completeness Metrics of DW (Kumar et al., 2013) for the Hospital Requirements Model

Requirements Completeness Metrics		Calculated Values
Requirements Decomposition metrics	NSA	7
	NGH	3
	NSG	2
	NDH	3
	NSD	3
	NI	5
Specification to Be Completed metrics	NCGTBC	2
	NSGTBC	1
	NCDTBC	2
	NSDTBC	0

serves a variety of functions, such as the complex goals (CG) broken till simple goals (SG) and offering data to make simple decisions.

b. **NGH Metric:** It will total up all the CG that the agent has attained, which will eventually result in goal hierarchies. As a result of Figure 21.8, we first determine the full values of CG = 3, that the hospital requirements model must satisfy. Additionally, all three CGs will result in goal hierarchies i.e., quality and safety audit, provide care to patients, and reduce total cost, meaning that the NGH metric will also be considered as 3. The "Arrange Funds" goal was taken by "regulatory agency" (agent) and will not be taken into account in this case.

c. **NSG Metric:** It determines how many SGs are involved in each hierarchical goal. Therefore, based on Figure 21.8, the NSG measure will facilitate more patients and increase the quality of doctor and staff in accordance with customer feedback will be 2.

d. **NDH Metric:** It compiles all complex decisions made by agents for carrying out the SG that produced a decisional hierarchy. As a result, the NDH metric value from Figure 21.8 will be 3, i.e., provide better infrastructure facilities, provide cleanliness and hygiene, and provide good services to the patients.

e. **NSD Metric:** It computes all simple decisions made by agents to achieve the goal of each decisional hierarchy. As a result, the NSD metric in Figure 21.8 equals 3, i.e., offer online services, provide subsidized treatment to SWS patients, and provide special services to international patients.

f. **NI Metric:** It determines how many different types of information the agents need to support all of their decisions to achieve their goals. Therefore, based on Figure 21.8, the NI metric value is 5, which are services in demand department wise, details of medical history and financial background of patients by registration number, number of potential patients service wise, details of medical history by registration number, and details of immigration of patients by registration number.

2. **Specification to Be Completed Metrics:** In this metrics, we have four metrics that contain all the goals and decisions that need to be completed and the detailed explanation is as follows:

   a. **NCGTBC:** The CG for all unfinished hierarchical goals is calculated. Since just one of the three CGs (*provide care to patients*) has been completed according to Figure 21.8 and the other two (*quality and safety audit* and *reduce total cost*) have not, the NCGTBC metric = 2.

   b. **NSGTBC:** It determines the NSG value for which suggestions for decisions have not yet been made. Therefore, based on Figure 21.8, the NSGTBC metric's value will be 1, i.e., *increase quality of doctor and staff.*

   c. **NCDTBC:** It calculates all difficult decisions in hierarchical decisions. Therefore, based on Figure 21.8, two complex decisions have not yet been made, i.e., *provide better infrastructure facilities* and *provide cleanliness and hygiene.*

   d. **NSDTBC:** It summarizes the value of NSD for which the agents have not yet determined any information. Therefore, based on Figure 21.8, the simple decision remains unfinished. Consequently, the NSDTBC metric has a value of 0.

21.8   LESSON LEARNT

As shown in Figures 21.6 to 21.8, we successfully incorporated the suggested metrics into the hospital requirements model of DW. The metrics have provided a quantitative evaluation of quality.

We have gathered data for the completeness metrics and explained in Section 21.7 the calculated values are depicted in Table 21.1 for the DW hospital requirements model. The ensuing observations are noteworthy in our opinion, as follows:

1. The hospital requirements model will be improved over time, and metrics will once more be gathered to track the development of RE models. Thus, during the RE phase of the DW life cycle, measurements must be gathered at regular intervals. The value of each STBC metric should be 0 to denote that RE models are 100% complete.

2. It is important to gather the information from time to time and validate its importance in terms of the decisions and the goals.

21.9   CONCLUSION AND FUTURE SCOPE

The following are the conclusions made from this chapter, which included numerous measures for ensuring the quality of the DW requirements model:

1. The hospital requirements model is created based on the AGDI model of DW.
2. The completeness metrics were successfully applied to the early and late DW requirements models for hospitals.
3. The suggested metrics for requirements completeness help analysts in objectively tracking the development of the DW requirements definition.
4. The conceptual design phase will make use of the ensured quality, i.e., complete hospital requirements models that have been developed to produce a complete DW conceptual model, which is not related to this chapter.
5. The AGDI model can further help the hospital management to keep track of the requirements and to achieve the goals made by the management of the hospital.

In upcoming research, we'll concentrate on the following:

1. In the near future, the impact of RE measurements on the metrics of DW conceptual models will also be examined.
2. The requirements management tool must support the collection of suggested metrics. Soon, work will begin on developing the tool.

REFERENCES

Amalfitano, D., De Simone, V., Scala, S., & Fasolino, A.R. (2020). A Model-Driven Engineering Approach for Supporting Questionnaire-based Gap Analysis Processes Through Application Lifecycle Management Systems. *Software Quality Journal*, 1–31.

Amaral, V., & Mernik, M. (2016). Special issue on quality in model-driven engineering. *Software Quality Journal*, 24(3), 597–599.

Ballard, C., Herreman, D., Schau, D., Bell, R., Kim, E., & Valencic, A. (1998). *Data modeling techniques for data warehousing* (p. 25). San Jose: IBM Corporation International Technical Support Organization.

Berenbach, B., & Borotto, G., (2006). Metrics for Model Driven Requirements Development. In *Proceedings of the 28th International Conference on Software Engineering* (pp. 445–451).

Bonifati, A., Cattaneo, F., Ceri, S., Fuggetta, A., & Paraboschi, S. (2001). Designing data marts for data warehouses. *ACM Transactions on Software Engineering and Methodology*, 10(4), 452–483.

Bouzeghoub, M., & Kedad, Z. (2002). Quality in Data Warehousing. *Information and Database Quality*, 163–198.

Bresciani, P. & Donzelli, P. (2003). REF: A practical agent-based requirement engineering framework. In *International conference on conceptual modeling* (pp. 217–228). Berlin, Heidelberg: Springer.

Bubenko, J.A. (1980). Information modeling in the context of system development. In *Proceedings of IFIP Congress 1980* (pp. 395–411).

Busborg, F., Christiansen, J.B., & Jensen, L. (1998a). *A method for data warehouse development, Dat5 report, Part I, CS department*. Aalborg University.

Busborg, F., Christiansen, J.B., & Jensen L. (1998b). *Data warehouse modelling: The NyKredir case study, Dat5 report/Part II, CS department university*. Aalborg University.

Böhnlein, M., & Ulbrich-vom Ende, A. (2000). Business process oriented development of data warehouse structures. In *Data warehousing 2000* (pp. 3–21). Heidelberg: Physica.

Castro, J., Kolp, M., & Mylopoulos, J. (2002). Towards requirements-driven information systems engineering: The Tropos project. *Information Systems*, *27*(6), 365–389.

Chung, K.L. (1993). Representing and Using Non-Functional Requirements for Information System Development: A Process-Oriented Approach. *Ph.D. Thesis*, also 169 *Tech. Rpt. DKBS-TR-93-1*, Department of Computer Science, University of Toronto.

Dardenne, A., Lamsweerde, A., & Fickas, S. (1993). Goal directed requirements acquisition. *Science of Computer Programming*, *20*(1-2), 3–50. Elsevier.

Debevoise, T. (1999). The Data Warehouse Method: Integrated Data Warehouse Support Environments. *(No Title)*.

Dubois, E. (1989). A logic of action for supporting goal-oriented elaborations of requirements. *ACM Sigsoft Software Engineering Notes*, *14*(3), 160–168.

Dubois, E., Hagelstein, J., Lahou, E., Ponsaert, F., & Ri-faut, A. (1986). A knowledge representation language for requirements engineering. *Proceedings of the IEEE*, *74*(10), 1431–1444.

English, L. P. (1996). Redefining information management: IM as an effective business enabler. *Information Systems Management*, *13*(1), 65–67.

Feather, M.S. (1987). Language support for the specification and development of composite systems. *ACM Transactions on Programming Languages and Systems*, *9*(2), 198–234.

Fenton, N.E., & Neil, M. (2000). Software metrics: Roadmap. In A. Finkelstein (Ed.), *Future of software engineering*. ACM Press.

Fickas, S., & Helm, R. (1992). Knowledge representation and reasoning in the design of composite systems. *IEEE Transactions on Software Engineering*, *18*(6), 470–482.

Frendi, M., & Salinesi, C. (2003). Requirements Engineering for Data Warehousing. In *Proc. of the 9th International Workshop on Requirements Engineering: Foundations of Software Quality*, Klagenfurt/Velden, Austria, (pp. 75–82).

Giorgini, P., Rizzi, S., & Garzetti, M. (2008) .GRAnD: A goal-oriented approach to requirement analysis in data warehouses. *Decision Support Systems*, *45*(1), 4–21. Elsevier.

Golfarelli, M. (2009). Data Warehouse Life Cycle and Design.

Golfarelli, M., & Rizzi, S. (1998, November). A methodological framework for data warehouse design. In *Proceedings of the 1st ACM international workshop on Data warehousing and OLAP* (pp. 3–9).

Golfarelli, M., & Rizzi, S. (1999) Designing the data warehouse: Key steps and crucial issues. *Journal of Computer Science and Information Management*, *2*(3), 88–100.

Golfarelli, M., Maio, D., & Rizzi, S. (1998a). The dimensional fact model: A conceptual model for data warehouses. *International Journal of Cooperative Information Systems*, *7*(02n03), 215–247.

Golfarelli, M., Maio, D., & Rizzi, S. (1998b, January). Conceptual Design of Data Warehouses From E/R Schemes. In *Proceedings of the thirty-first Hawaii International Conference on System Sciences* (Vol. 7, pp. 334–343). IEEE.

Greenspan, S., Mylopoulos, J., & Borgida, A. (1994). On Formal Requirements Modeling Languages: RML Revisited. In *Proceedings of 16th International Conference on Software Engineering (ICSE'1994)* (pp. 135–147). IEEE Computer Society.

Gupta, A., & Gupta, C. (2019). A novel collaborative requirement prioritization approach to handle priority vagueness and inter-relationships. *Journal of King Saud University-Computer and Information Sciences*.

Husemann, B., Lechtenborger, J., & Vossen, G. (2000). Conceptual Data Warehouse Design. In *International Workshop on Design and Management of Data Warehouses*, Sweden (pp. 3–9).

Hüsemann, B., Lechtenbörger, J., & Vossen, G. (2000). *Conceptual data warehouse design* (Vol. 168). Universität Münster. Angewandte Mathematik und Informatik.

Inmon, W.H. (1991). *Building the data warehouse.* 1st Edition, Wiley.

Inmon, W.H. (1996). The data warehouse and data mining. *Communications of the ACM, 39*(11), 49–51.

Inmon, W.H. (2005). *Building the data warehouse.* John Wiley & Sons.

ISO/IEC, "ISO/IEC 9126-1. (2001). *Software engineering: Product quality-quality model.*Geneva: International Standards Organization.

ISO/IEC 25010. (2010). *Systems and Software Engineering-Systems and Software Quality Requirements and Evaluation (SQuaRE)-System and software quality models.* Geneva: International Standards Organization.

Jackson, M. (1995). *Software requirements & specifications- a lexicon of practice, principles and prejudices,* 1st edition. ACM Press, Addison-Wesley.

Jarke, M., Lenzerini, M., Vassiliou, Y., & Vassiliadis, P. (2002). *Fundamentals of data warehouses.* Springer Science & Business Media.

Jeusfeld, M.A., Quix, C., & Jarke, M. (1998). Design and analysis of quality information for data warehouses. In *Conceptual Modeling–ER'98: 17th International Conference on Conceptual Modeling,* Singapore, November 16–19, 1998. Proceedings 17 (pp. 349–362). Springer Berlin Heidelberg.

Kimball, R., & Ross, M. (2002). *The data warehouse lifecycle toolkit,* 2nd ed. New York: John Wiley & Sons.

Kumar, M. (2015) Validation of Data Warehouse Requirements-Model Traceability Metrics Using a Formal Framework. In *2015 2nd International Conference on Computing for Sustainable Global Development (INDIACom),* New Delhi, India (pp. 216–221). IEEE.

Kumar, M., Gosain, A., & Singh, Y. (2009). Agent oriented requirements engineering for a data warehouse. *ACM SIGSOFT Software Engineering Notes, 34*(5), 1–4.

Kumar, M., Gosain, A., & Singh, Y. (2010). Stakeholders driven requirements engineering approach for data warehouse development. *Journal of Information Processing Systems, 6*(3), 385–402.

Kumar, M., Gosain, A., & Singh, Y. (2011). Quality-oriented requirements engineering for a data warehouse. *ACM SIGSOFT Software Engineering Notes, 36*(5), 1–4.

Kumar, M., Gosain, A., & Singh, Y. (2012). Quality-oriented requirements engineering approach for data warehouse. *International Journal of Computational Systems Engineering, 1*(2), 127–138.

Kumar, M., Gosain, A., & Singh, Y. (2013). On completeness and traceability metrics for data warehouse requirements engineering. *International Journal of Computational Systems Engineering, 1*(4), 229–237.

Kumar, M., Gosain, A., & Singh, Y. (2016). A novel requirements engineering approach for designing data warehouses. *International Journal of System Assurance Engineering and Management, 7*(1), 205–221.

Loucopoulos, P., Louridas, P., Kavakli, V., Filippidou, D., & Tzanaki, A. (1996). The Enterprise Requirements Analysis (ERA) Approach. *MIST, Technical Report, ISE Group.*

Macaulay, L. (1996). *Requirements engineering.* London: Springer-Verlag.

Mazón, J.N., Pardillo, J., & Trujillo, J. (2007). A model-driven goal-oriented requirement engineering approach for data warehouses. In *International conference on conceptual modeling* (pp. 255–264). Berlin, Heidelberg: Springer.

Moody, D.L., & Simsion, G.C. (1995). Justifying investment in information resource management. *Australasian Journal of Information Systems, 3*(1), 25–37.

Mylopoulos, J., Chung, L., & Yu, E. (1994). From object-oriented to goal oriented requirements analysis. *Communications of the ACM, 42*(1), 31–37.

Mylopoulos, J., Chung, L., Liao, S., Wang, H., & Yu, E.S.K. (2001). Exploring alternatives during requirements analysis. *IEEE Software*, 2–6.

Nuseibeh, B., & Easterbrook, S. (2000) Requirements Engineering: a roadmap. In *Proceedings of the conference on the future of software engineering (ICSE'2000)* (pp. 35–46). ACM.

Paim, F.R., & Castro, J.B. (2003). DWARF: An Approach for Requirements Definition and Management of Data Warehouse System. In *Proc. of 11th IEEE International Requirements Engineering Conference*, California, USA.

Prakash, N., & Gosain, A. (2003). Requirements Driven Data Warehouse Development. In *Proc. of 15th Conference on Advanced Information Systems Engineering (CAiSE'2003)*, Austria (pp. 13–16).

Prakash, N., & Gosain, A. (2008). An approach to engineering the requirements of data warehouses. *Requirements Engineering*, *13*(1), 49–72.

Schiefer, J., List, B., & Bruckner, R. (2002). A holistic approach for managing requirements of data warehouse systems. In *AMCIS 2002 Proceedings*, Dallas, U.S. (p.13).

Serrano, M., Trujillo, J., Calero, C., & Piattini, M. (2007). Metrics for data warehouse conceptual models understandability. *Information and Software Technology*, *49*(8), 851–870.

Serrano, M.A., Calero, C., Sahraoui, H.A., & Piattini, M. (2008). Empirical studies to assess the understandability of data warehouse schemas using structural metrics. *Software Quality Journal*, *16*, 79–106.

Singh, T., & Kumar, M. (2020, December). Empirical validation of requirements traceability metrics for requirements model of data warehouse using SVM. In *2020 IEEE 17th India Council International Conference (INDICON)* (pp. 1–5). IEEE.

Singh, T., & Kumar, M. (2021a, May). Formally investigating traceability metrics of data warehouse requirements model using Briand's framework. In *2021 5th International Conference on Intelligent Computing and Control Systems (ICICCS)* (pp. 1203–1209). IEEE.

Singh, T., & Kumar, M. (2021b). Empirical study to predict the understandability of requirements schemas of data warehouse using requirements metrics. *International Journal of Intelligent Engineering Informatics*, *9*(4), 329–354.

Singh, T., & Kumar, M. (2022a). Investigating requirements completeness metrics for requirements schemas using requirements engineering approach of data warehouse: A formal and empirical validation. *Arabian Journal for Science and Engineering*, *47*(8), 9527–9546.

Singh, T., & Kumar, M. (2022b). Theoretical validation of data warehouse requirements metrics based on agent goal decision information model using Zuse's framework. In *Communication and Intelligent Systems: Proceedings of ICCIS 2021* (pp. 107–118). Singapore: Springer Nature Singapore.

Williams, D.W., Hall, T., & Kennedy, M. (1999). A framework for improving the requirements of engineering process management. *Software Quality Journal*, *8*(2), 133–147.

Winter, R., & Strauch, B. (2003, January). A Method for Demand-Driven Information Requirements Analysis in Data Warehousing Projects. In *36th Annual Hawaii International Conference on System Sciences, 2003. Proceedings of the* (pp. 9–pp). IEEE.

Winter, R., & Strauch, B. (2004). Information Requirements Engineering for Data Warehouse Systems. In *Proceedings of the 2004 ACM symposium on Applied computing* (pp. 1359–1365).

Yu, E., & Mylopoulos, J. (1998). Why Goal-Oriented Requirements Engineering. In *Proceedings of the 4th International Workshop on Requirements Engineering: Foundations of Software Quality* (Vol. 15, pp. 15–22).

Yu, E.S.K. (1995). Modelling Strategic Relationships for Process Reengineering. *Ph.D. thesis*, Dept. of Computer Science, University of Toronto.

Yu, E.S.K. (1997). Towards Modeling and Reasoning Support for Early-Phase Requirements Engineering. In *Proceedings of IEEE International Symposium on Requirements Engineering* (pp. 226–235).

Yu, E.S.K., & Mylopoulos, J. (1994). Understanding Why in Requirements Engineering with an Example. In *Proceedings of Workshop on System Requirements: Analysis, Management, and Exploitation*.

Zave, P. (1997). Classification of research efforts in requirements engineering. *ACM Computing Surveys*, *29*(4), 315–321.

22 Paving the Way for Healthcare with AI, ML, and DL

Opportunities, Challenges, and Open Issues

Mitali Chugh and Neeraj Chugh

22.1 INTRODUCTION

The current healthcare landscape is characterized by a rapidly growing and aging population, a rising rate of chronic diseases and healthcare costs, and a shortage of healthcare specialists. The transformed healthcare landscape is steered by several aspects such as developments in technology, advancing patient expectations, demographic changes, and increasing healthcare costs. These reasons are putting the burden on healthcare systems worldwide to enhance care quality, raise efficacy, and lower costs. Technological improvements in medical research are granting new alternatives for diagnosis and treating diseases (Vyas et al., 2023).

However, the healthcare industry is also confronting several challenges, such as division of care, restricted access to care in rural areas, and deficiency of interoperability among different healthcare systems. This needs an amendment from a reactive to a proactive approach to healthcare, with the prominence on prevention and early interference. From this view, there is a growing implication in the application of AI, ML, and DL in healthcare to address some of these challenges. These technologies have the potential to transform healthcare delivery by enabling personalized, data-driven decision making, improving diagnostic accuracy, and reducing healthcare costs (Esteva et al., 2019; Rajkomar et al., 2019; Rasheed et al., 2022).

AI describes the development of computer systems that can execute chores that characteristically need human intelligence, such as decision making, problem solving, and pattern recognition. ML is a subset of AI that is algorithm-centric and uses statistical models that support computers to learn from and make data-centric projections or decisions. On the other hand, DL is a subset of ML that uses artificial neural networks with multiple layers to process and extract high-level features from complex data.

DOI: 10.1201/9781003346289-22

Voluminous patient data can be analyzed including electronic health records, genetic information, and medical literature, to support clinical decision making, identify risk factors, and personalize treatment plans. With the assistance of AI/ML and DL, medical data can be organized, standardized, and made easily accessible. This improves the efficiency of healthcare delivery, enhances the accuracy of diagnoses, and enables healthcare professionals to provide personalized care to patients. By alleviating the burden of manual data management, clinicians are empowered to make informed decisions and deliver timely and effective treatments.

AI, ML, and DL algorithms can help predict disease outcomes, recommend optimal treatment options, and support patient monitoring and management. These technologies are also promising in remote patient monitoring, telemedicine, and precision medicine and treatment optimization, improving access to care, reducing costs, and enhancing patient experiences. There are various AI, ML, and DL tools used in healthcare that have shown promising results in improving patient care and healthcare outcomes. Here are a few examples:

- Computer-aided diagnosis (CAD): AI-based CAD systems assist radiologists in analyzing medical images such as X-rays, MRIs, and CT scans. These tools use ML algorithms to detect and highlight abnormalities or potential signs of diseases, aiding in accurate diagnosis and reducing interpretation errors.
- Natural language processing (NLP): NLP techniques are used to extract relevant information from unstructured medical texts, such as clinical notes, research papers, and electronic health records (EHRs). NLP algorithms can help convert free-text data into structured formats, enabling better data analysis, decision support, and clinical research.
- Predictive analytics: ML models are employed to analyze patient data and identify patterns that can predict disease progression, readmission rates, or treatment outcomes. By leveraging historical patient data, these models help healthcare professionals make proactive decisions and personalize treatment plans for better patient management.
- Virtual assistants and chatbots: AI-powered virtual assistants and chatbots are designed to interact with patients, answer their questions, and provide basic medical advice. These tools use natural language understanding and ML algorithms to provide personalized responses and support triage processes, improving patient engagement and accessibility to healthcare information (Vyas, S. 2023).
- Genomic analysis: DL algorithms are utilized to analyze vast genomic data sets, identifying genetic markers associated with diseases, drug responses, and risk factors. This enables precision medicine approaches, allowing tailored treatments and personalized healthcare based on an individual's genetic profile.
- Robotics and surgical assistance: AI and ML algorithms are integrated into robotic systems used in surgical procedures. These technologies assist surgeons in performing complex operations with precision, enhancing surgical outcomes, and reducing the risk of human error.

Subsequently seeing all these contexts, it is difficult to get a complete insight into healthcare disciplines using ML and DL algorithms. Nonetheless, several reviews on AI-based healthcare have been published in the recent few years (Kumar et al., 2023; Loh et al., 2022; Rasheed et al., 2022). In many scenarios, the contexts are limited, such as the automatic detection of cardiovascular disease in computed tomography angiography acquisitions (Jin et al., 2021) or the image analysis for certain illnesses.

This study aims to offer a comprehensive introduction to healthcare, focusing on the latest machine learning (ML) and deep learning (DL) techniques. With the global shift towards digitalization in healthcare and the continuous advancements in AI, there is a significant need to understand the implications and nuances of ML and DL in this field. Thus, comprehensively covering the aspects of these technologies in the healthcare domain is essential to bridge this knowledge gap. This chapter seeks to fill that gap by providing a comprehensive study that delves into the opportunities, challenges, and open issues of AI/ML/DL in healthcare services. By doing so, it aims to benefit both the research community and policymakers involved in the healthcare industry worldwide (Vyas, S., & Gupta, S. 2023).

22.2 OPPORTUNITIES IN HEALTHCARE WITH AI, ML, AND DL

AI, ML, and DL have emerged as promising tools to improve healthcare services. As seen in Figure 22.1, disruptive technologies have the potential to revolutionize the healthcare industry and improve patient outcomes.

This section presents the opportunities offered by employing AI, ML, and DL in healthcare.

22.2.1 MEDICAL IMAGE ANALYSIS AND DIAGNOSTICS

Medical image analysis using AI, ML, and DL techniques supports interpreting medical images for diagnostics to improve efficiency and speed in diverse medical

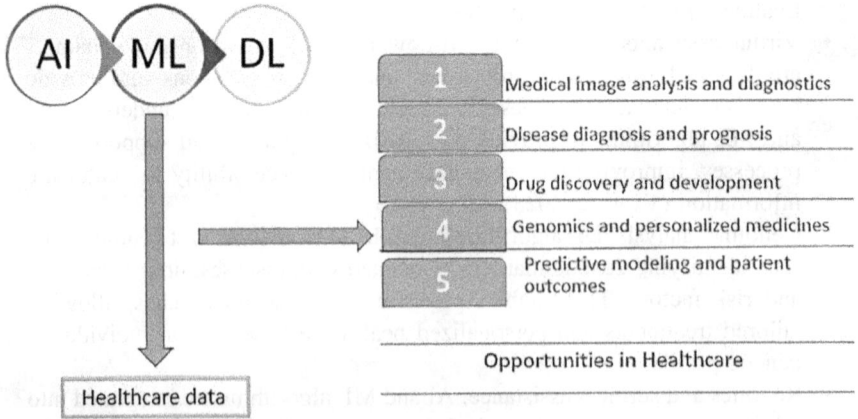

FIGURE 22.1 Opportunities in healthcare.

cases. Medical imaging comprises several modalities such as X-ray, computed tomography (CT), magnetic resonance imaging (MRI), ultrasound, positron emission tomography (PET), and more. Diverse images are generated using these modalities that require specific techniques for analysis.

Medical images are pre-processed to reduce noise, normalized, and enhanced before the application of AI, ML, or DL algorithms. This facilitates the optimization of the quality and consistency of images, aiding precise analysis. Next, the image segmentation that separates the relevant regions in an image follows. AI techniques, such as deep learning, can automatically segment organs, tumors, blood vessels, or other anatomical structures, assisting in diagnosis and treatment planning. Followed image segmentation is lesion detection and classification, where AI, ML, and DL algorithms can detect and classify abnormalities, lesions, or pathologies in medical images such as tumors, nodules, microcalcifications, etc. These algorithms can aid radiologists in early detection and accurate diagnosis. In addition, AI techniques enable a quantitative analysis to monitor disease progression, assess treatment response, or aid in surgical planning. Medical image analysis and diagnostics using AI, ML, and DL have the potential to enhance the accuracy, efficiency, and speed of diagnosis, leading to improved patient outcomes, reduced costs, better utilization of healthcare resources, and support in clinical decision-making.

Specifically, DL is widely used for medical imaging, and the most prevalent DL technique convolutional Neural Networks (CNNs) is used for image classification, object detection, and segmentation (Li et al., 2023). During the pandemic, AI-assisted CT imaging analysis has been used for COVID screening; the proposed system consisted of classification and segmentation and saved about 30%–40% of the detection time for physicians and promoted the performance of COVID-19 detection (Wang et al., 2021). Microsoft's Project Inner Eye utilizes computer vision and ML to distinguish between tumors and healthy anatomy using 3D radiological images that facilitate medical practitioners in radiotherapy and surgical planning. SubtleMR harnesses AI, ML, and DL to generate better quality medical images for radiologists that make it easier to complete exams, lowering the time it takes to get care and diagnoses by patients.

However, medical imaging requires labeled data sets for training algorithms by expert radiologists, which can be time-consuming and resource-intensive.

As technology advances and data sets grow, medical image analysis and diagnostics are expected to play an increasingly crucial role in healthcare, empowering clinicians with advanced tools to improve diagnosis, treatment, and patient care.

22.2.2 Disease Diagnosis and Prognosis

Disease diagnosis and prognosis using AI, ML, and DL offer vast opportunities for improving healthcare outcomes, optimizing resource utilization, and enabling personalized treatment approaches. As these technologies continue to evolve and integrate with clinical practice, they have the potential to transform disease management and improve patient care across various medical specialties.

AI algorithms are trained for disease pattern identification in medical images, such as radiological scans, pathological slides, or dermatological images. By leveraging ML and DL techniques, these algorithms can precisely find and classify diseases, assisting in the prompt diagnosis of diseases like cancer, cardiovascular diseases, or neurological disorders. AI, ML, and DL models can be used to develop clinical decision support systems that assist healthcare professionals in diagnosing diseases. These systems analyze patient data and provide real-time recommendations and insights based on learned patterns, helping doctors make informed decisions and improving diagnostic accuracy.

AI, ML, and DL techniques can build prognostic models that predict disease outcomes, treatment response, or survival rates based on patient characteristics, biomarkers, and historical data. These models can aid in treatment planning and help patients and healthcare providers make more informed decisions. Automated Diagnosis: AI algorithms can be trained to recognize disease patterns in medical images, such as radiological scans, pathology slides, or dermatological images. By leveraging ML and DL techniques, these algorithms can accurately detect and classify diseases, aiding in the early diagnosis of conditions like cancer, cardiovascular diseases, or neurological disorders.

Recent research has shown the exploitation of these sophisticated techniques in treating a wide range of diseases as cancer (Huang et al., 2020), interstitial lung disease (Dack et al., 2023), heart (Kapoor et al., 2023), etc. Also, examples of AI, ML, and DL usage in the industry include Tempus in Chicago, Illinois, which uses AI-powered algorithms for diagnostic biomarking. Utilizing IBM's Watson AI technology, Pfizer employs ML for immuno-oncology research to analyze massive patient data sets to foster more impactful immuno-oncological treatments for patients.

22.2.3 DRUG DISCOVERY AND DEVELOPMENT

By leveraging AI, ML, and DL techniques, the drug discovery and development process can be accelerated, optimized, and made more cost-effective. These techniques can contribute to finding possible drug targets, augmenting drug discovery pipelines, and predicting the efficiency of the latest treatments. In addition, these technologies enable precision medicine approaches.

AI, ML, and DL can aid in target identification and validation by analyzing massive volumes of biological and genetic data to recognize proteins, receptors, or pathways that play a crucial role in disease development (Vamathevan et al., 2019). This enables researchers to focus their efforts on targets with higher therapeutic potential. These techniques predict the binding affinity and interactions between molecules, such as small compounds and target proteins, and hence accelerate the identification of principal compounds and lessen the number of molecules that need to be experimentally tested. Also, by analyzing huge data sets of the medical arena comprising clinical information these techniques can reveal formerly unidentified relationships relating to drugs and diseases. This re-purposing approach offers a cost-effective and time-efficient strategy to find new indications for approved or investigational drugs.

These techniques based on large data set analysis help guide medicinal chemists in designing compounds with improved drug-like properties, reducing the failure rate during preclinical and clinical stages. ADME-Tox (absorption, distribution, metabolism, excretion, and toxicity) properties are critical factors in drug development. AI, ML, and DL algorithms can predict these properties based on chemical structure and other relevant parameters. This helps prioritize and select drug candidates with a higher likelihood of success and fewer adverse effects.

DL techniques, such as generative models and generative adversarial networks (GANs), can generate novel molecular structures with desired properties (Martinelli, 2022). These models learn from large chemical databases and can propose new compounds that have the potential to become drug candidates, expanding chemical space exploration.

Insitro is a San Francisco–based company that specializes in using machine learning and computational biology to improve the drug discovery process. By integrating these technologies, Insitro aims to accelerate drug development and reduce the cost associated with bringing new therapies to market.

One of the key ways Insitro utilizes ML is to analyze large biological data sets and uncover patterns and trends that may be difficult for humans to identify. This includes identifying new disease subtypes, which can inform the development of more targeted and personalized therapies. For instance, Insitro efficiently analyzes voluminous data to present insights for the identification of potential drug options to optimize the drug development process. Thus, researchers and developers can gain new insights into diseases and develop more successful medications for patients.

22.2.4 GENOMICS AND PERSONALIZED MEDICINE

Genomics is the study of an individual's complete set of genes (genome), including all their DNA and genetic material. With advancements in DNA sequencing technologies, it has become more accessible and affordable to obtain an individual's genomic information.

Personalized medicine, also known as *precision medicine*, considers an individual's unique genetic, environmental, and lifestyle factors to guide medical decisions and treatments. It aims to provide targeted interventions that are specifically tailored to an individual's needs, increasing treatment effectiveness, and reducing adverse effects.

Genomics and personalized medicine are related domains that help to understand an individual's genetic makeup to adapt healthcare choices and treatments (Ching et al., 2018). The techniques of AI/ML and DL are employed for the identification of genes associated with diseases, patterns of gene expressions, or genetic markers that support disease prognosis (Caudai et al., 2021). Moreover, genomic data is also helpful when assessing the risk of developing specific diseases. For personalized risk assessment, genomic data having specific genetic variants or biomarkers for diseases like cardiovascular diseases, cancer, etc. are used. This also facilitates proactive screening, monitoring, and formulation of prevention strategies for diverse diseases. By finding genetic mutations directing disease development,

22.4 INTEGRATION OF AI, ML, AND DL INTO EXISTING HEALTHCARE SYSTEMS

The integration of AI, ML, and DL into existing healthcare systems involves thorough planning, association, and conformance to privacy and ethical standards. The state-of-art healthcare systems generate voluminous data such as electronic health records, lab reports, image analysis data, etc. Thus, this needs sophisticated data integration and interoperability for effective access and analysis of data. AI, ML, and DL algorithms frequently need important computational resources and processing power. Guaranteeing that the healthcare infrastructure can uphold the computational needs of these techniques is vital. This may include hardware upgradation, employing cloud-based solutions, or applying distributed computing structures to handle the enhanced computational requirements.

Moreover, AI, ML, and DL algorithms should impeccably integrate into clinical workflows to boost their effect. The designed interfaces should be user-friendly and integrate intuitive workflows for facilitating the healthcare professionals' interaction with the systems and interpretation of generated outputs. Integration should include addressing the specific needs and preferences of healthcare stakeholders to confirm smooth adoption and acceptance. Developers, researchers, and industry professionals must collaborate to speed up the integration of AI, ML, and DL technologies into healthcare systems. Collaborations can accelerate access to innovative technologies and knowledge exchange to enhance the implementation process.

22.5 OPEN ISSUES IN HEALTHCARE WITH AI, ML, AND DL

AI, ML, and DL have already created an impact in the healthcare industry. Yet a few open issues are to be addressed before clinical practice implementation. Prior to incorporating AI, ML, and DL into clinical practice, the points are contemplated.

22.5.1 DATA QUALITY AND SHARING

To ensure accuracy in AI/ML and DL models, the data quality and addressing the biases in healthcare data are significant that require regular data audits, validation procedures, and bias detection techniques for disparate outcomes for different patient populations. Also, in large-scale AI implementation, data sharing is necessary. However, among the stakeholders, a few are hesitant in data sharing due to various reasons: privacy and security of personal or commercial data. Thus, regulatory frameworks and guidelines need to evolve for getting a comprehensive understanding of big data and AI-related technologies. For all real-world scenarios, AI-related techniques are not required. AI-, ML-, and DL-based models are specifically helpful when the data set is huge, parameters are hard to predict, and drawing inferences from the results is time-consuming.

Further, data collection and usage should be a transparent process within AI, ML, and DL systems. Organizations should reveal their data handling methods, algorithms used, and the role and constraints of AI-driven insights.

22.5.2 Regulatory Frameworks and Guidelines

Establishing clear data governance policies and frameworks is essential to address issues of data ownership, stewardship, and responsible use. Collaboration between healthcare organizations, researchers, policymakers, and regulatory bodies can help define guidelines for data sharing, access, and usage while respecting patient rights and organizational responsibilities.

22.5.3 Collaboration between Healthcare Stakeholders and Technology Developers

Collaboration among stakeholders is crucial to establishing industry-wide standards and best practices for responsible data handling and governance in AI, ML, and DL in healthcare. Engaging with patients, researchers, technology developers, and regulatory bodies can foster a collective understanding of responsible data practices and promote responsible innovation.

Some research in this area has been proposed, including a wide range of interested parties—clinicians, policymakers, data scientists, ML researchers, hospital staff, and so on in the process of creating new ML/DL methods. By bringing together clinicians and AI researchers, for example, healthcare service providers will be able to pool their knowledge for the greater good of patient care.

22.5.4 Overcoming Barriers to Adoption and Implementation

To overcome barriers to the adoption and implementation of AI, ML, and DL systems, healthcare professionals, administrators, and policymakers must be trained and educated so that their confidence and knowledge are enhanced about using the systems. The best practices and effective case studies must be shared to promote awareness and adoption of these systems. Also, conducting thorough evaluations, clinical trials, and validation studies are necessary to build confidence and address concerns. Furthermore, effective change management strategies, clear communication, involvement of stakeholders, and explaining tangible benefits can assist and overcome resistance and foster a culture of innovation to adopt the AI, ML, and DL systems in healthcare.

22.6 CONCLUSION

The integration of AI/ML/DL systems supports an increasing number of medical tasks. This needs the coevolution of these systems with healthcare systems, and this will govern the performance degree at which it will influence the landscape of medical practice, incorporating disease detection and treatment. This chapter comprehensively presents the opportunities and challenges offered when incorporating the AI/ML/DL systems into the healthcare landscape that includes patient care, diagnostics, treatment, and healthcare management.

From medical image analysis and disease diagnosis to drug discovery, genomics, personalized medicine, and predictive modeling, AI, ML, and DL are transforming different healthcare aspects. In this chapter, the current state of AI technologies and various applications that offer various prospects to enhance effectiveness, precision, and patient outcomes are discussed (Vyas, S., & Gupta, S., 2022). The integration of AI, ML, and DL can facilitate healthcare practitioners and professionals to make more precise diagnoses and develop tailored treatment approaches. In addition, the new therapies development and healthcare research advancement are also supported. However, the challenges of AI/ML/DL integration along with ethical considerations, data quality, algorithm transparency, and validation need to be addressed.

This chapter is an effort to explore and address the challenges and opportunities of integrating AI, ML, and DL for personalized, accessible, efficient, and effective healthcare. By incorporating these technologies reliably and with a patient-centered approach, we can pave the way for a future where AI, ML, and DL play a transformative role in improving health outcomes and enhancing the overall healthcare experience.

REFERENCES

Albahri, A.S., Duhaim, A.M., Fadhel, M.A., Alnoor, A., Baqer, N.S., Alzubaidi, L., Albahri, O.S., Alamoodi, A.H., Bai, J., Salhi, A., Santamaría, J., Ouyang, C., Gupta, A., Gu, Y., & Deveci, M. (2023). A systematic review of trustworthy and explainable artificial intelligence in healthcare: Assessment of quality, bias risk, and data fusion. *Information Fusion*, *96*(January), 156–191. 10.1016/j.inffus.2023.03.008

Balasubramaniam, N., Kauppinen, M., Rannisto, A., Hiekkanen, K., & Kujala, S. (2023). Transparency and explainability of AI systems: From ethical guidelines to requirements. *Information and Software Technology*, *159*(July 2022), 107197. 10.1016/j.infsof.2023.107197

Caudai, C., Galizia, A., Geraci, F., Le Pera, L., Morea, V., Salerno, E., Via, A., & Colombo, T. (2021). AI applications in functional genomics. *Computational and Structural Biotechnology Journal*, *19*, 5762–5790. 10.1016/j.csbj.2021.10.009

Ching, T., Himmelstein, D.S., Beaulieu-Jones, B.K., Kalinin, A.A., Do, B.T., Way, G.P., Ferrero, E., Agapow, P.M., Zietz, M., Hoffman, M.M., Xie, W., Rosen, G.L., Lengerich, B.J., Israeli, J., Lanchantin, J., Woloszynek, S., Carpenter, A.E., Shrikumar, A., Xu, J., … Greene, C.S. (2018). Opportunities and Obstacles for Deep Learning in Biology and Medicine. In *Journal of the Royal Society Interface* (Vol. 15, Issue 141). 10.1098/rsif.2017.0387

Dack, E., Christe, A., Fontanellaz, M., Brigato, L., Heverhagen, J.T., Peters, A.A., Huber, A.T., Hoppe, H., Mougiakakou, S., & Ebner, L. (2023). Artificial Intelligence and Interstitial Lung Disease: Diagnosis and Prognosis. *Investigative Radiology*, *00*(00), 1–8. 10.1097/RLI.0000000000000974

Esteva, A., Robicquet, A., Ramsundar, B., Kuleshov, V., DePristo, M., Chou, K., Cui, C., Corrado, G., Thrun, S., & Dean, J. (2019). A guide to deep learning in healthcare. *Nature Medicine*, *25*(1), 24–29. 10.1038/s41591-018-0316-z

Farah, L., Murris, J.M., Borget, I., Guilloux, A., Martelli, N.M., & Katsahian, S.I.M. (2023). Line Farah, PharmD; Juliette M. Murris, MSc; Isabelle Borget, PhD, PharmD; Agathe Guilloux, PhD; Nicolas M. Martelli, PhD, PharmD; and Sandrine I.M. Katsahian, MD, PhD. *Mayo Clinic Proceedings*, *1*(2), 120–138.

González-Gonzalo, C., Thee, E.F., Klaver, C.C.W., Lee, A.Y., Schlingemann, R.O., Tufail, A., Verbraak, F., & Sánchez, C.I. (2022). Trustworthy AI: Closing the gap between development and integration of AI systems in ophthalmic practice. *Progress in Retinal and Eye Research*, *90*(December 2021). 10.1016/j.preteyeres.2021.101034

Gupta, S., Sharma, H.K., & Kapoor, M. (2023). Digital Medical Records (DMR) security and privacy challenges in smart healthcare system. In: *Blockchain for Secure Healthcare Using Internet of Medical Things (IoMT)*. Cham: Springer. 10.1007/978-3-031-18896-1_6

Huang, S., Yang, J., Fong, S., & Zhao, Q. (2020). Artificial intelligence in cancer diagnosis and prognosis: Opportunities and challenges. *Cancer Letters*, *471*, 61–71. 10.1016/j.canlet.2019.12.007

Jin, Y., Pepe, A., Li, J., Gsaxner, C., Zhao, F., Kleesiek, J., Frangi, A.F., & Egger, J. (2021). *AI-based Aortic Vessel Tree Segmentation for Cardiovascular Diseases Treatment: Status Quo*. http://arxiv.org/abs/2108.02998

Kapoor, A., Kapoor, S., Upreti, K., Singh, P., Kapoor, S., Alam, M.S., & Nasir, M.S. (2023). Cardiovascular Disease Prognosis and Analysis Using Machine Learning Techniques. In *Advanced Communication and Intelligent Systems: First International Conference, ICACIS 2022*, 180–194.

Kumar, P., Chauhan, S., & Awasthi, L.K. (2023). Artificial Intelligence in Healthcare: Review, Ethics, Trust Challenges & Future Research Directions. *Engineering Applications of Artificial Intelligence*, *120*(December 2022), 105894. 10.1016/j.engappai.2023.105894

Li, N., Fan, L., Xu, H., Zhang, X., Bai, Z., Li, M., Xiong, S., Jiang, L., Yang, J., Chen, S., & Qiao Y.C.B. (2023). An AI-Aided Diagnostic Framework for Hematologic Neoplasms Based on Morphologic Features and Medical Expertise. *Laboratory Investigation*, *103*(4), 100055.

Loh, H.W., Ooi, C.P., Seoni, S., Barua, P.D., Molinari, F., & Acharya, U.R. (2022). Application of explainable artificial intelligence for healthcare: A systematic review of the last decade (2011–2022). *Computer Methods and Programs in Biomedicine*, *226*, 107161. 10.1016/j.cmpb.2022.107161

Martinelli, D.D. (2022). Generative machine learning for de novo drug discovery: A systematic review. *Computers in Biology and Medicine*, *145*(January), 105403. 10.1016/j.compbiomed.2022.105403

Rajkomar, A., Dean, J., & Kohane, I. (2019). Machine learning in medicine. *New England Journal of Medicine*, *380*(14), 1347–1358. 10.1056/nejmra1814259

Rasheed, K., Qayyum, A., Ghaly, M., Al-Fuqaha, A., Razi, A., & Qadir, J. (2022). Explainable, trustworthy, and ethical machine learning for healthcare: A survey. *Computers in Biology and Medicine*, *149*(August), 106043. 10.1016/j.compbiomed.2022.106043

Vamathevan, J., Clark, D., Czodrowski, P., Dunham, I., Ferran, E., George Lee, B.L. et al. (2019). Applications of machine learning in drug discovery and development. *Nature Reviews Drug Discovery*, *186*, 463–477.

Vyas, S. (2023). Extended reality and edge AI for healthcare 4.0: Systematic study. In *Extended reality for healthcare systems* (pp. 229–240). Academic Press.

Vyas, S., & Gupta, S. (2022). Case study on state-of-the-art wellness and health tracker devices. In *Handbook of research on lifestyle sustainability and management solutions using AI, big data analytics, and visualization* (pp. 325–337). IGI Global.

Vyas, S., & Gupta, S. (2023). WBAN-based remote monitoring system utilising machine learning for healthcare services. *International Journal of System of Systems Engineering*, *13*(1), 100–108.

Vyas, S., Bhargava, D., & Khan, S. (2023). Healthcare 4.0: A Systematic Review and Its Impact Over Conventional Healthcare System. *Artificial Intelligence for Health 4.0: Challenges and Applications*, 1–17.

Vyas, S., Gupta, S., & Shukla, V.K. (2023, March). Towards edge AI and varied approaches of digital wellness in healthcare administration: A study. In *2023 International Conference on Computational Intelligence and Knowledge Economy (ICCIKE)* (pp. 186–190). IEEE.

Wang, B., Jin, S., Yan, Q., Xu, H., Luo, C., Wei, L., Zhao, W., Hou, X., Ma, W., Xu, Z., Zheng, Z., Sun, W., Lan, L., Zhang, W., Mu, X., Shi, C., Wang, Z., Lee, J., Jin, Z., … Dong, J. (2021). AI-assisted CT imaging analysis for COVID-19 screening: Building and deploying a medical AI system. *Applied Soft Computing*, *98*. 10.1016/j.asoc.2020.106897

Wysocki, O., Davies, J.K., Vigo, M., Armstrong, A.C., Landers, D., Lee, R., & Freitas, A. (2023). Assessing the communication gap between AI models and healthcare professionals: Explainability, utility and trust in AI-driven clinical decision-making. *Artificial Intelligence*, *316*, 103839. 10.1016/j.artint.2022.103839

Index

For Product Safety Concerns and Information please contact our EU
representative GPSR@taylorandfrancis.com
Taylor & Francis Verlag GmbH, Kaufingerstraße 24, 80331 München, Germany